GENDERING GREEN CRIMINOLOGY

Edited by
Emma Milne, Pamela Davies, James Heydon,
Kay Peggs and Tanya Wyatt

BRISTOL
UNIVERSITY
PRESS

First published in Great Britain in 2023 by

Bristol University Press
University of Bristol
1–9 Old Park Hill
Bristol
BS2 8BB
UK
t: +44 (0)117 374 6645
e: bup-info@bristol.ac.uk

Details of international sales and distribution partners are available at bristoluniversitypress.co.uk

British Library Cataloguing in Publication Data
A catalogue record for this book is available from the British Library

ISBN 978-1-5292-2961-5 hardcover
ISBN 978-1-5292-2963-9 ePub
ISBN 978-1-5292-2964-6 ePdf

Cover design: Liam Roberts
Front cover image: iStock/FangXiaNuo
Bristol University Press uses environmentally responsible print partners.
Printed and bound in Great Britain by CPI Group (UK) Ltd, Croydon, CR0 4YY

Contents

List of Figures and Tables   v

Notes on Contributors   vi

Acknowledgements   xii

Foreword by Ragnhild Sollund   xiii

1   Why Gendering Green Criminology Matters   1
*Emma Milne, Pamela Davies, James Heydon, Kay Peggs
and Tanya Wyatt*

**PART I   Gendered Nature of Green Crimes and
Environmental Harm**

2   Eco-Feminism and the Gendering Green Criminology Project   17
Pamela Davies

3   New Directions Please! Veganising Green Criminology   34
Kay Peggs

4   Men and the Climate Crisis: Why Masculinities Matter for   53
Green Criminology
Stephen R. Burrell

5   Reconceptualising Gendered Dimensions of Illegal Wildlife   72
Trade in Sub-Saharan Africa through Legal, Policy and
Programmatic Means
Helen U. Agu, Josiah C. Ogbuka and Meredith L. Gore

6   The Attitudes of People with Different Gender Identities   97
and Different Perceptions of Gender Roles towards
Nonhuman Animals and Their Welfare
Aphra Hope-Forest, Ekaterina Gladkova and Tanya Wyatt

PART II Gendered Impacts and Victimisation

7   Queering Green Criminology: The Impacts of Zoonotic   121
Diseases on the LGBTQ Community
Laurence Pedroni and Benja Kromash

8   Women and the Structural Violence of 'Fast-Fashion'   148
    Global Production: Victimisation, Poorcide and
    Environmental Harms
    Sandya Hewamanne and Nigel South

9   Green Victims of the International Waste Industry: An   170
    Analysis from a Gender Perspective
    María-Ángeles Fuentes-Loureiro

10 The Green Road Project and Women's Green   187
    Victimisation in Turkey
    Halil Ibrahim Bahar

11 'Daughters of Dust': An Eco-Feminist Analysis of   205
    Debt-for-Nature Swaps and Underage Marriage in Indonesia
    Delon Alain Omrow

PART III Resistance

12 Women's Experiences of Environmental Harm in   229
    Colombia: Learning from Black, Decolonial and
    Indigenous Communitarian Feminisms
    Daniela Suárez Vargas and Rachel Killean

13 Vegan Feminism Then and Now: Women's Resistance to   251
    Legalised Speciesism across Three Waves of Activism
    Corey Lee Wrenn and Lynda M. Korimboccus

14 'To Preserve and Promote': Gendering Harm in Green   267
    Cultural Criminology
    Angeline Marie Letourneau

15 David and Goliath: Exploring the Male Burdens of   289
    Patriarchal Capitalism
    Rob White

Index   304

List of Figures and Tables

Figures

14.1   A rider and his horse take a tumble in the saddle bronc event   274
       at Calgary Stampede
14.2   Cattle grazing on the foothills of Chief Mountain in southern   279
       Alberta

Tables

5.1   Summary of country-specific assessments of the state of   83
       gender and illegal wildlife trade in national wildlife legal and
       regulatory frameworks of some member states of the sub-
       Saharan African region
6.1   The pre-defined gender identities and their definitions, as   103
       used in the survey
6.2   Non-cisgender identities   105
6.3   Gender identity and ownership of a companion animal   106
6.4   Gender identity and awareness of status dogs/dog fighting   107
6.5   View of gender roles and pet ownership   107
6.6   View of gender roles and awareness of status dogs   108
15.1   Gender-based vulnerabilities in disaster responses   295

Notes on Contributors

Helen U. Agu is Senior Lecturer in Environmental, Climate Change and Conservation Law, Department of International and Comparative Law, University of Nigeria, Enugu Campus. She is an African Futures fellow of the Alliance for Africa Partnership at Michigan State University, United States and a Research Fellow of the Raoul Wallenberg Institute for Humanitarian Law (Lund, Sweden) at the Centre for Human Rights, Addis Ababa University, Ethiopia. Her research interests centre on gendered dimensions of wildlife crime, climate change and conservation. She holds a BSc in agricultural education, bachelor's, master's and doctoral degrees in environmental law and policy from the University of Nigeria. She has published many journal articles and book chapters and co-edited the volume *Women and Wildlife Trafficking* (Wiley-Blackwell, 2022).

Halil Ibrahim Bahar graduated from the Police Academy in 1988 in Turkey. He earned his MA and PhD from the Department of Criminology, Leicester University, UK. From 1995 to 2015, Dr Bahar taught sociology, criminology and victimology at the Faculty of Security Sciences and the Institute for Security Sciences at the Police Academy. Currently, Dr Bahar is a freelance researcher in Ankara. His research interests include the sociology of institutions, green crime, victimology, urbanisation and crime. Some of his publications have appeared in *International Criminology*, *International Journal for Crime, Justice and Social Democracy* and *The Journal of Population, Space and Place*.

Stephen R. Burrell is Assistant Professor (Research) in the Department of Sociology, Durham University, UK. He is currently undertaking a Leverhulme Early Fellowship on masculinity and violence in the climate crisis, and engaging men and boys in caring for the environment. Stephen is Deputy Director of Durham's Centre for Research into Violence and Abuse, where he is part of a team conducting research on men, masculinities and social change (for more information see www.mmasc.org.uk). Together with Sandy Ruxton, Stephen co-hosts a podcast called 'Now and Men: Current Conversations on Men's Lives'.

Pamela Davies is Professor of Criminology and Head of the Criminology subject group in the Department of Social Sciences at Northumbria University, UK. Pamela's research interests coalesce around gender, crime, harm and victimisation, and the tensions around social and environmental justice. She has edited numerous textbooks and scholarly volumes and has published widely on the subject of victimisation and social harm. She is the series editor of the Palgrave Macmillan 'Victims and Victimology' book series (with Associate Professor Tyrone Kirchengast, University of New South Wales, Sydney).

María-Ángeles Fuentes-Loureiro holds a PhD in law from the University of A Coruña, Spain, where she presented her dissertation on environmental criminality in the field of waste and currently works as Lecturer in Criminal Law. In her research career, she has studied crimes against the environment and green criminology issues. Other secondary research lines are animal protection and corruption in the environmental field, urban planning and public procurement. She has presented her work at more than 20 conferences, and has published various book chapters, scientific reports and journal articles in Spain, Italy and the United States.

Ekaterina Gladkova is Lecturer in Sociology and Criminology at Northumbria University, UK. She is interested in critical understandings of interrelations/interconnections between the environment, human and nonhuman. Ekaterina has published in journals such as *Critical Criminology, Journal of White Collar and Corporate Crime, International Journal for Crime, Justice and Social Democracy* and *Polar Research.*

Meredith L. Gore is Associate Professor of Human Dimensions of Global Environmental Change in the Department of Geographical Sciences at University of Maryland, College Park, United States. Her research focuses on human–environment relationships; she has expertise in conservation criminology, community-based conservation, science diplomacy, and gender in wildlife. She is the editor and author of *Conservation Criminology* (Wiley-Blackwell, 2017) and co-editor of *Women and Wildlife Trafficking* (Wiley-Blackwell, 2022). Meredith held a post-doc at Michigan State University, earned her PhD at Cornell University, MA at the George Washington University and BA at Brandeis University.

Sandya Hewamanne is Professor of Anthropology and Director of the Center for Global South Studies at the University of Essex, UK. She is the author of *Stitching Identities in a Free Trade Zone: Gender and Politics in Sri Lanka* (University of Pennsylvania Press, 2008); *Sri Lanka's Global Factory Workers: (Un)Disciplined Desires and Sexual Struggles in a Post-Colonial*

Society (Routledge, 2016); *Re-stitching Identities in Rural Sri Lanka: Gender, Neoliberalism and the Politics of Contentment* (University of Pennsylvania Press, 2020) and the co-editor of *The Political Economy of Post-COVID Life and Work in the Global South: Pandemic and Precarity* (Springer, 2022). She is the Founder, Director of IMPACT-Global Work, a non-profit which connects academics and activists to initiate positive policy changes for workers in the Global South.

James Heydon is Assistant Professor of Criminology in the School of Sociology and Social Policy at the University of Nottingham, UK. James's research interests centre on crimes of the powerful more broadly and environmental harm specifically. In particular, he is interested in the effectiveness of formal and informal social controls on the behaviour of individuals, states and corporate actors. This includes the more traditional 'top-down' elements, such as law and regulation, but also those deemed 'bottom-up', such as the application of deviance labels by 'non-elite' groups.

Aphra Hope-Forest (she/her) was carrying out a research project at Northumbria University when she collected these data. Her research interests included gender differences in attitudes to nonhuman animals, and how students with different learning styles create their preferred learning environments. She received a BSc in Zoology from Cardiff University and is now working for the Ministry of Justice.

Rachel Killean is Senior Lecturer at Sydney Law School and a member of the Sydney Institute of Criminology, the Sydney Southeast Asia Centre, and the Sydney Environment Institute, Australia. Prior to joining Sydney Law School, she was Senior Lecturer at the Queen's University Belfast School of Law. Dr Killean's research centres responses to violence, with a focus on transitional justice, victims' rights, sexual and gender-based violence, and harms perpetrated against the natural world.

Lynda M. Korimboccus is Lecturer in Sociology at West Lothian College, Scotland, and a PhD sociology student at the University of East Anglia, UK, investigating the lived experiences of young vegan children in Scottish education. An advocate for equity and justice, Lynda is an activist scholar and a committed ethical vegan and grassroots campaigner since 1999. She is Editor-in-Chief of the *Student Journal of Vegan Sociology*, a volunteer writer for Faunalytics, and a member of both the Researcher and Education Networks of the Vegan Society.

Benja Kromash (he/him) is originally from Chicago, United States, where he first found his passions for environmental justice, queer community and racial equity. Benja completed his undergraduate degree at Macalester College prior to serving as a Peace Corps Volunteer in Paraguay. Following his work on environmental education and conservation in Paraguay, Benja completed his master's degree in Sociology at Colorado State University. Today, Benja lives in Denver, Colorado, where he works at the non-profit Volunteers for Outdoor Colorado. In his spare time, he enjoys pottery, hiking and cheering on local drag queens.

Angeline Marie Letourneau is a PhD candidate in environmental sociology at the University of Alberta whose research focuses on gendering the sociology of climate change. Angeline's work examines the affective influence of identities under threat in the age of environmental and climate crisis, and the contribution of these identity processes to widening cultural divisions and political polarisation. Her research is inspired by the possibility of synergies despite differences. Angeline's prior experience in the energy industry and non-profit landscape informs her attention to purposeful intervention and actionable outputs. Her work has been published in such journals as *Environmental Sociology* and *Climatic Change*.

Emma Milne is Associate Professor of Criminal Law and Criminal Justice at Durham University. She is a socio-legal scholar and feminist criminologist. Emma's research is interdisciplinary, focusing on criminal law and justice responses to infant killing and foetal harm. The wider context of Emma's work is social controls and regulations of women, notably of pregnancy, sex and motherhood. Emma's monograph *Criminal Justice Responses to Maternal Filicide: Judging the Failed Mother* was published in 2021 (Emerald Publishing Limited). She co-authored *Sex and Crime* (SAGE, 2020), and co-edited *Women and the Criminal Justice System: Failing Victims and Offenders?* (Palgrave, 2018).

Josiah C. Ogbuka is Research Fellow and Lecturer at the Institute of Maritime Studies, University of Nigeria, Enugu Campus, Nigeria. His areas of research interest include marine pollution management, marine biogeochemistry, marine governance and marine spatial planning, wildlife conservation, circular economy and sustainability, climate change and environmental changes, and has published in some of these areas. He holds a BSc in education/biology, and MSc in marine environmental protection (UK) and PhD (forthcoming).

Delon Alain Omrow is Instructor in the Faculty of Social Science and Humanities at Ontario Tech University, Canada. He earned his undergraduate and graduate degrees in criminology from York University and the University of Toronto, respectively. He completed his PhD at York

University, specialising in green criminology, environmental justice and biosecurity. He is currently completing a postdoc at Ontario Tech University, exploring what he refers to as racialised ecologies and the androcentric-anthropocentric symbiosis of trauma; and the way the lived experiences of the disempowered and marginalised are articulated through contemporary environmental discourse.

Laurence Pedroni (he/they) is originally from San Jose, California, where he found his passion for challenging systems of oppression and liberation ideologies. Laurence completed his degrees in psychology and justice studies before completing his master's degree in justice studies at San Jose State University. He continues to teach at San Jose State University. He is currently completing his PhD at Colorado State University while living in Denver, Colorado, where he spends most of his time walking his German Shepherd, Artemis, and practising martial arts.

Kay Peggs is Professor of Criminology and Sociology at Kingston University, UK and is a Fellow of the Oxford Centre for Animal Ethics. Her current research examines the persistence of complex inequalities associated with species. Kay has published widely on this theme and her books include *Animals and Sociology* (Palgrave, 2012), *Experiments, Animal Bodies and Human Values* (Routledge, forthcoming) and co-authored *(Not) Consuming Animals: Ethics, Environment and Lifestyle Choices* (Routledge, forthcoming). Kay is also a research methods specialist, and is co-editor of two major four-volume sets, *Observation Methods* (SAGE, 2013) and *Critical Social Research Ethics* (SAGE, 2018).

Nigel South is Emeritus Professor of Sociology at the University of Essex, UK. He has made various contributions to the development of 'green criminology' including co-authoring *Green Cultural Criminology* (Routledge, 2014) and *Water, Crime and Security in the Twenty-First Century* (Palgrave, 2018), as well as co-editing the *International Handbook of Green Criminology* (2nd edn, Routledge, 2020). In 2022 he received the 'Outstanding Achievement Award' from the British Society of Criminology and in 2013 received a 'Lifetime Achievement Award' from the American Society of Criminology, Division on Critical Criminology and Social Justice.

Daniela Suárez Vargas is a PhD researcher in the School of Law at Queen's University Belfast, UK. She is a Northern Bridge Consortium doctoral scholar, funded by the Arts and Humanities Research Council of the UK. Daniela is a qualified lawyer from Universidad del Rosario, Colombia. She holds an LLM from Queen's University Belfast and a postgraduate certificate in gender, conflict, and transitional justice from Ulster University. Daniela's doctoral project examines narratives of victimhood in transitional justice, focusing on the case of sexual violence within armed groups in Colombia. Her research interests include transitional justice, international criminal law, human rights and gender.

Rob White is Emeritus Distinguished Professor of Criminology at the University of Tasmania, Australia. He has published extensively in the areas of youth studies, criminology and eco-justice. Among his recent books are *Advanced Introduction to Applied Green Criminology* (Edward Elgar, 2023), *Theorising Green Criminology* (Routledge, 2022) and *Critical Forensic Studies* (with Roberta Julian and Loene Howes) (Routledge, 2022). He is Editor-in-Chief of *Forensic Science International: Animals and Environments*.

Corey Lee Wrenn is Lecturer in Sociology at the University of Kent, UK, and specialises in vegan studies and the politics of animal rights mobilisation. She is the author of *A Rational Approach to Animal Rights* (Palgrave, 2016), *Piecemeal Protest: Animal Rights in the Age of Nonprofits* (University of Michigan Press, 2019) and *Animals in Irish Society* (SUNY Press, 2021). She has served as the Chair of the Animals and Society Section of the American Sociological Association (2018–21) and the Research Advisory Committee of the Vegan Society.

Tanya Wyatt was Professor of Criminology at Northumbria University. She is a green criminologist specialising in wildlife crime and trafficking and nonhuman animal abuse and harm. Her research explores the intersections with organised crime, corporate crime and corruption. Her most recent books are *Is CITES Protecting Wildlife? Assessing Implementation and Compliance* (Earthscan, 2021) and *Wildlife Trafficking: A Deconstruction of the Crime, Victims and Offenders* (Palgrave Macmillan, 2021).

Acknowledgements

We would like to extend our thanks to the British Society of Criminology Green Criminology Network and the Women, Crime and Criminal Justice Network for organising and hosting the conference that inspired this book. The half-day conference, held online in November 2020, provided a space for discussion of important issues relating to the gendering of green criminology, many of which are central to chapters in this collection.

Thanks to anonymous reviewers who provided us with some helpful comments and feedback.

We would also like to thank Rebecca Tomlinson from Bristol University Press, who believed in and encouraged this project from the start.

Foreword

Ragnhild Sollund

Green criminology is an urgent field of criminology that has expanded tremendously alongside the growing nature crisis caused by human-induced climate change and dramatic species loss due to anthropogenic action, or lack of action. This in turn has led criminologists all over the world to engage in research on different forms of harms and crimes committed against nature and nonhuman animals. Cooperation between scholars in the Global North and the Global South has formed part of this transnational endeavour to address environmental crimes from non-colonialist, non-anthropocentric point of departures, bridging people, knowledge and scholarship. This book is an excellent example of the international character and fusion of different disciplines and fields of green criminology. Most importantly, this book presents us with the gendered aspects of green crimes and environmental harms and the ways in which neither offending nor victimisation of such harms are gender neutral, but caused by overarching structures, such as colonialism, capitalism and patriarchy.

While feminist theorisation has been important in critical criminology generally, in green criminology, unfortunately, it has been more absent and while intersectional theorisation swiped sociology and other social sciences in the beginning of this century, green criminology seemed generally unaffected by this theoretical approach. Applying intersectional theory methodologically is not an easy task due to its complexity (McCall, 2005). For example, when does one aspect of a person's identity, which renders the person vulnerable to oppression, end and another begin? What factors intersect where, when, how and why? These questions, I think, are questions many feminist social scientists have grappled with since intersectionality was introduced as a theoretical concept. Nonetheless, the underlying basis of this approach; that multiple features and mechanisms that shape different aspects of oppression and victimisation intersect, for example, class, ethnicity, gender, and not the least species affiliation, are important to keep in mind, and that is one important endeavour of this book.

Prior to intersectionality theory and perhaps more familiar to green criminology, eco-feminism appeared as an interdisciplinary feminist approach to the natural world, nature ethics and nonhuman animals. Eco-feminism (for example, Donovan and Adams, 1996; Kheel, 2007) as well as intersectionality are important in the analysis of offending and victimisation in regard to environmental crimes, harms and nonhuman victimisation. These perspectives are also vital in relation to activism and protest, so well exemplified in Part III of this book.

We are all part of deeply embedded social structures. Some of these are oppressive, and this makes most of us both complicit in and victims of various forms of environmental harm. It is urgent that our role, both as victims and perpetrators, is identified and to provide a theoretical and empirical foundation for how to deal with and counteract such harms. This book is essential and a welcome contribution to green criminology in this regard.

References

Donovan, J. and Adams, C.J. (eds) (1996) *Beyond Animal Rights: A Feminist Caring Ethics for the Treatment of Animals*. New York: Continuum.

Kheel, M. (2007) *Nature Ethics: An Ecofeminist Perspective*. Plymouth: Rowman & Littlefield.

McCall, L. (2005) The complexity of intersectionality. *Signs: Journal of Women in Culture and Society*, 30(3), 1771–800.

Why Gendering Green Criminology Matters

Emma Milne, Pamela Davies, James Heydon,
Kay Peggs and Tanya Wyatt

Introduction

Our ambition for this book is to bring together feminist and green criminology for the first time in a scholarly volume where all contributions are devoted to the project of gendering green criminology. The editorial team is comprised of experts in gender and crime and in green criminology/environmental harm. The idea for the edited collection, and some of the chapters, arose from a conference organised by the editors through the 'Green Criminology' and 'Women, Crime and Criminal Justice' Networks of the British Society of Criminology. That conference inspired us to expand the discussion and scope of inquiry into the gendering of green criminology.

   As a collective of scholars, we cannot help but observe how research in the green criminological field has proliferated. There is a growing body of theoretical thinking and imaginative and robust research arising out of thick descriptive and in-depth narrative accounts, ethnographies and visual methodologies. Much of this newfound knowledge and emerging qualitative data flows from a variety of sources in support of the often distinct, but nevertheless complex, developing picture of the patterns to environmental harms and green crimes and victimisations. The gendered nature of these patterns is especially evident, concerning and often exacerbated by a range of factors. Our starting point – gendering the problems – demands that we properly situate our thinking within a broader intersectional framework where these complex patterns are excavated early in the life of this new direction for green criminology. Consequently, we are confident that this book, *Gendering Green Criminology*, is a timely publication. Our theoretical perspective thus starts out as feminist and green, and this informs our trilogy of aims. First, while the contents of the volume inevitably present compelling evidence attesting to the gender patterning to, and gendered nature of, green crimes and environmental harms, our aim goes well beyond illustrating these features. Second, we illustrate the gendered impacts of these problems and the gendered nature of harm and victimisation caused and experienced in different contexts and in different parts of the globe. Our ambition extends further so that, third, we examine the gendered nature of resistance and aftermath recovery, thus allowing us to offer an informed critical understanding and appreciation of how to ameliorate the harms currently being experienced. As an editorial team, our theoretical perspective has sharpened as we have engaged in this project, and this introduction expands upon the scaffold of concepts and perspectives which has allowed this to become more sophisticated.

In the remainder of this first and introductory chapter we begin to draw together feminist and green criminology. In doing so we set the scene for expanding our theoretical positioning – which is one of our ambitions for the volume. This chapter closes with an outline of how the contributors' chapters are positioned within the volume together with a brief overview of their substantive contents.

Feminist and green criminology

Since its inception in the 1990s, green criminology has done much to draw attention to a range of environmental harms, green crimes and other forms of understudied victimisation. Echoing its parent discipline, the sub-field is characteristically 'parasitic' (Cohen, 1998: 4). Its focus ranges across philosophies of environment and it borrows perspectives, theories and concepts from other disciplines to understand more traditionally anthropocentric issues, like the effect of environmental degradation on the health of human animals (Hall, 2014) or cultural existence of human groups (Crook et al, 2018). At the same time, it is also attentive to the more eco-centric experiences of nonhuman species, such as those who are free-living, implicated in agriculture, or 'domesticated' (Sollund, 2012; Beirne, 2014; Maher, 2021). There is also a burgeoning body of green criminological work exploring the harms associated with climate change-related ecological disorganisation (White, 2012; Lynch et al, 2013), and the multiplicity of injustices involved with local opposition to environmentally harmful practices (Gladkova, 2020). Taken together, in its relatively short existence green criminology has been largely preoccupied with identifying and examining the understudied spaces between environmental harm and human/nonhuman victimisation.

It is in this context that the sub-field's relative inattention towards gendered aspects of green crime can be understood. In its effort to alert wider criminological scholarship to environmental issues, repurpose existing empirical and theoretical contributions to less determinedly anthropocentric phenomena (Brisman, 2014), and challenge the long-held ambivalence of its parent discipline towards the difficulties faced by the nonhuman, green criminology has yet to fully explore the specific intersections of focus in this book. This is not to say such a project does not exist, but that it is incipient. Several insightful works illustrate the potential for scholarship operating in this space, ranging from the gendered dynamics of environmental exposure (Lynch, 2018), intersections between sexism and speciesism (Beirne, 2020), and feminist analyses of nonhuman animal abuse (Taylor and Fitzgerald, 2018), among others. Here, that enterprise is advanced by establishing a touchstone for work in this space, and by providing a platform for researchers exploring the gendered aspects of green criminology into the future.

Despite the relatively small amount of empirical green criminological research focusing on gender, studies have shown – and the chapters of our edited collection demonstrate – that, like other forms of crime, green crimes and harms are also gendered in nature. Men are often perpetrators (particularly of corporate environmental offences), and women often disproportionately bear the brunt of environmental damage (as do children and nonhuman animals). In acknowledging women's increased risk of victimisation, we are not disregarding the fact that women are also perpetrators. As Robert Agnew (2013) has illustrated, everyday ecocide – those routine behaviours that harm the environment – are committed by all genders and, in this way, everyone contributes in some way to environmental degradation. What our edited collection reinforces is the need to further unpack the nuances and intersections of gender in the environmental sphere.

Recognising that every country is experiencing the drastic effects of climate change, 'Climate Action' is the thirteenth goal of the 17 Sustainable Development Goals (SDGs), adopted by the United Nations in 2015 as a universal call to action to end poverty, protect the planet, and ensure all people enjoy peace and prosperity. However, the effects of climate change extend to all species. Subsequent commentators have evaluated the SDGs ethically, in terms of their anthropocentric frame, concluding that 'the SDGs should therefore be reconsidered so that they take non-human animals into direct consideration' (Torpman and Röcklinsberg, 2021: 1). From the global to the national to the local, our major public sector institutions and non-governmental organisations are facing up to the environmental challenges of the 21st century. In England and Wales, a recent independent strategic review of policing headlines the environmental crisis as one of the three major transformations alongside the technological revolution and social change:

Human activity is transforming the natural environment in a way that poses new risks to public safety. Global warming is leading to more frequent and more intense extreme weather events such as floods, droughts, storms, heat waves and heavy rainfall. By 2050 climate change will force more than 143 million people in sub-Saharan Africa, South Asia and Latin America from their homes, with significant consequences for international migration. In the UK climate change is also generating an increase in political protest and the police increasingly find themselves having to manage the public order implications. (Police Foundation, 2022: 9)

Green criminologists have contributed to this high-level recognition of how human activity is resulting in the degradation of diverse species, the environment and the planet, and have undoubtedly shaped the pledges, strategies and goals around sustainability. Green criminology has matured to the extent that it is uniquely well-positioned to offer sophisticated evidence-based knowledge on how best to address the various environmental harms as they relate to humans and nonhumans alike. Gaining increasingly urgent relevance in a world where environmental concerns are coming to occupy social, political and economic agendas, green criminology is coming of age. However, one of the major areas where attention is both lacking and urgently needed is at the intersections between green criminology, feminism and gender-focused analysis of crime, victimisation and harm. This edited collection seeks to contribute to our understanding of these important intersections by addressing questions such as: How can feminism and a critical gender approach engage meaningfully with environmental harm? Why should those interested in environmental harm also be concerned with gender? What are the implications of gender on environmental harms? And, in what ways can a gender-focused lens provide greater insight into the nature and impact of gendered harms? Presenting a dialogue that connects international scholarship from a diverse range of disciplines, this edited collection makes a significant contribution to our understanding of these intersections such as, but not limited to, theoretical and empirical explorations of:

• gender and perpetrators of environmental harms;
• gender and victims of environmental harms;
• eco-feminism and justice;
• feminist approaches to environmental harm;
• gendered nature of structural violence and environmental harm;
• interactions between gender and environmental harms;
• gender, environmental catastrophe, resilience and recovery; and
• gender, intersectionality and species.

Lessons from feminist criminology

Carol Smart's (1977) book, *Women, Crime and Criminology: A Feminist Critique*, is widely acknowledged as one of the founding criminological texts that introduces gender as a concept significant to criminology. Prior to Smart's book, studies had mostly focused on the experiences of men who commit crime and who had involvement in the criminal justice system. However, the focus on men's actions and experiences and the state responses to their crimes failed to consider men 'as men'. Gender was absent from the analysis. 'Deviants', 'criminals', 'defendants', 'prisoners' were the focal point of the studies. If the demographic features of the studied population group were contemplated, men were the yardstick. As Catherine A. MacKinnon (1987) suggests, gender-neutrality simply equates to the male standard where masculinity and maleness are the yardsticks against which judgements of others are made. The *maleness* of 'deviants', 'criminals', 'defendants' and 'prisoners' was not thought about. In this regard, '[t]he history of criminology has been largely a history of men but one in which the "maleness" of crime, crime control and crime analysis has been so taken for granted as to be rendered invisible' (Smith, 2005: 346).

Smart (1977) and subsequent second-wave feminist theorists offered a clear challenge to the 'malestream' approach by foregrounding the experiences of women, specifically in relation to their victimisation by men (see, for example, Russell, 1975; Dobash and Dobash, 1979; Kelly, 1988; Stanko, 1990). However, feminist criminologists wanted to do more than simply add women into the equation. Previous studies that incorporated women into their analysis of criminality often assumed that women's motivations, experiences and their criminal justice outcomes were on a par with the experiences of men (for example, Pollak, 1961). Such studies often drew on negative stereotypes of women to explain criminality, for example the 'chivalry' proposition', which is characterised by categorisation based in 'the instigative female offender, the sexualized female offender, and the protected female offender' (Anderson, 1976: 350). Theories were developed and refined based on men, which were then assumed to be applicable to women. In contrast, feminist criminologists set out to understand women's involvement in crime and criminal justice from women's perspectives, so developing theoretical approaches that go beyond simply adding data about women to 'monosexual criminological theories' or uncritically building on existing literature that is inherently blind to the dynamics of gender (Chesney-Lind, 1986: 84). The result was a far richer understanding of the causes of women's offending, the impact on their victims, the state responses and the consequences of different forms of punishment, such as imprisonment.

We do not have to look far to see the importance of making women visible in debates and research related to green crimes and environmental harms. For example, UN Women (2022) makes it clear that women are not only more affected by climate change today than men, but that research demonstrates the greater involvement of women in national parliaments leads to the adoption of more stringent climate change policies. Moreover, higher percentages of women on corporate boards positively correlates with the disclosure of carbon emissions information.

Significantly, feminist thinkers have promoted the notion that a gendered analysis does not simply mean an analysis of women's experiences and behaviours. Contrary to popular discourse, men are also gendered. A key aspect of feminist criminological work has been to ask about men as men, 'not as working-class, not as migrants, not as underprivileged individuals but as men, considering what induces them to commit crime' (Grosz, 1987: 5).

Raewyn Connell's (1987, 1995) examination of the concept of masculinity has been incredibly influential for criminology, as well as for other disciplines. As Connell argues, masculinity is negotiated and practised in varying ways in different contexts: there is not one masculinity, but multiple. Hegemonic masculinity is cited as the normative standard of heterosexual masculinity, which endorses the ideology of male dominance over women, children and other men who present different forms of masculinity. Connell argues that hegemonic masculinity has a key role to play in legitimating patriarchy and the subjugation of women. Crime, and perhaps specifically violence, has been identified as one way for men to practice 'doing' their gender. As the work of James W. Messerschmidt (1993) illustrated, committing acts of violence allowed young men with fewer social and economic resources to demonstrate their masculinity. Research, such as that conducted by Ragnhild Sollund (2020: 105), also indicates that men's involvement in the trafficking of nonhuman animals may also be explained through hegemonic masculinity: 'The hunting and killing of nonhuman animals and the collecting of nonhuman animals' parts may represent one of the last bastions of men to exercise "traditional, hegemonic masculinity"'. Domination of nonhuman animals has a gendered element that needs to be explored further in criminology. For example, masculinity is associated with eating nonhuman animals, with hunting nonhuman animals, and with violence towards companion animals (see Hunnicut, 2020).

In discussing a gendered analysis of crime and harm, whether that be in the field of green criminology or more generally, we would be remiss if we did not acknowledge the continued debate about the nature of and distinction between sex and gender. Second-wave feminists, such as Simone de Beauvoir (1949) and Ann Oakley (2015 [1972]), clearly distinguished between women's biological function (their sex) and the social expectations ladened upon them (their gender). They noted that women's positions in the world mostly aligned with domestic labour within a heterosexual relationship, arguing this is a consequence of the social construction of gender. Few chief executive officers of global conglomerates are women not because women are physically and mentally less capable of completing such a role, compared to men, but because the socialisation of women and gendered expectations of how they pursue their lives generally bar them from such opportunities. Feminist theorisation of sex and gender has developed since de Beauvoir and Oakley were writing, and different ideas have progressed. Constant development in the area of feminist thinking, across the broad spectrum of feminisms, continue to make the debates imaginative and dynamic. Readers will likely see different feminist influences present in chapters in this volume. Engagement with how sex and/or gender manifest and the impact upon crime and harm continues to be an important facet of any discussion about gendering green criminology.

Theoretical inspirations

In this volume, theoretical contributions will draw on insights from feminist influenced scholarship including intersectionalities theorising, eco-feminism and environmental jurisprudence. Kimberlé Crenshaw (1989) used the term 'intersectionality' to illustrate the interplay between law, race and racial power, and in particular to describe the multidimensionality of the Black woman's experiences. Intersectionality describes how an individual's different identities – such as class, race and gender – intersect and are interdependent systems of discrimination or disadvantage. Class-race-gender has become a shorthand expression for theorising social structure, class-race-gender and other intersections, and for linking social movement politics and struggles with changed consciousness and analysis of structure and process (Daly, 1993). Kathleen Daly later extended Candace West and colleagues' ideas about 'doing-gender' to 'doing race' and 'class' (Daly, 1994, 1997) so that a new understanding of 'difference' is that it is viewed as an ongoing interactional accomplishment (Daly, 1997). Race and class are thus similarly ongoing methodical and situated accomplishments (West and Zimmermann, 1987; Lorber, 1994; Simpson and Elis, 1996; West and Fenstermaker, 1995). While a range of feminist voices have informed the study of crime and victimisation, socialist feminism in particular dwells on the interplay between patriarchy and capitalism, insisting that the intersectionalities of class-race-sex-gender-age be accounted for in the search for social justice (Davies, 2007). Intersectionalities thinking has permeated feminist criminology and, together with the influential work of Connell (1995) and Messerschmidt (1995, 1997) – who is credited with introducing the construct of doing gender into criminology – these theoretical developments and trajectories are expanding to contend with the agendas of green criminology.

Several of the chapters in the volume align with these useful and sophisticated theoretical formulations and constructs. Early in the volume

Pamela Davies elaborates on the origins and influences of intersectionalities thinking as she expands on the broader historical and theoretical trajectories informing our project of gendering green criminology in her chapter entitled 'Eco-Feminism and the Gendering Green Criminology Project'. The theoretical formulations discussed in that chapter and other more recent innovations that are subsumed within green criminology warrant wider exposure, moving these theoretical insights from the periphery to the mainstream. While Davies's chapter specifically addresses some of these important developments the chapter does not exhaust the theoretical insights that are emerging. Other chapters in the volume variously illustrate how new empirical evidence is being revealed from research that builds on these theoretical traditions. Theory building is evident and refined theoretical formulations are emerging as we have witnessed in our editorial journey on this project.

Ambitions of the volume

As we note in our introductory paragraph, our ambition for this book is to bring together feminist and green criminology for the first time in a scholarly volume where all contributions are devoted to the project of gendering green criminology. The original research that will be reported on will of course foreground gender, and this will feature headline catastrophe violence as well as everyday, systematic, systemic and structural violence, abuse and harms. The gendered nature of resilience and recovery struggles as well as the implications of this in terms of promoting sustainable pathways to change policy and practices are part of this project of drawing together feminist and green criminology. The book as a whole offers fresh theoretical insights emerging from this merger and presents new findings derived from scholarship that has been advancing since the mid 1990s, and from more recent PhD and early career research inquiries. All these new developments are brought together in this volume to illustrate the journey of green criminology and its exposure of environmental harms and criminality through a gendered lens. We are proud to present research from scholars based around the world, making this volume a benchmark text for this area of study around the globe.

The contents

Following on from this introductory chapter the book is sub-divided into three parts. In some respects, our decision to divide the book into parts is a heuristic organising device. However, we feel that there is sufficient commonality of content to warrant these three umbrella headings. Part I foregrounds five chapters (Chapters 2–6) that variously illustrate the gendered nature of green crimes and environmental harm. Each chapter goes beyond this remit and there are features in each contribution that inevitably fit equally well under one or more of the other parts of the book. Part II features a run of five chapters under the heading 'Gendered Impacts and Victimisation'. Part III of the volume is entitled 'Resistance'; this part of the book includes four chapters which in the main foreground the gendered nature of resilience and recovery struggles, and the implications of this in terms of promoting sustainable pathways to change policy and practice.

Part I: Gendered nature of green crimes and environmental harm

Part I begins with Chapter 2, 'Eco-Feminism and the Gendering Green Criminology Project', where Pamela Davies expands upon the theoretical aspirations of the introduction. Foregrounding ecological (eco-)feminism, the chapter assesses the extent to which green criminology is gendered by reviewing contemporary scholarship in the discipline. Considering the enduring strengths, critiques and limitations of eco-feminism throughout, the chapter develops a framework for embedding a gendered approach to the green criminological and victimological project into the future. Chapter 3, 'New Directions Please! Veganising Green Criminology' calls on criminology to move away from its anthropocentric focus and argues for green criminology to take an ethical veganism approach. Here, Kay Peggs advocates opposition to speciesism and the commodity status of nonhuman animals. In so doing, she advances our understanding of what gendering green criminology could entail, by drawing on early feminist criminological thought and by embracing a non-speciesist intersectional approach. In Chapter 4, 'Men and the Climate Crisis: Why Masculinities Matter for Green Criminology', Stephen R. Burrell draws attention to the disproportionate role of men and masculinities in contributing to the climate crisis. The chapter illustrates why the study of gender and masculinities should be central to green criminology more broadly, and to efforts aimed at understanding and tackling the climate crisis more specifically.

Chapter 5 sees Helen U. Agu, Josiah C. Ogbuka and Meredith L. Gore consider the gendered dimension of the illegal wildlife trade and trafficking in Nigeria. Drawing on a critical review of existing literature, and interviews with wildlife conservationists in the public and private spheres, this chapter considers the gendered nature of this crime. The chapter assesses the effectiveness of wildlife policy and legal frameworks in achieving a sustainable wildlife economy. In Chapter 6, 'The Attitudes of People with Different Gender Identities and Different Perceptions of Gender Roles towards Non-human Animals and their Welfare', Aphra Hope-Forest, Ekaterina Gladkova and Tanya Wyatt explore the attitudes of people with a range of gender identities – and different perceptions of gender roles – towards nonhuman animals. Moving beyond binary gender investigations, the inclusive approach adopted here contributes a detailed understanding of gender identity and gender roles-related attitudes towards nonhuman animal welfare.

Part II: Gendered impacts and victimisation

The second part of the book begins with Chapter 7, 'Queering Green Criminology: The Impacts of Zoonotic Diseases on the LGBTQ Community'. In this chapter Laurence Pedroni and Benja Kromash use an approach informed by eco-feminism and necrocapitalism to analyse how green crimes disproportionately harm members of LGBT+[1] communities. Drawing links between structures of patriarchal domination, pandemics and the state, this chapter calls for an understanding of environmental justice informed by eco-feminism to counter the subjugation of nature and LGBT+ communities. In Chapter 8, 'Women and the Structural Violence of "Fast-Fashion" Global Production: Victimisation, Poorcide and Environmental Harms', Sandya Hewamanne and Nigel South explore how private corporations and the demand for cheap goods by Western consumers results in environmental damage that furthers the oppression and marginalisation of women in the Global South. Focusing on global assembly line workers in Bangladesh and Sri Lanka, this chapter argues for the foregrounding of gender in policies aimed at preventing or reversing environmental damage, and for a feminist political economy of environmental damage and victimology.

María-Ángeles Fuentes-Loureiro uses a gender perspective to examine green victimisation in the international waste industry in Chapter 9, 'Green Victims of the International Waste Industry'. Through an analysis of secondary sources, this chapter explores various environmental and health effects of the industry – including cancers and infertility – to consider its differential impact on men and women. In Chapter 10, 'The Green Road Project and Women's Green Victimisation in Turkey', Halil Ibrahim Bahar uses the treadmill of production theory to examine the green victimisation of women in the East Black Sea Region of Turkey. Using data from online ethnographic research, this chapter argues that plans to build a 1,600-mile 'green highway' will cause a myriad of environmental consequences that will disproportionately effect women. Drawing attention to patriarchal power structures, the chapter highlights the role of regional and national resistance networks in opposing such projects and challenging gendered patterns of victimisation. In Chapter 11, '"Daughters of Dust": An Eco-Feminist Analysis of Debt-for-Nature Swaps and Underage Marriage in Indonesia', Delon Alain Omrow uses eco-feminism to explore the parallels between debt-for-nature swaps and 'money marriages' in African countries. Drawing attention to an 'androcentric-anthropocentric symbiosis of oppression' this chapter explores the 'logic of domination' at the core of these activities.

Part III: Resistance

The final part of the book begins with Chapter 12, 'Women's Experiences of Environmental Harm in Colombia: Learning from Black, Decolonial and Indigenous Communitarian Feminisms'. Here, Daniela Suárez Vargas and Rachel Killean bring green criminology into conversation with decolonial feminism to explore women's diverse and nuanced experiences of environmental crime in Colombia. Resisting framings of Colombian women as agentless victims in need of saving, this chapter highlights the phenomenon of 'complex' victimhood while exploring women's use of agency and resistance in the face of environmental and gendered harms. In Chapter 13, 'Vegan Feminism Then and Now: Women's Resistance to Legalised Speciesism Across Three Waves of Activism', Corey Lee Wrenn and Lynda M. Korimboccus examine the history of women's activism in solidarity with other nonhuman animals. They do so by spotlighting the efforts of three women across the three waves of vegan feminism in the West, exploring how the centuries-old campaign to liberate nonhuman animals has been buoyed by several generations of female activists.

In Chapter 14, '"To Preserve and Promote": Gendering Harm in Green Cultural Criminology', Angeline Marie Letourneau examines how gendered representations of resource development and environmental resistance in Alberta, Canada, affect individual and collective behaviour, and how these characterisations shape attitudes and policies. Working in the tradition of cultural criminology, this chapter goes on to explore how a critical gender approach can illuminate and challenge the hegemony of resource development and fossil fuels both culturally and politically. The final chapter (Chapter 15), 'David and Goliath: Exploring the Male Burdens of Patriarchal Capitalism', differentiates between those fighting against the structural power and institutional weight of patriarchal capitalism, the 'Davids', and those at the helm of the existing global capitalism power structure, the 'Goliaths'. In doing so, Rob White explores the structural and oppressive burdens experienced by men in combating ecological destruction and, in particular, contesting the pressures and limits of patriarchal capitalism. We decided to close the volume with this chapter because discussions about gender often use 'gender' as an abbreviation for 'women'. Including men as a focus of the discussion raises issues about gender and power and, as White argues, in order to address green harms and crimes we need more men who are 'fighting against the structural power and institutional weight of patriarchal capitalism' (White, Chapter 15, this volume).

Taken together, these chapters seek to explore and clarify the value of bringing gender into green criminology. We hope the volume will lead to further discussion in the field, and offer a corrective to the relative inattention given towards gendered aspects of environmental harms and crimes.

Note

[1] We have purposefully not standardised the use of this term across the chapters as we wish to reflect the political intentions of the chapter authors.

References

Agnew, R. (2013) The ordinary acts that contribute to ecocide: A criminological analysis. In N. South and A. Brisman (eds) *Routledge International Handbook of Green Criminology*. London: Routledge, pp 58–72.

Anderson, E.A. (1976) The 'chivalrous' treatment of the female offender in the arms of the criminal justice system: A review of the literature. *Social Problems*, 23(3), 350–7.

Beirne, P. (2014) Theriocide: Naming animal killing. *International Journal for Crime, Justice and Social Democracy*, 3(2), 49–66.

Beirne, P. (2020) Animals, women and terms of abuse: Towards a cultural etymology of con(e)y, cunny, cunt and c★nt. *Critical Criminology*, 28, 327–49.

Brisman, A. (2014) Of theory and meaning in green criminology. *International Journal for Crime, Justice and Social Democracy*, 3(2), 21–34.

Chesney-Lind, M. (1986) 'Women and crime': The female offender. *Signs*, 12(1), 78–96.

Cohen, S. (1998) *Against Criminology*. New Brunswick: Transaction Publishers.

Connell, R.W. (1987) *Gender and Power*. Cambridge: Polity.

Connell, R.W. (1995) *Masculinities*. Berkeley: University of California Press.

Crenshaw, K. (1989) Demarginalizing the intersection of race and sex: A Black feminist critique of antidiscrimination doctrine, feminist theory and antiracist politics. *University of Chicago Legal Forum*, 8, 139–67.

Crook, M., Short, D. and South, N. (2018) Ecocide, genocide, capitalism and colonialism: Consequences for indigenous peoples and glocal ecosystems environments. *Theoretical Criminology*, 22(3), 298–317.

Daly, K. (1993) Class-race-gender: Sloganeering in search of meaning. *Social Justice*, 20(1–2), 56–71.

Daly, K. (1994) *Gender, Crime, and Punishment*. London: Yale University Press.

Daly, K. (1997) Different ways of conceptualising sex/gender in feminist theory and their implications for criminology. *Theoretical Criminology*, 1(1), 25–51.

Davies, P. (2007) Lessons from the gender agenda. In S. Walklate (ed) *Handbook on Victims and Victimology*. Cullompton: Willan Publishing, pp 175–201.

De Beauvoir, S. (1949) *The Second Sex*. Harmondsworth: Penguin.

Dobash, R.E. and Dobash, R. (1979) *Violence against Wives: A Case against the Patriarchy*. New York: Free Press.

Gladkova, E. (2020) Farming intensification and environmental justice in Northern Ireland. *Critical Criminology*, 28(3), 445–61.

Grosz, E. (1987) Feminist theory and the challenge to knowledge. *Women's Studies International Forum*, 10(5), 208–17.

Hall, M. (2014) Environmental harm and environmental victims: Scoping out a 'green victimology'. *International Review of Victimology*, 20(1), 129–43.

Hunnicutt, G. (2019) *Gender Violence in Ecofeminist Perspective*. London: Routledge.

Kelly, L. (1988) *Surviving Sexual Violence*. Cambridge: Polity.

Lorber, J. (1994) *Paradoxes of Gender*. New Haven: Yale University Press.

Lynch, M.J. (2018) Acknowledging female victims of green crimes: Environmental exposure of women to industrial pollutants. *Feminist Criminology*, 13(4), 404–27.

Lynch, M., Long, M., Barrett, K. and Stretesky, P. (2013) Is it a crime to produce ecological disorganization? Why green criminology and political economy matter in the analysis of global ecological harms. *The British Journal of Criminology*, 53(6), 997–1016.

MacKinnon, C. (1987) *Feminism Unmodified: Discourses on Life and Law*. Cambridge, MA: Harvard University Press.

Maher, J. (2021) Documenting harm to the voiceless: Research on animal welfare. In P. Davies, P. Leighton and T. Wyatt (eds) *The Palgrave Handbook of Social Harm*. London: Palgrave Macmillan, pp 167–95.

Messerschmidt, J.W. (1993) *Masculinities and Crime: Critique and Reconceptualization of Theory*. Lanham: Rowman & Littlefield.

Messerschmidt, J.W. (1995) From patriarchy to gender: Feminist theory, criminology and the challenge of diversity. In N. Rafter and F. Heidensohn (eds) *International Feminist Perspectives in Criminology*. Maidenhead: Open University Press, pp 167–88.

Messerschmidt, J.W. (1997) *Crime as Structured Action*. Thousand Oaks: SAGE.

Oakley, A. (2015 [1972]) *Sex, Gender and Society*. Farnham: Ashgate.

Police Foundation (2022) *A New Mode of Protection: Redesigning Policing and Public Safety for the 21st Century*. The Police Foundation. Available at: www.policingreview.org.uk/wp-content/uploads/srpew_final_report.pdf (accessed 21 October 2022).

Pollak, O. (1961) *The Criminality of Women*. Philadelphia: University of Pennsylvania Press.

Russell, D.E.H. (1975) *The Politics of Rape, the Victim's Perspective*. New York: Stein and Day.

Simpson, S.S. and Elis, L. (1996) Theoretical perspectives on the corporate victimisation of women. In E. Szockyi and J.G. Fox (eds) *Corporate Victimisation of Women*. Boston: Northeastern University Press, pp 32–58.

Smart, C. (1977) *Women, Crime and Criminology: A Feminist Critique*. London: Routledge and Kegan Paul.

Smith, C. (2005) Gender and crime. In C. Hale, K. Hayward, A. Wahidin and E. Wincup (eds) *Criminology*. Oxford: Oxford University Press, pp 345–66.

Sollund, R. (2012) Speciesism as doxic practice versus valuing difference and plurality. In R. Ellefsen, R.A. Sollund and G. Larsen (eds) *Eco-global Crimes: Contemporary Problems and Future Challenges*. Farnham: Ashgate, pp 91–113.

Sollund, R. (2020) Wildlife crime: A crime of hegemonic masculinity? *Social Sciences*, 9(6), 93–109.

Stanko, E.A. (1990) *Everyday Violence: How Women and Men Experience Sexual and Physical Danger*. London: Pandora.

Taylor, N. and Fitzgerald, A. (2018) Understanding animal (ab)use: Green criminological contributions, missed opportunities and a way forward. *Theoretical Criminology*, 22(3), 402–25.

Torpman, O. and Röcklinsberg, H. (2021) Reinterpreting the SDGs: Taking animals into direct consideration. *Sustainability*, 13(2), 1–11.

UN Women (2022) Explainer: Why women need to be at the heart of climate action. *UN Women*, 1 March. Available at: www.unwomen.org/en/news-stories/explainer/2022/03/explainer-why-women-need-to-be-at-the-heart-of-climate-action (accessed 19 October 2022).

West, C. and Fenstermaker, S. (1995) Doing difference. *Gender and Society*, 9(1), 8–37.

West, C. and Zimmermann, D.H. (1987) Doing gender. *Gender and Society*, 1(2), 125–51.

White, R. (2012) *Climate Change from a Criminological Perspective*. New York: Springer.

PART I

Gendered Nature of Green Crimes and Environmental Harm

2

Eco-Feminism and the Gendering Green Criminology Project

Pamela Davies

Introduction

This chapter follows up on some of the observations made in the introduction to this volume where we briefly reflect on contributions to theoretical approaches notably by eco-feminists. The chapter focuses very specifically on these latter contributions and in the following pages I seek to advance and develop our theoretical and conceptual understanding of green criminology using eco-feminism as the springboard and benchmark for assessing the extent to which green criminology is gendered and for developing a framework for embedding a gendered approach to the green criminological and victimological project. Part of the task that is set for this chapter, therefore, is to help with the review of determining what has been achieved so far, and part is about stepping up and forward to thoroughly gender the green criminological project drawing on appropriate theoretical traditions and trajectories.

Ecological (eco-)feminism holds out for a gender-sensitive form of justice into the 21st century. In lay terms, eco-feminism is a feminist approach to analysing the relationships between humans and the natural world. The foundation of eco-feminism is the relationship between women, the earth and environmentalism. If we take this as the starting point, we can begin to ask some important questions about green criminology. A modern definition of *eco-justice* is one that encompasses environmental justice (relating to humans and their relationship with specific environments), ecological justice (referring primarily to the health and wellbeing of eco-systems) and species justice (involving rights and biodiversity concerns as these pertain to nonhuman animals and plants) (White, 2013). This definition is also a useful starting point, and conjoining this with our interest in exploring eco-feminism we can illuminate some important features about the state of green criminology knowledge more broadly. For example, we can expose what we know and do not know about who and what perpetrates crime and harm and who and what can be and are victims. The latter thread – the victimological – has been especially underdeveloped and until very recently knowledge of the gendered nature of victimisation has been scant. Broader questions about the nature and extent of victimisation have, however, surfaced such that scholars are now proposing that victimisation extends beyond humans to nonhuman species, that inanimate objects can be a victim, that multi- and poly-species victimisation is experienced. In this volume, most are persuaded that there is no objective standard of victimisation at the same time as we argue there is a gender bias to these questions of who, what and where victimisation occurs. All of these questions have a global dimension to them and the 'where' question certainly transcends cartographical boundaries that are place- or land-centric. These are fundamental questions about crime and victimisation in the 21st century that demand ever more sophisticated and sensitive theoretical formulations to aid understanding and to ensure policy and practice development make good sense and are likely to have impact.

As the editors also note in the introduction to this volume, it is an unfortunate aspect of some of the gender work that has previously been completed in green criminology (and other areas of criminology) that it does not draw more thoroughly on the rich theoretical work of feminist criminology. Consequently, the gender analysis in some of the existing work is not, we feel, as rich and strong as it could be. While we do not propose that eco-feminism is the only lens through which gender and green/environmental crime/harm can be or has been viewed, there is merit in pausing to dwell on eco-feminism in the way proposed in this chapter. The chapter is therefore organised around the theoretical contribution of eco-feminism.

The chapter begins in earnest by stressing the import of robust theoretical foundations to scholarship in criminology and in green criminology in particular moving into a fuller explication of the hallmarks of eco-feminism in the first substantive section of the chapter. Here, key contributions from founding members of the eco-feminist approach are outlined and the various trajectories of their arguments are considered. In doing so, the chapter tunes in to the interesting history of eco-feminism as well as insights from more recent environmental justice/environmental racism contributions from outside of criminology all under the heading 'eco-feminism as a benchmark'. The chapter then examines contemporary scholarship within green criminology and identifies the extent, variety and strength of the eco-feminist theoretical underpinnings to that work. The third part of the chapter considers eco-feminism and intersectionalities. Overall the chapter considers the enduring strength of eco-feminism as well as critiques and limitations of it, reflecting on why the influence of gendered thought is not more embedded in green criminology. The chapter draws to a close about our learning from and beyond eco-feminism.

A note on the importance of theory

Ideologies and their respective accompanying theoretical perspectives capture our beliefs about how social change might be brought about. Our worldview is formed according to how our environmental, social, economic and political beliefs come together. In criminology there has tended to be a very select range of ideologies dominating the agenda, several of which, as illustrated in this volume, are problematic in terms of furthering the aims of the green criminological project. The 'good theory' test demands that for maximum effectiveness, theories must make sense (logical consistency), explain as much crime as possible (scope) and be as concise as possible (parsimony). Most importantly, the theory must be true or correct (validity) (Akers, 2012). The merits and hallmarks of a good theory also include the ability to provide a novel interpretation to an old problem and will attend to both epistemological and methodological considerations (Kings, 2017).

I have recently argued the case that ideology and theory – and indeed ethics (following Raymen, 2021) – matter in the context of social harm and victimisation (Davies, 2021), suggesting an ideology that is ethically wedded to human and nonhuman species and environmental betterment be embedded in a sustainable and future proofed criminology. Such theory can satisfy the demands of academia for sound scholarship based on sound theorising that will have real-world applicability and utility with the policy pathways illuminated by it leading to sustainable and harmonious planetary living. In the remainder of this chapter, it is argued that, in order to gender and further the green criminological project, we may draw more systematically on an intellectual tradition that has thus far been marginalised in academic criminology.

Eco-feminism as a benchmark

Having reminded ourselves of the import of robust theoretical foundations to scholarship in criminology and, in suggesting ways of challenging dominant discourse by turning to strong traditions of scholarship that exist in other disciplines, I now consider one such neglected intellectual tradition and delve into a fuller explication of the hallmarks and history of eco-feminism. This leads into discussion of eco-feminism and its relationship to and impact on theorising environmental, ecological and species justice and environmental racism.

The founders, hallmarks and trajectories of eco-feminism

Eco-feminism – a term credited to Françoise d'Eaubonne – entered our vocabulary in the early 1970s as feminist, peace, health and ecology movements found common ground. Eco-feminism emerged as a form of political activism and an academic critique of society–nature relations and an ideological prescription for a more harmonious relationship between society and nonhuman nature. A range of different eco-feminist positions within this normative philosophy for saving planet Earth unfolded. Although lacking a uniform trajectory, the contributions are more or less all on a continuum of thought which opposes dominant Western scientific, dualistic philosophising and critiques the twin oppressions of the domination of women and nature. Some key hallmarks of eco-feminism began to emerge in scholarly English language publications in earnest throughout the 1980s as eco-feminists articulated their arguments in a range of book-length monographs sponsored by UK and US publishers, chapters in edited volumes solely on the subject of eco-feminism and in journal articles where authors heralded from, and readership was steered towards, philosophers, theologians, ethicists, sociologists and social and cultural scientists. Among the most recognised English language scholars and most widely cited are the works of Carol Adams (1993; Adams and Gruen, 2021), Françoise d'Eaubonne (1980), Susan Griffin (1978, 1989), Petra Kelly (1984, 1988, 1994), Ynestra King (1989, 1990), Carolyn Merchant (1980, 1990, 1992, 1996), Mary Mellor (1992a, 1992b, 1992c, 1997), Val Plumwood (1986), Ariel Salleh (1982, 1984, 1995, 1997) and Vandana Shiva (1989; Mies and Shiva, 1993), among others.

It is never wise and rarely possible to reduce grand, ambitious and rich theory-building traditions to a short descriptive outline and being selective in picking out key founders to the eco-feminist approach will inevitably mean there are omissions. However, in order to elaborate on the brief definition of eco-feminism provided in the introduction to this chapter most of the authors' work referred to in the previous paragraph will be drawn on. Many of those cited mobilised as activists and academics. In this chapter, the emphasis is on their documented theoretical insights. To launch in to a deeper illustration of what eco-feminism is and what it addresses as a movement, a direct quote is presented from the preface of Mary Mellor's *Feminism & Ecology* (1997). Later in the chapter a second direct quote is provided from the first page of Adams and Gruen's second edition (2021) of *Ecofeminism Feminist Intersections with Other Animals and the Earth*, and some comparisons are made in the wake of a three-decade time lapse in the scholarship. In 1997, by which time eco-feminism was already very much established as a body of theoretically important work, Mellor held that:

> Eco-feminism brings together the analysis of the ecological consequences of human 'progress' from the green movement, and the feminist critique of women's disproportionate responsibility for the costs and consequences of human embodiment, to show how relations of inequality within the human community are reflected in destructive relations between humanity and the non-human world. (Mellor, 1997: viii)

In her first book, Mellor (1992a) had clearly articulated the fundamental shift required for a sustainable nature stewardship future from men to women, from rich to poor, from North to South, from nature exploitation to nature stewardship. At this point in time Mellor (1992b, 1992c) had also already acknowledged the dilemmas of essentialism and materialism thrown up by the politics of eco-feminism. Those who hold the view that women have different essential nature than men align to essentialist positions. The dilemma of materialism refers to women's relationship to social and physical materialist production. These dilemmas tended to separate two broad schools of thought within eco-feminism. Cultural eco-feminists valorise and celebrate women and nature. They analyse the environmental problems from within their critique of patriarchy. Socialist eco-feminists (eco-socialists) analyse and address the dialectic contradiction between production and reproduction. Mellor's position is aligned to the latter where she uses the concept of human embodiment and embeddedness to illustrate women's dilemma of materialism in a sexed and gendered society. Mellor resists categorising contributions to eco-feminism (1997), preferring instead to explore themes evident within its broad church. Nevertheless, she does articulate two shades to green thinking. One incorporates 'light green' or 'shallow ecology' thinkers who propose humanity can use its technological ingenuity to overcome or adapt to the bounding conditions imposed by ecosystems. The other is 'dark green' or 'deep ecology' thinking where humanity must fundamentally rethink its relationship to the natural world. Mies and Shiva's (1993) position, which is especially sensitive to rural women's lives, particularly Indian women and the violence of ethnic chauvinisms, typifies the latter tradition in an appeal for a collective resistance and shift to earth-centred politics and economics (Shiva, 1989). Light/shallow and dark/deep ecology represent extremes within the spectrum of eco-feminist thought that Hallen (1994: 207) suggests offers ten distinct approaches (liberal eco-feminists, radical eco-feminists, cultural eco-feminists, social eco-feminists, socialist eco-feminists, ecological eco-feminists, deep ecological eco-feminists, critical or transformative eco-feminists, aboriginal or native eco-feminists, eco-feminists of the Third World), all of which are alluded to within this depiction of the hallmarks of eco-feminism.

Plumwood's (1986) overview and discussion of positions and arguments relating to eco-feminism remains a useful touchstone. She classified proponents of eco-feminism into three groups. One group problematises a set of sexist dualisms concerning women and nature. The subjugation of women and sexism tends to be the spur to the unfolding arguments and Ruether (1975, 1989) is often cited in respect of solutions to gender inequalities and ecological degradation. In 1989, Plant and Collard separately made their own mark on our understanding of eco-feminism and the pain and death of animals in their examinations of the various interpretations of the emergence of the nature/culture dualism and how the dominant masculine imprint towards nature led to pollution, exploitation and extinction. Merchant's (1980, 1990, 1992, 1996) position is not dissimilar to that of Ynestra King (1989, 1990) who famously articulates the link between the eco-feminist movement and radical political activism. Petra Kelly is the obvious candidate to illustrate the latter point (Kelly, 1984, 1988, 1994). Kelly wrote about the interrelationship between feminism and ecology and her work featured cultural and spiritual overtones. As an interesting aside, Kelly was a pioneer founder of the West German Green Party, instrumental in placing feminism at the heart of green politics, at least temporarily in the 1980s. Illustrative of a second group who explore the convergence of a feminist and ecological metaphysics, Merchant explains her own environmental ethic as a partnership that goes beyond egocentric and homocentric ethics. Her partnership ethic of earthcare 'treats humans (including male partners and female partners) as equals in personal, household, and political relations and humans as equal partners with (rather than controlled-by or dominant-over) nonhuman nature' (Merchant, 1996: 8). A third set of scholarship gives prominence to 'difference' and different experiences among women. Salleh (1982, 1984), for example, initially focused on challenging the polarity of the masculine dualism theses and argued for a privileged epistemological perspective for women that simultaneously acknowledges class and race dimensions. Salleh's (1995, 1997) resistance to deep ecology lies in its uncritical acceptance of male sexist values and the exploitative social relations that flow between men and women prompting her towards a more critical neo-Marxist form of eco-feminism which foregrounds material oppression in her later work.

Two unique humanitarian/emancipation eco-feminist approaches complete Hallen's (1994) list of ten varieties of eco-feminism: Aboriginal or native eco-feminists who live close to nature, nurturing sacred lands and reconsecrating degraded spaces and eco-feminists of the 'Third World' who criticise maldevelopment in the 'First World' and show us how women of colour may be in a privileged position because their minds are not yet colonised and because they do not profit from the oppression of others (Hallen, 1994: 207).

Mellor and Salleh in particular developed debates within eco-feminism into the 1990s and, in their ever-sophisticated refinements and articulations of their position, they find common ground. Space precludes a deep and thorough exposition of each of these rich and important bodies of work. This brief overview serves only to illustrate that in the early proliferation of ideas around eco-feminism there were variations on the theme, often all being 'lumped together under the "ecofeminism" label' (Plumwood, 1986: 120), yet all of which heralded different promises (Plant, 1989). Plumwood remained optimistic about the eco-feminist perspective, proclaiming in 1986 that eco-feminism was the most exciting and potentially radical set of ideas. Plumwood's own position emerged more clearly in 1993 and veered towards the radical end of the eco-feminist spectrum in also drawing on critical theory. She drew on the concepts of the relational and ecological 'self' to move beyond dualistic logic recognising both difference and continuity between people and nature and between self and others.

The story of eco-feminism continued to develop and thrive and there remains much evidence of this healthy thriving in several disciplinary areas much of which has overt crime, harm and victimisation and thus criminological significance (note the activist work of Carol Adams in the 1980s challenging racism, domestic violence and violence to animals and in later publications [Adams, 1994; Adams and Fortune, 1995]) though within criminology the embracing of eco-feminist approaches is not so apparent. The seeping of eco-feminist thought is latent within green criminology. This is discussed further in the second part of the chapter.

As promised earlier, a second direct quote is now introduced from a more recent articulation of the aims of eco-feminism. Adams and Gruen (2021: 1) propose that '[e]cofeminism addresses the various ways that sexism, heteronormativity, racism, colonialism, and ableism are informed by and support speciesism and how analysing the ways these forces intersect can produce less violent, more just practices'. There are similarities and subtle but important differences between this articulation of eco-feminism and that quoted from Mellor (1997). Two subtle differences are no doubt due to the quarter of a century passage of time between 1997 and 2021. One difference is that the second quote makes it apparent that intersectionalities thinking has permeated feminist approaches to social problems. A second difference is that the more recent articulation is more specific with reference to speciesism. Although the impacts of harms on nonhuman species are implicit in both quotes, the second flags the problem of binarism in this regard. That said, the overall ambition evident in both articulations is remarkably similar, showing that the original formulation has stood the test of time.

Eco-feminism, green criminology and green victimology

I have suggested that the seeping of eco-feminist approaches into green criminology is latent. In this second part of the chapter, I examine contemporary green scholarship and evidence of eco-feminist influences. Thus, here I begin to unpick the submerged and potential connectivity between eco-feminism and green criminology. In order to do so, it is useful to outline some additional concepts including green harms and prominent scholarship workstreams within green criminology and victimology. Efforts to locate women in the environmental debate – eco-feminist contributions – are noted as this unfolds.

As is evident throughout this volume, green criminology often uses the shorthand term 'green harms'. Green harms include legal as well as illegal environmental harms. In acknowledging the former – legal environmental harms – green criminology is keenly aware that some of the most ecologically harmful and destructive practices (such as clear felling of old growth forests) may be entirely legal (Vegh Weis and White, 2020). Other examples of green harm include harms against human and nonhuman bodies such as animal rights and animal welfare abuses including those arising from industrial and unhygienic farming and concentrated feeding of livestock – cows, pigs, sheep – and poultry. State-corporate land grabs, through to the victimisation of Indigenous peoples, environmental injustices and the consequences wrought by extractive industries (Davies and Wyatt, 2021) are other examples where often the harm is against the non-speaking (animals, plants, soil, rocks, minerals, water, air) or those rendered voiceless (oppressed and supressed cultures and communities) (Davies and Wyatt, 2021). For many criminologists, the starting point for their inquiries is crime, indeed as already noted, this approach is a dominant tradition in criminology. An alternative starting point identified by Hillyard and colleagues (Hillyard et al, 2004; Hillyard and Tombs, 2004, 2021) is to use 'harm' as the starting point for alternative narratives that capture a fuller range of harmful practices and experiences to the narrow ones found within criminology. As Leighton and Wyatt (2021: 2) have observed, 'crime as predominantly a social construct of the powerful creates its own harms that are largely discriminatory towards certain human populations'. Such thinking is in the tradition of a more critical approach to the study of criminology. A harms-based, or zemiological, perspective argues that much harm is equally destructive and violent and sometimes fatally harmful and seriously maims (Box, 1983). Thus non-criminal, legal harms feature in the research endeavours of scholars wedded to harms-based inquiry (see, for example, Hillyard and Tombs, 2004; Pemberton, 2015; Kotze, 2018; Lloyd, 2018, 2021; Stephens-Griffin and Griffin, 2021; Tombs, 2022). Few of those just cited would consider themselves green criminologists. Much green criminology scholarship, however, does follow in this tradition where an expansive definition of *crime* is adopted. Hence the concept of 'green harms' is often employed. Pockets of green criminology dwell not only on the harms and destructions that are *perpetrated* but also on the nature, extent and experience of *victimisation*.

Following in the tradition of the social harm approach, Flynn and Hall (2017) have proposed the case for a victimology of nonhuman animal harms. Their argument for a nonhuman animal victimology arises out of their previous separate endeavours to pioneer the case for a victimological perspective in environmental harm (Hall, 2013, 2014) and a social harm lens for considering environmental harm and wildlife trafficking in particular (Wellsmith, 2010, 2012). Though their argument for a nonhuman animal victimology is compelling, at the point of their proposal there is no recognition of the gendered features and dynamics of such harms. Sollund (2012), on the other hand, has sought to understand crimes through an eco-feminist perspective and does so through a focus on the victimisation of women, children and nonhuman species through trafficking and trade.

Accepting there are both criminological and victimological dimensions to green criminology, what are and where are the traces of eco-feminism? How have women been located within the environmental debate? Within green criminology, there are pockets of inspiration in this regard, many of whom are contributors to this volume who cite the work of like-minded colleagues' work too. Thus, the body of work that dwells on the harms and destructions that are perpetrated is complemented by an eco-justice perspective that foregrounds victimhood. An eco-justice perspective frames victimhood in terms of the particular subject or object that is harmed (White, 2013). Nonhuman environmental entities, for example, include rivers, mountains, birds, flowers, forests, and so on. They also include humans, specific ecosystems and/or animals and/or plants, all of which endure suffering (Vegh–Weis and White, 2020). The scholars referred to in this paragraph are illustrative of a small caucus of green criminologists who may yet emerge as pioneers of the eco-feminist influence in our discipline.

As noted in the introduction to this chapter, some gender work evident in green criminology (and other areas of criminology) fails to adequately draw on the rich theoretical work within feminist criminology or from eco-feminism. While it is disappointing that it may still be necessary to remind one another, as scholars passing on new research discoveries and theory-building ourselves, that gender matters, it is not enough to simply add women into the mix to make up for the previous neglect of women in our research agenda. Much important work has materialised through applying a gender lens to the climate change debate. Within such work women have not only been revealed as highly vulnerable to and victims of climate change and environmental degradation, but also women are revealed as capable of contributing solutions that increase women's adaptive capacity (see, for example, the edited volume by Terry [2009] on climate change and gendered justice in the context of development studies, and the earlier works of Enarson and Morrow [1998] on the gendered terrain of disaster). Our gendering of the green criminology project calls for a research agenda that does this from a criminological starting point and from a platform of robust contemporary theorising such that our activism and academic approaches are in tune and become ever more sophisticated in preventing the need for such adaptive strategies. There are traces of such endeavours emerging within green criminology over the last 15 years (see, for example, Wachholz, 2007).

Eco-feminism, green criminology and activism

Before moving in to the third section of the chapter, and in light of the positive and theoretically promising pockets of scholarship alluded to earlier, which capitalise (albeit implicitly rather than explicitly) on the strong foundations of eco-feminism, I pause to flag and celebrate the impact of the activist and academic movements linked to eco-feminism and environmental concerns. In the same way that the early founders of eco-feminism were often joined at the hip, activist-academics, so too are several colleagues who work in the field of green criminology today.

Violence Against Women and Girls has long been recognised as a global issue albeit suffered in very local and often hidden places. Recently scholars have begun to argue that there is cause to be concerned about increased violence in response to the strains of living through climate change, global warming and environmental degradation. Such strains precipitate social conflict, irritability, anger, and aggression and violent crime including homicide and gendered violence such as rape and domestic abuse (Agnew, 2011; Heckenberg and Johnston, 2012).

While recognising that women play an important role in response to climate change, the United Nations acknowledges:

> Women commonly face higher risks and greater burdens from the impact of climate change in situations of poverty, and the majority of the world's poor are women. Women's unequal participation in decision-making processes and labour markets compound inequalities and often prevent women from fully contributing to climate-related planning, policy-making and implementation. (UNFCCC, nd)

The Intergovernmental Panel on Climate Change (IPCC, 2007) also acknowledge women's particular vulnerability to increased climate variability. Poor women in urban areas suffer disproportionate health problems due to heat effects, diseases and water shortages, whereas rural women in developing countries are noted as one of the most vulnerable groups (IPCC, 2007).

As noted earlier, eco-feminism surfaced as a body of work throughout the 1980s when its interdisciplinary foundations were evident. The tradition of crossing and pushing disciplinary boundaries continues to be evident in more recent searches of literature pertaining to gender, victimisation, environmental problems and climate change. These robust theoretical foundations developed in parallel with political activism undoubtedly contributed to the cited official recognitions of the stubborn problems of climate change, environmental degradation and violence against women and girls as intersecting social concerns.

Eco-feminism and intersectionalities

Power, the concept of intersectionality and the theoretical construct *class-race-gender* are significant formulations in the advancement of eco-feminism and potentially for gendering the green project. There are many different facets to the concept of power and criminologically the word conjures up several additional concepts, including control, dominance, rule, supremacy, and so forth. The construction of problems in society, representations of and the visibility of problems, solutions and responses to such problems are all mediated through a range of mechanisms, organisational, institutional and structural processes, cultural, relational and situational contexts. In several green crime and harm contexts, it is evident that gender matters when it comes to exposing the lived experience of suffering and harm, and the burden of reconstructing and repairing the damage and harm done. In the context of green crime, harm and victimisation, those contributing to this volume contend that gender ought to be front and centre stage – omnipresent – where we know the vectors of power intersect. The theoretical construct *intersectionalities* is identified by some as having the potential to harness the experiences and phenomenon that occur at these crossroads where various dimensions of power collide. The seeds of a critical race and intersectional theorising are evident in the early work of Carol Adams and the effects on women and the environment of the interconnectedness of race, class, gender, disability, sexuality, caste, religion and age have been foregrounded in the wake of eco-feminist and intersectionalities thinking emerging since the early 1990s.

A core component of critical race theory, the term intersectionality was coined by Kimberlé Crenshaw in 1989 in her exposé of the interplay between law, race and racial power in American society. She used the term to describe the multidimensionality of the Black woman's experience. Intersectionality describes how an individual's different identities – such as gender identity, race, class, and so on – intersect and overlap to create compound, interdependent systems of discrimination or disadvantage. For example, a white woman may experience misogyny and a Black man may experience racism, but a Black woman will experience both misogyny *and* racism. Gender and race are but two social categories which provide a basis for shared identity and shared experience including experiences of oppression and subjugation. Iris Marion Young (1997) argued that essential to redressing structural inequalities, and indeed what she explains as the five faces of oppression – exploitation, marginalisation, powerlessness, cultural domination, violence (Young, 2005) – is that we appreciate social group distinctions. Crenshaw and Young's thinking is representative in illustrating the interplay between class and race and how interlocking systems define women of colour's lives. Their work, and that of their contemporaries and countless other feminist thinkers in their wake, has prompted much theoretical development around the concept of intersectionality.

In bringing to a close this extended outline and critical review of the contribution of eco-feminist theorising, we have arrived at a position where the theoretical strength of eco-feminism is indebted to critical race theory and intersectionalities thinking for sound theory-building. According to Kings, eco-feminism is a continually evolving academic/activist tradition which has, since the late 1980s,

> been taking into account the interconnected nature of social categories such as gender, race, class, sexuality, caste, species, religion, nationality, dis/ability, and issues such as colonialism. It has also challenged anthropocentric modes of thought, by incorporating both species and the natural environment into the ongoing debate concerning the workings of social categorization and identity construction. (Kings, 2017: 71)

As the theoretical construct of intersectionalities moves firmly into the application phase, Kings's celebration of the enthusiastic adoption of intersectionalities in the feminist community concludes this warrants eco-feminism continuing to work within an intersectional framework. The reasons for this include the mass market appeal and intersectionalities' ability to provide a novel interpretation to an old problem as well as the steadfast consideration of epistemological and methodological matters. Kings also observes, however, that much environmental and climate research remains ignorant to issues of gender, class, race, caste and sexuality, in spite of research clearly demonstrating the increased vulnerability of women to environmental threats. Others are arriving at parallel conclusions regarding the value of intersectional approaches. Vegh Weis (2019), for example, examines how crime control in Argentina is being intensively used to the detriment of Indigenous peoples claiming their rights, which constitutes a form of *over-criminalisation*. Simultaneously, the criminal justice system is not being used to prosecute green harms perpetrated by corporations nor the unlawful use of force against native peoples by law enforcement agencies, a form of *under-criminalisation*. However, here too the gendered nature of these controls in the Argentinean criminal justice system which targets the most vulnerable peoples while failing to provide environmental protections is not foregrounded.

Conclusion

The questions posed in the introduction to this chapter were posed from a specific starting point – eco-feminism – and through a particular lens – criminological and victimological. It was noted that all of the questions posed have a global dimension to them and the 'where' question certainly transcends cartographical boundaries that are place- or land-centric. A thoroughly gendered green project automatically and always already recognises the gendered interplay between land, air, sea, space, the Anthropocene. In focusing on eco-feminism and the green criminology project, this chapter has traced our theoretical and conceptual understanding of green criminology using eco-feminism as the springboard and benchmark for assessing the extent to which green criminology is gendered and for developing a framework for embedding a gendered approach to the global green criminological and victimological project into the future. The chapter has outlined the hallmarks of eco-feminism, the significant contributions made by the founders of the theoretical arm of it and the main commonalities and points of divergence among these feminist approaches. The chapter has also traced insights from more recent environmental justice/environmental racism contributions from outside of criminology and contemporary scholarship within green criminology where the strength of the eco-feminist approach is embraced. In terms of the wider green project, however, it is important to put this applause of eco-feminist theorising achievement in perspective. Gendered thought has not percolated into the mainstream, and it is disappointing to the editorial and contributor team of this text that feminist-influenced thought has not been more prominent feature in green criminology. Adams and Gruen's (2021) recent second edited collection, *Ecofeminism Feminist Intersections with Other Animals and the Earth*, consisting of 13 chapters by philosophers, artists, activists, sociologists and political scientists, suggests eco-feminism may be making a comeback though the absence of contributions from those identifying as criminologists is noted and the readership and uptake of the ideas within that book by those within and studying criminology, green crime and harm is yet to be determined. To end on a more optimistic note, however, overall, in having considered the gendering of the green criminology project through the lens of eco-feminist approaches, the chapter ends by suggesting that eco-feminism continues to hold promise for a gender-sensitive form of justice into the 21st century.

References

Adams, C.J. (1993) *Ecofeminism and the Sacred*. New York: Continuum.
Adams, C.J. (1994) *Woman-Battering*. Minneapolis: Fortress Press.
Adams, C.J. and Fortune, M. (eds) (1995) *Violence Against Women and Children: A Christian Theological Sourcebook*. London: Bloomsbury.

Adams, C.J. and Gruen, L. (eds) (2021) *Ecofeminism Feminist Intersections with Other Animals and the Earth*, 2nd edn. London: Bloomsbury.

Agnew, R. (2011) Dire forecast: A theoretical model of the impact of climate change on crime. *Theoretical Criminology*, 16(1), 21–42.

Akers, R.L. (2012) *Criminological Theories*, 2nd edn. Abingdon: Routledge.

Box, S. (1983) *Power, Crime and Mystification*. London: Tavistock Publications.

Crenshaw, K. (1989) Demarginalizing the intersection of race and sex: A black feminist critique of antidiscrimination doctrine, feminist theory and antiracist politics. *The University of Chicago Legal Forum*, 140, 139–67.

d'Eaubonne, F. (1980) Le Feminisme ou le mort. In E. Marks and I. de Courtivron (eds) *New French Feminisms: An Anthology*. Amherst: University of Massachusetts Press, pp 64–7.

Davies, P. (2021) Why social harm matters: Five reasons from a feminist influenced perspective. In P. Davies, P. Leighton and T. Wyatt (eds) *The Palgrave Handbook of Social Harm*. London: Palgrave Macmillan, pp 453–73.

Davies, P. and Wyatt, T. (2021) *Crime and Power*. London: Palgrave Macmillan.

Enarson, E. and Morrow, B. (eds) (1998) *The Gendered Terrain of Disaster*. Westport: Praeger.

Flynn, M. and Hall, M.D. (2017) The case for a victimology of nonhuman animal harms. *Contemporary Justice Review*, 20(3), 299–318.

Griffin, S. (1978) *Women and Nature: The Roaring Inside Her*. New York: Harper & Row.

Griffin, S. (1989) Remembering who we are: The meaning of ecofeminism. In J. Plant (ed) *Healing the Wounds: The Promise of Ecofeminism*. London: Green Print, pp 7–15.

Hall, M. (2013) *Victims of Environmental Harm: Rights, Recognition and Redress under National and International Law*. London: Routledge.

Hall, M. (2014) Environmental harm and environmental victims: Scoping out a 'green victimology'. *International Review of Victimology*, 20(1), 129–43.

Hallen, P. (1994) Re-awakening the exotic: Why the conservation movement needs eco-feminism. *Habitat Australia: The Magazine of the Australian Conservation Foundation*, 22(1).

Heckenberg, D. and Johnston, I. (2012) Climate change, gender and natural disasters: Social differences and environment related victimisation. In R White (ed) *Climate Change from a Criminological Perspective*. New York: Springer, pp 149–71.

Hillyard, P. and Tombs, S. (2004a) Beyond criminology? In P. Hillyard, C. Pantazis, S. Tombs and D. Gordon (eds) *Beyond Criminology: Taking Harm Seriously*. London: Pluto Press, pp 10–29.

Hillyard, P. and Tombs, S. (2004b) Beyond criminology? In P. Davies, P. Leighton and T. Wyatt (2021) (eds) *The Palgrave Handbook of Social Harm*. London: Palgrave Macmillan, pp 11–36.

Hillyard, P. and Tombs, S. (2021) Beyond criminology? In P. Davies, P.S. Leighton and T. Wyatt (eds) *The Palgrave Handbook of Social Harm*. London: Palgrave Macmillan, pp 11–36.

Hillyard, P., Pantazis, C., Tombs, S. and Gordon, D. (eds) (2004) *Beyond Criminology: Taking Harm Seriously*. London: Pluto Press.

IPCC (2007) Synthesis Report. Contribution of Working Groups I, II and III to the Fourth Assessment Report of the Intergovernmental Panel on Climate Change Geneva: IPCC.

Kelly, P. (1984) *Fighting for Hope*. London: Chatto & Windus.

Kelly, P. (1988) Towards a green Europe and a green world. In F. Dodds (ed) *Into the 21st Century*. London: Green Print, pp 39–50.

Kelly, P. (1994) *Thinking Green!* Berkeley: Parallax Press.

King, Y. (1989) The Ecology of Feminism and the Feminism of Ecology. In J. Plant (ed) *Healing the Wounds: The Promise of Ecofeminism*. London: Green Print, pp 18–28.

King, Y. (1990) Healing the wounds: Feminism, ecology and nature/culture dualism. In I. Diamond and G.F. Orenstein (eds) *Reweaving the World*. San Francisco: Sierra Book Clubs, pp 106–21.

Kings, A.E. (2017) Intersectionality and the changing face of ecofeminism. *Ethics & the Environment*, 22(1), 63–87.

Kotze, J. (2018) *The Harms of Work: An Ultra-Realist Account of the Service Economy*. Bristol: Bristol University Press.

Leighton, P. and Wyatt, P. (2021) The case for studying social harm. In P. Davies, P. Leighton and T. Wyatt (eds) *The Palgrave Handbook of Social Harm*. London: Palgrave Macmillan, pp 1–8.

Lloyd, A. (2018) *The Harms of Work: An Ultra-Realist Account of the Service Economy*. Bristol: Bristol University Press.

Lloyd, A. (2021) Work-based harm. In P. Davies, P. Leighton and T. Wyatt (eds) *The Palgrave Handbook of Social Harm*. London: Palgrave Macmillan, pp 227–49.

Mellor, M. (1992a) *Breaking the Boundaries: Towards a Feminist Green Socialism*. London: Virago.

Mellor, M. (1992b) Ecofeminism and ecosocialism: Dilemmas of essentialism and materialism. *Capitalism, Nature, Socialism*, 3(2), 1–20.

Mellor, M. (1992c) Green politics: Ecofeminist, ecofeminine or ecomasculine? *Environmental Politics*, 1(2), 229–51.

Mellor, M. (1997) *Feminism & Ecology*. London: Polity

Merchant, C. (1980) *The Death of Nature: Women, Ecology, and the Scientific Revolution*. New York: Harper & Row.

Merchant, C. (1990) Ecofeminsim and feminist theory. In I. Diamond and G.F. Orenstein (eds) *Reweaving the World*. San Francisco: Sierra Book Clubs, pp 12–21.

Merchant, C. (1992) *Radical Ecology: The Search for a Livable World*. London: Routledge.

Merchant, C. (1996) *Earthcare: Women and the Environment*. London: Routledge.

Mies, M. and Shiva, V. (1993) *Ecofeminsim*. London: Zed Books.

Pemberton, S. (2015) *Harmful Societies*. Bristol: Policy Press.

Plant, J. (ed) (1989) *Healing the Wounds: The Promise of Ecofeminism*. London: Green Print.

Plumwood, V. (1986) Ecofeminsim: An overview and discussion of positions and arguments. *Australasian Journal of Philosophy*, 64(suppl), 120–38.

Plumwood, V. (1993) *Feminism and the Mastery of Nature*. London: Routledge.

Raymen, T. (2021) The assumption of harmlessness. In P. Davies, P.S. Leighton and T. Wyatt (eds) *The Palgrave Handbook of Social Harm*. London: Palgrave Macmillan, pp 59–88.

Ruether, R.R. (1975) *New Woman, New Earth*. New York: Seabury Press.

Ruether, R.R. (1989) Toward an eco-feminist theory of nature. In J. Plant (ed) *Healing the Wounds: The Promise of Ecofeminism*. London: Green Print, pp 235–48.

Salleh, A. (1982) On the dialectics of signifying practice. *Thesis Eleven*, 5–6(1), 72–84.

Salleh, A. (1984) Deeper than deep ecology. *Environmental Ethics*, 6, 331–41.

Salleh, A. (1995) Nature, women, labour, capital: Living the deepest contradiction. *Capital, Nature, Society*, 6(1), 21–39.

Salleh, A. (1997) *Ecofeminism as Politics: Nature, Marx and the Postmodern*. London: Zed Books.

Shiva, V. (1989) *Staying Alive: Women, Ecology and Development*. London: Zed Books.

Sollund, R. (2012) The victimization of women, children and non-human species through trafficking and trade: Crimes understood through an ecofeminist perspective. In N. South and A. Brisman (eds) *Routledge International Handbook of Green Criminology*. London: Routledge, pp 317–30.

Stephens-Griffin, N. and Griffin, N. (2021) Beyond meat? Taking violence against non-human animals seriously as a form of social harm. In P. Davies, P. Leighton and T. Wyatt (eds) *The Palgrave Handbook of Social Harm*. London: Palgrave Macmillan, pp 281–310.

Terry, G. (ed) (2009) *Climate Change and Gender Justice*. Warwickshire: Practical Action Publishing.

Tombs, S. (2022) Consumption, crime and harm at home: Regulating for what and whom? In P. Davies and M. Rowe (eds) *A Criminology of the Domestic*. London: Routledge, pp 50–68.

UNFCCC (nd) United Nations Climate Change Quarterly Report. Available at: https://unfccc.int/gender (accessed 7 June 2022).

Vegh Weis, V. (2019) Towards a critical green southern criminology: An analysis of criminal selectivity, indigenous peoples and green harms in Argentina. *The International Journal for Crime, Justice and Social Democracy*, 8(3), 38–55.

Vegh Weis, V. and White, R. (2020) Environmental victims and climate change activists. In P. Davies and J. Tapley (eds) *Victimology Research, Policy and Activism*. London: Palgrave Macmillan, pp 301–19.

Wachholz, S. (2007) 'At risk': Climate change and its bearing on women's vulnerability to male violence. In P. Beirne and N. South (eds) *Issues in Green Criminology. Confronting Harms against Environments, Humanity and Other Animals*. Cullompton: Willan, pp 161–85.

Wellsmith, M. (2010) The applicability of crime prevention to problems of environmental harm: A consideration of illicit trade in endangered species. In R. White (ed) *Global Environmental Harm: Criminological Perspectives*. Cullompton: Willan Publishing, pp 132–49.

Wellsmith, M. (2012) Preventing wildlife crime. *Criminal Justice Matters*, 90(1), 18–19.

White, R. (2013) *Environmental Harm: An Eco-Justice Perspective*. Bristol: Policy Press.

Young, I.M. (1997) *Intersecting Voices: Dilemmas of Gender, Political Philosophy, and Policy*. Princeton: Princeton University Press.

Young, I.M. (2005) Five faces of oppression. In A.E. Cudd and R.O. Andreasen (eds) *Feminist Theory: A Philosophical Anthology*. Oxford: Blackwell, pp 91–104.

3

New Directions Please! Veganising Green Criminology

Kay Peggs

Introduction

This chapter calls for a new direction in green criminology – an ethical veganising of the field. Ethical veganism opposes speciesism and the commodity status of nonhuman animals. Through the lens of ethical veganism, this chapter offers a critical consideration of gendering green criminology by drawing in issues associated with species and with human–nonhuman animal entanglements. I seek to address (the occlusion of) the enormity of nonhuman animals' victimisation by proposing a veganising of the field of green criminology within the context of a mainstreaming of non-anthropocentric, and preferably non-speciesist, criminology. Although some green criminologists have overlooked or marginalised the position of nonhuman animals, especially nonhuman animals who live outside of what is usually conceived to be 'the environment', green criminology can encourage thinking about and alleviating the human exploitation of nonhuman animals within and beyond the bounds of 'the environment'. As well, criminology that is not 'green' has the tools and perspectives to study more effectively harms and crimes to all species, thus shifting away from the largely anthropocentric focus that is currently evident.

To reflect on how to fulfil these goals, where appropriate this chapter draws on lessons from feminist criminological thought and from intersectional analysis. Because the chapter embraces a non-speciesist intersectional approach, speciesism and the place of species within intersectionality and within green criminology are discussed at length. As veganism is the driver for this chapter, there is a section on what constitutes veganism. By employing both intersectionality (Hill Collins and Bilge, 2016) and veganism as forms of critical praxis the chapter concludes with a call to veganise and gender green criminology via activism, with the aim of changing the world for the better (see Belknap, 2015). I begin with a discussion about speciesism and its place within multiple axes of inequalities, inequalities that are at the heart of much criminological theorising and research (Heimer, 2019).

Speciesism, intersectionality and criminology

Although generally marginalised in the social sciences, 'species is a foundational identifier of difference' (Hovorka, 2012: 876). Species classifications are based in essentialist assumptions about purported natural and intrinsic properties that are considered to be impervious to change (for discussion see de Vel-Palumbo et al, 2019). The classification 'human' not 'animal' is the foundation of the idea of the superiority of 'humans' relative to all other animals. Indeed, humans often are not considered to be animals at all – rather the term 'animal' is applied 'to the whole animal kingdom with the exception of human' (Derrida, 2008: 41). Rene Descartes's dualistic notion that 'the mind is distinct from and superior to matter' (Sutcliffe, 1968: 19) is at the heart of this classification. Regrettably, and with appalling consequences for nonhuman animals, Cartesian science recognised only humans as having conscious minds (Rowlands, 2002: 3), which permitted 'humans to act as though other animals are simply "inanimate" resources to be used as we like' (Boss and Boss, 1994: 120). Consequently, Cartesian scientists legally performed experiments on living nonhuman animals who were fully awake throughout (Rowlands, 2002: 3). Although the Cartesian idea that nonhuman animals are nonconscious has been challenged fundamentally by 'the advance of ethology' (Midgley, 2002 [1979]: 138), still nonhuman animals are used as experimental subjects (Rowlands, 2002) and concepts such as consciousness 'are often defined on the basis of what humans do' (Hauser, 2000: xviii). Notions of the 'natural' uniqueness of humans (Laland and Seed, 2021) are fundamental to these definitions, and humans take this as evidence of human superiority (Peggs, 2009). However, the grounds for human uniqueness are questioned increasingly. For instance, studies in genetics show that most genes are shared by all animals (human and nonhuman), and for this reason species are convenient classifications instead of natural types (Fuller, 2006: 29). Nevertheless, conceptualisations of human uniqueness and superiority are used to justify human gains via nonhuman animals' costs (Peggs, 2009). Justifiably, essentialist notions of genetic predispositions to criminality among humans have been criticised by many criminologists (for example, see Rose, 2000). However, aside from a few exceptions found within green criminology (for example, Beirne, 1999; Sollund, 2011, 2021b), essentialist notions of 'species 'are rarely discussed in criminology, let alone in a critical way. This is problematical for criminology not least because species are ranked classifications that are sources of inequalities and discriminations and '[t]he study of inequalities undergirds much of criminology' (Heimer, 2019: 377). Conventional criminology is missing analysis of an important area of oppression and injustice.

The overarching label 'animal' as it is applied to all animals who are not human is based in lazy assumptions about the 'natural' shared characteristics of the category 'animal'. There are, after all, between 2 million and 50 million species of nonhuman animals on this planet (Factmonster, 2017). Discourses about innate differences between humans and all those labelled 'animal' obscure the significance of power in hierarchical, anthropocentric, insider(human)/outsider(animal) classifications, which mask the heterogeneity of the living (Peggs, 2009: 87). In this regard, Jacques Derrida notes that '[t]here is no animal in the general singular, separated from man [*sic*] by a single indivisible limit' (2008: 47). Additionally, this constructed 'indivisible limit' is an uneven and messy threshold because, for example, some nonhuman animal species are judged more like the human species than others. Great apes such as gorillas and orang-utans are often viewed as being 'honorary humans' (Midgley, 2004: 147) and in Spain in 2008 were granted legal personhood (Strickland, 2008). More recently, in India in 2013 dolphins were recognised officially as nonhuman persons 'whose rights to life and liberty must be respected' (Coelho, 2018). Added to such recognitions, within the category 'animal' there is a ranked 'sociozoologic scale' (Arluke and Sanders, 1996), which influences how humans treat and view nonhuman animals in terms of their social, cultural and moral value to humans. The classification of some nonhuman animal species as 'pests' and others as 'pets' is one example. Nevertheless, essentialist discourses assume that the categorisation human/animal is universal and natural, and this occludes the social construction of the classification, the nonsensical aggregation of billions of nonhuman animals into one category, and the power relations that serve this classification. In short, humans have classified all nonhuman animal species as being 'animals' and have classified themselves as hierarchically superior 'humans'. Speciesism is the reason and speciesism is the result.

Speciesism is the perspective that allows 'the unjustified disadvantageous consideration or treatment of those who are not classified as belonging to one or more particular species' (Horta, 2010: 247). This description does not capture the usual specificity of speciesism. Typically, speciesism is anthropocentric. Being based in a belief in human superiority, it promotes the interests of humans over the greater interests of members of all other species (Ryder, 1983 [1975]). This leads to the oppression, harm, exploitation and killing of other species of animals (Singer, 2019). Speciesism as an ideology works as 'a set of socially shared beliefs that legitimates an existing or desired social order' (Nibert, 2002: 8). Like racism and sexism, speciesism has social structural and economic causes, it is institutionally based, and it supports oppressive social arrangements (Nibert, 2002: 10). However, while speciesism relates to groups, 'speciesism is not suffered by species as such, but by their individual members' (Horta, 2010: 250). For example, each one of the billions of individual nonhuman animals who are incarcerated in the 'animal-industrial complex' (Noske, 1989), which converts them into commodities for human consumption, 'suffer every waking minute they are alive' (Regan, 2004: 89). However, as Piers Beirne makes clear, 'as a discipline criminology has a considerable way to travel before it arrives at a well-rounded research programme in human-animal relationships' (2011: 355). Although criminologists (are beginning to) recognise that there are multiple axes of difference and discrimination (Paik, 2017), aside from work within green criminology, criminology as a discipline has largely left unrecognised species as an axis of diversity, inequality and oppression. By adopting a broader intersectional approach, criminology could 'further highlight oppressive power relations, invisible and marginalised "others", and shared locations in social hierarchies' (Hovorka, 2012: 875).

In the 20th century, feminist critiques of criminology propelled criminological research and theorising in the direction of intersectionality (Hill Collins and Bilge, 2016: 10). Writers such as Meda Chesney-Lind (1973) and Frances Heidensohn (1985) were asking the same question as Carol Smart (1976: 178), 'Where are the women [in criminology]?' Unfortunately, on the rare occasions when criminology had not completely excluded women, the study of 'female criminality leaves much to be desired' (Smart, 1976: 176). Such work, Smart argued, either relies on 'a determinate model of female behaviour', as seen in the work of Cesare Lombroso (Smart, 1976: 176), or women are included in a tokenistic way, warranting only a minor reference (Smart, 1976: 177). Although feminist criminology challenged the omission and the misrepresentation of women in criminology (Chesney-Lind, 2006: 7), Kimberley Cook laments that 'there remain missed opportunities within criminology to analyze gender and crime' (2016: 335). Given the focus of this chapter, there is no room here to detail the issues regarding feminist theory.[1] However, feminist theorising opened the way for an intersectional approach, which enables criminology to 'engage more critically in how the justice system embodies, perpetuates, and transforms existing social inequalities such as "gender"' (Paik, 2017: 4), sexuality (Copson and Boukli, 2020), 'race' (Vegh Weis and León, 2021) and class (Lynch, 2015), and the various way in which these intersect (Paik, 2017: 5). What is disappointing is that species is rarely included when intersectionality is discussed (Deckha, 2009: 249). But, as Maneesha Deckha (2009: 249) explains, '[o]ur identities and experiences are not just gendered or racialized, but are also determined by our species status and the fact that we are culturally marked as human … [our] … experiences of gender, race, sexuality, ability, etc., are often based on and take shape through speciesist ideas of humanness vis-a-vis animality'. Deckha concludes that intersectionality needs to 'reach across the species divide to consider species as a force of social construction, experience formation, and source of difference' and '[j]ust as feminism has turned toward intersectionality, intersectionality itself must now turn toward posthumanism and integrate species into its analysis' (2009: 267). This leads me to ask an amended version of Smart's (1976) question: 'Where are the nonhuman animals in criminology?' Corresponding with Smart's (1976) reading of the problem of the misrepresentation of women in criminology that preceded feminist critique, preceding the work of green criminologists, nonhuman animals had not been completely excluded from criminology. However, the inclusion was, and often remains, problematical. For example, Beirne notes that 'when animals do appear in criminology, they are almost always, passive insentient objects acted on by humans' (2009: 3). Moreover, conventional criminology does not question the legal status of nonhuman animals as 'the property of human masters' (Beirne, 2009: 5), and the property-status of nonhuman animals differs according to species.

An intersectional approach is vital when considering inequalities associated with species. Although the division human/'animal' discriminates against all nonhuman animals, among nonhuman animals there are overlapping and interdependent systems of anthropocentric discrimination and disadvantage that relate to their species and that intersect with other discriminations. Alice Hovorka (2012: 876) comments that '[i]ntersections of specifically gender and species have been theorized and investigated by, for example, ecofeminists, who have explored the shared oppressions of women and animals'. For instance, Carol Adams (1990) notes that nonhuman animals who are eaten as 'meat' and women generally are reduced to their body parts – breast, thigh, leg. Animal geographers[2] also have investigated 'how speciesism closely reflects other Western-based hierarchies', in terms of differences and inequalities both 'between-species' and 'within-species '(Hovorka, 2019: 876). The title of Melanie Joy's (2011) book, *Why Do We Love Dogs, Eats Pigs, and Wear Cows*, underlines in just ten words the differences and inequalities between species. Taking the corporate agricultural sector as an example, all nonhuman animals who are bred for human food are 'bred-disabled' (Taylor, 2017), such as being bred to grow too quickly, to put on excessive weight, to produce much more milk than they should, or to produce large numbers of eggs. These inequalities intersect with ageism and sexism. For example, in the 'egg industry' it is female chickens who are exploited for their eggs, 'male chicks … have no economic value' so are killed at birth (Harris, 2017: 70). In the 'dairy industry' it is female cows who suffer the agony and distress of constant milking. Male calves are taken from their mothers within a few days of their birth, so their mother's milk can be used for humans. Then they are either 'fattened' for their flesh or are killed immediately (Harris, 2017: 70). Once the female cow is older and has been drained of her milk and the hen ages so she can no longer lay as many eggs as she used to, they will be killed too (Harris, 2017: 70). Beirne and colleagues argue that 'if the killing of an animal by a human is as harmful to her as homicide is to a human, then the proper naming of such a death [is] "theriocide"' (2018: 5). But even if this term were applied, in speciesist societies, '[b]ecause the life of a human is almost always valued more highly than the life of an animal, homicide draws more attention than theriocide' (Beirne, 2014: 61). Conventional criminology does little to draw attention to 3 billion (Zampa, 2018) theriocides that are committed daily for food alone.

(Green) criminology, speciesism and animals

Conventional criminology focuses largely on the perpetrators and victims of illegal actions – that is, crimes as defined by criminal law (see Newburn, 2017). Because most nonhuman animal abuse is legal, billions of nonhuman animal victims are ignored (Nibert, 2013). Legally, vast numbers of nonhuman animals are exploited and killed in experiments, are seized from their own families to become companions to humans, have their liberty stolen by being incarcerated in zoos and in agribusiness, are used as living donors of secretions such as milk and eggs, and are victims of theriocides for 'meat' and other products such as bags, shoes and clothing. Because none of these usually involves a crime (Beirne, 1999: 128), it is not possible to study them using the 'conventional criminology lens' (Sollund, 2021a: 313). But even when the abuse is illegal, nonhuman animals are not usually regarded legally to be *victims* of crimes (Ritvo, 1987). Derrida (1992: 18) summarises this position: 'An animal can be made to suffer, but we would never say, in a sense considered proper, that it [*sic*] is a wronged subject, the victim of a crime, of a murder, of a rape or a theft, of a perjury.' He stresses that when legal actions are brought against humans who inflict suffering on nonhuman animals, 'these are considered to be either archaisms or still marginal and rare phenomena not constitutive of our culture' (Derrida, 1992: 18). Conventional victimologists seem to agree because they have largely recognised only humans as victims of crimes (for discussion see Flynn and Hall, 2017).[3] Despite its 'long-standing criminalization', criminologists have said very little indeed about nonhuman animal abuse (Beirne, 1999, 2011). It seems that the illegal abuse of nonhuman animals is of little interest to conventional criminologists, and the legal abuse of nonhuman animals is of practically no interest at all.

Green criminology offers a critical alternative to conventional criminology, because it aims to study 'environmental crimes and harms affecting human and non-human life, ecosystems and the biosphere' (Brisman and South,

2018: 1). The remit is wide because green criminological researchers 'consider not just harms to the environment, but also the links between green crimes and other forms of crime, including organized crime's movement into the illegal trade in wildlife' (Nurse, 2016: 2). As the reference to 'wildlife' shows, green criminologists include harms and crimes that are perpetrated against nonhuman animal species. For example, in his book *An Introduction to Green Criminology and Environmental Justice*, Angus Nurse (2016) devotes a chapter to species justice and animal rights and another to 'wildlife' crime. Nurse notes there is increasing interest in issues associated with the abuse of nonhuman animals in criminal law and argues 'there is evidence that criminal justice systems are increasingly embracing the importance of animal abuse as a specific area of crime and within contemporary discourse on applying a green perspective to crime and justice' (2016: 42). This is welcome news but given that the 'frame of reference' of green criminology is 'based in nature, the environment, or natural ecology' (Lynch and Stretesky, 2016: 6), I wonder how possible or appropriate it is for green criminology to study the myriad, often occluded, harms that are inflicted by humans on nonhuman animals outside the spaces that are conventionally considered to be 'the environment'. Although it is without question problematical to conceive of the 'environment and society' as 'distinct analytical categories' (Lockie, 2015: 139), Bill McClanahan observes that green criminology's 'framing of "nature" still often clings to the narrow binary of nature/culture and the ontological dualism of man-as-subject/nature-as-object' (2020: 645). This framing means that the abuse of nonhuman animals in the 'cultural context' continues to be overlooked, ignored, occluded.

Using the label 'green criminology' when considering crimes and harms towards nonhuman animals gives the impression that only those crimes and harms that are perpetrated in 'the environment' are within its remit. Nik Taylor and Amy Fitzgerald's (2018) review of the field of green criminology provides some evidence for this. They found that:

> [W]hile the amount of consideration given to nonhuman animals by green criminologists has increased dramatically over the years, much of this work has focused on crimes and harms against wild [*sic*] animals (for example, 'wildlife poaching', 'trafficking'), comparatively less attention has been paid to so-called 'domesticated animals' or to larger questions of species justice. (Taylor and Fitzgerald, 2018: 402)

Their review causes me to worry that harms perpetrated by humans in the 'cultural sphere' on 'an enormous proportion of nonhuman animals who are entangled with humans' would be 'at best marginalised when thinking about "green" issues' (Peggs, 2023). The harms and crimes that occur in the urban, in the home, in the factory, in the laboratory, in the stadium, and in myriad other spaces would be ignored. While being mindful of McClanahan's (2020) concern that the 'nature/culture' binary remains evident in green criminological scholarship, I am concerned as well about the consequences, if green criminology were to become the overall and only area of criminological scholarship in which violence to nonhuman animals is discussed. Nonhuman animals could return to being seen as parts of 'nature' only in criminological theorising and research. Such a move would obscure, once more, the central place of nonhuman animals in the realm of what Clifton Bryant (1979) referred to as 'social enterprise' or 'society'. As well, it would marginalise all the work that has been done in the 40 years or so since then in fields such as geography, sociology and politics, to theorise and research the consequences of human violence to nonhuman animals outside as well as inside 'the environment'. Furthermore, the compartmentalising of violence to nonhuman animals into green criminology would entrench rather than remove the anthropocentric and speciesist assumptions about the appropriate focus of conventional criminology; it would remain anthropocentric.

Conventional criminology is extending beyond a focus on illegal activities to include research and theorising about legal harms (Newburn, 2017). Given that most nonhuman animal abuse is legal this move could presage a less anthropocentric focus in conventional criminology where study of harms to currently excluded beings, such as nonhuman animals, would be included. Conventional criminological recognition that nonhuman animals are victims of humans' criminal and legal behaviours could follow. This would lead to a critical address of the anthropocentric hierarchy, discussed by Ragnhild Sollund, which classifies 'who legitimately has the right to claim victimhood, who is ascribed victimhood, and for whom this is not accepted' (2021b: Summary: para 1). A critical approach to the belief that nonhuman animals can be the property of humans, recognition that harms to nonhuman animals are as damaging as they are to humans and integrating species into intersectional analysis could advance a nonspeciesist approach to criminology. Where does veganism fit in?

Veganisms

It is quite difficult to discuss veganism in relation to criminology because there is little engagement with veganism in the field. Given that conventional criminology leaves unquestioned the legal status of nonhuman animals as the property of humans, unfortunately it is not surprising that there is an absence of discussion of veganism as an address of violence towards nonhuman animals. Coming to vegans themselves, although ethical veganism is a 'protected philosophical belief under employment law' (McKeown and Dunn, 2021: 207), there are incidents of hate crimes (in all but name) towards vegans (for example, see Nachiappan, 2020), the vegaphobic[4] news media in the UK often makes unwarranted links between veganism and criminality (for example, see Robertson, 2018) and there is discernible criminalisation of veganism regarding laws such as 'ecoterrorism' (for example, see Griffin, 2017), there is little discussion of veganism in conventional criminology. This omission is in the context of growing numbers of vegans (for example, see Wunsch, 2021) and increasing discussion of veganism in academic work (for example, see Yilmaz, 2019).

In much academic work on veganism, what veganism means is assumed rather than specified (North et al, 2021). Broadly, veganism 'eschews the consumption of nonhuman animals' (Peggs, 2020). This avoidance is crucial. Humans' entanglements with nonhuman animals are 'foremost organized around humans' consumption and "needs"' (Sollund, 2011: 437). Given that many of us live in consumer societies that require us to be 'consumer[s]-by-vocation' (Bauman, 2007: 55), abstaining from the consumption of nonhuman animals is not easy to achieve. Nonhuman animal exploitation is a central feature of human consumption; it is often occluded though it is present almost everywhere (Peggs, 2012). Consequently, it is very difficult to avoid entirely. For this reason, all vegans are aspiring vegans (Castricano and Simonsen, 2016) because it is likely that we consume nonhuman animals in one way or another.

'Consumption' is difficult to define. It is so vast and diverse that it is hard to incorporate 'such variety within a single interpretative framework' (Brewer and Trentmann, 2006: 60). Still, an essential feature of consumption is the 'using up' of things, utilities, services (Lawson, 2009: 2) and beings. A society defined by consumption ' "uses up" on a systematic and industrial scale' (Lawson, 2009: 2). For example, almost 3 billion terrestrial and aquatic nonhuman animals are killed every day across the world for food alone (Zampa, 2018). This killing is entirely legal. This 'using up' is on such an industrial scale that some nonhuman animal species are being eaten towards extinction.[5] I refer to eating here because the intentional avoidance of nonhuman animal-based food and drink is the signal behaviour that defines all vegans. Mock-vegan versions that renounce the consumption of nonhuman animal flesh, secretions and foods at specific times of day (for example, Bittman's *Vegan Before 6* [2019]), on specific days of the week (for example, Vegan Mondays, see *The Economist*, 2017), or that eschew only specific nonhuman animal food products (for example, 'pesco- vegans', who eat fishes [Urban Dictionary, 2008]) are not vegan. Even though following a vegan diet is fundamental to being a vegan, it is not the only criterion for all vegans. For this reason, it is more precise to refer to 'veganisms' (Jones, 2016).

One way to differentiate between veganisms involves distinguishing between 'lifestyle', 'environmental' and 'ethical' perspectives (Peggs, 2020).

Diet-only vegans who avoid ingesting nonhuman animals for health or other reasons associated with their individual self-interest can be categorised broadly as being 'lifestyle' vegans (Peggs, 2020). Their veganism focuses on their own health and wellbeing rather than on political protest (Gheihman, 2021). Drawing on Emile Durkheim's argument that '[m]orality begins only with disinterest, with attachment to something other than ourselves' (2005: 36), lifestyle veganism does not constitute a moral boycott. Rather it is based in an egoistic preoccupation with such matters as individual human health and identity. Close to lifestyle veganism is 'speciesist veganism', which is motivated by a desire to adopt a diet that contributes to 'human flourishing' (Holdier, 2016). It is not self-interest alone that motivates speciesist vegans, rather the concern is for all humans on the grounds that 'the current standard of industrialized animal husbandry leads to human suffering' (Holdier, 2016: 57). In his writing about this position, A.G. Holdier (2016) includes the financial corruption within agribusiness, the climate catastrophes that are exacerbated by agribusiness and the appalling conditions of humans who work in 'animal-processing plants'. Criminological research has found that 'slaughterhouse employment increases total arrest rates, arrests for violent crimes, arrests for rape, and arrests for other sex offenses in comparison with other industries' (Fitzgerald et al, 2009: 158). Thinking back to Durkheim, speciesist veganism seems to be a moral position because it entails concerns beyond the individual self, but it represents an anthropocentric morality that does not question the belief that 'all other beings are means to human ends' (Kopnina et al, 2018: 109).

Environmental veganism involves strategic visions and practices that aim to protect this planet. For many environmentalists, veganism 'is arguably the most impactful change humans can make to combat climate change' (Ghahari and McAdam, 2018: 1). For example, the scientist Joseph Poore argues that 'a vegan diet is probably the single biggest way to reduce your impact on planet Earth, not just greenhouse gases, but global acidifcation, eutrophication, land use and water use' (cited in Kortetmäki and Oksanen, 2020: 729). Again, the focus is mainly on diet because, as Deborah Kalte explains, although environmental vegans would avoid nonhuman animal food products they might 'buy … leather products instead of synthetic materials, which are considered to be more ecologically harmful' (2021: 818). Even though the scope of moral consideration of environmental veganism is beyond individual self-interest, it is nevertheless often, though not always, based in anthropocentric concerns about sustaining the planet for existing and future humans (for example, see Makkar and Ankers, 2014; Düwell et al, 2018). The speciesism is obvious (Faria and Paez, 2019). Although some environmentalists view the anthropocentric perspective as an impediment to sustainable development because 'it promotes dualisms, hierarchies, and the belief that humans are separate from nature' (Speed, 2006: 327), the notion of human distinctiveness and solidarity (for example, see Bielefeldt, 2021) still drives much environmental vegan arguments into speciesism.

Ethical veganism is non-anthropocentric. It opposes the commodity status of nonhuman animals, contests speciesism and refutes the notion that there can be a 'humane' use of nonhuman animals (Peggs, 2020). While being an ethical vegan might improve one's health and would likely involve strong concerns about the linkages between climate catastrophe, species loss, human suffering and the consumption of nonhuman animals, the emphasis for ethical vegans is on the elimination of the human oppression of nonhuman animals, with a rejection and eventual purging of speciesism. The focus is on justice for nonhuman animals and includes, though expands beyond, diet, by embracing all forms of using up. The following statement sums up ethical veganism rather well: 'To be vegan is to call for another world where one stands with animals while disrupting the current order of power, sovereignty, and authority that is built on the exploitation of animals, and Earth's others' (Schuster, 2016: 211). The veganising of green criminology would encourage a call for a better world based in justice for all animals.

Veganising and gendering green criminology: activism and change

Humans harm and kill billions of nonhuman animals every day. Consuming the results implicates humans who were not involved in the initial acts of violence in the illegal and legal suffering and theriocides of these nonhuman animals. Ethical veganism is a political position that stands against this. Veganism (and, I think, especially ethical veganism) is linked 'explicitly with its primary political end: to resist and overturn the oppression of non-human animals' (Cochrane and Cojocaru, 2023: 60). Alasdair Cochrane and Mara-Daria Cojocaru (2023) summarise the political dimension of veganism as being its address of 'routine harms created by social structures and systems for which members of a political community are responsible in virtue of their connection to them' and as 'a form of activism to be conducted collectively, in solidarity, with others' (2023: 60). Melanie Flynn argues that for vegan criminology to be realised it should take the form of a species justice perspective, which focuses not on reducing crime but on limiting harms, such as 'the harm, pain and suffering of, and even the interference with, individual nonhuman animals' (2021: para 9). This means replacing the assumption in criminology that nonhuman animals can be treated as the property of humans with recognition that nonhuman animals must be treated as the victims of various and diverse harms perpetrated by human actions, billions of which involve theriocides. Green criminology has already made progress in this regard, because it explores 'those harms against humanity, against the environment (including space) and against non-human animals committed both by powerful institutions … and also by ordinary people' (Beirne and South, 2007: xiii). I argue here that green criminology could develop further by adopting an ethical vegan approach, which would not only explore these harms but would contest the commodity status of nonhuman animals, oppose speciesism, and repudiate the idea that the human use of nonhuman animals can be 'humane'. This is more than an academic position; it is a political position that requires action and activism. Activism is not unusual in critical criminology.

In her presidential address to the American Society of Criminology in 2014, Joanne Belknap explained that what drove her to pursue a career as a criminologist was her 'desire to improve responses to injustices, on both small and large scales' (2015: 1). Belknap's definition of activist criminology is one in which criminologists engage in 'social and/or legal justice at individual, organizational, and/or policy levels, which goes beyond typical research, teaching, and service' (2015: 5). Criminologists who are activist academics can play a crucial role in opposing oppression (Arrigo and Bersot, 2016: 2). After all, 'the criminological gaze is more exposed to problems of power, stigmatization, and the context of values than any other area of social sciences' (Young, 2011: 19). For me, 'activism begins at home' (Peggs, 2023: 111). Opposing speciesism and the commodity status of nonhuman animals by being an ethical vegan is fundamental to being a non-speciesist activist and is essential to the project of green criminology precisely because green criminology extends beyond the limited anthropocentric view, to include nonhuman animals. However, the fact that ethical veganism is 'politically motivated and aims to induce change in society at large' (Kalte, 2021: 814) might estrange some criminologists. But if we are committed to intersectional theorising and research in criminology, we must recognise that an intersectional approach does not end with critical inquiry. As Deckha (2009) makes clear, intersectional analysis is an unfinished framework (also see Hill Collins and Bilge, 2016). The intersectional approach is also 'a form of critical praxis' in which research and practice go hand in hand (Hill Collins and Bilge, 2016: 5). Indeed, Patricia Hill Collins and Sirma Bilge emphasise 'both scholarship and practice are recursively linked, with practice being foundational to intersectional analysis' (2016: 5).

In 1986 in the journal *New Directions for Women*, Marie Shear remarked that '[f]eminism is the radical notion that women are people' (1986: 6). Like most fields is the social sciences, until the early 1970s criminology had focused mainly on men (Smart, 1976). Feminist criminology has sought to address the overall neglect of women in criminology. This provides lessons for green criminology. Green criminology has done surprisingly little to enhance our understanding of the ways in which green harms, crimes and victimisations are gendered. This edited book seeks to begin to address this omission. Although it has largely overlooked gender, green criminology has embraced the 'radical notion' that nonhuman animals are victims too. Nonhuman animals are victims of us all. I have used the following extract several times in my writing (for example, Peggs, 2023: 111). Unfortunately, it is as pertinent now as it was when David Nibert wrote it 20 years ago. Regarding sociologists, he writes:

> Members of the discipline, who like most other humans in society partake in the privileges derived from entangled oppressions – such as eating and drinking substances derived from the bodies of 'others', wearing their skin and hair, and enjoying the entertainment value their exploitation provides – can do so only by accepting the self-interested realities crafted by powerful agribusiness, pharmaceutical and other industries that rely on public acquiescence in oppressive social arrangements. Privilege is not so easy to give up. Silence, denial and substantial intellectual acrobatics are necessary for oppression in all forms to continue. (Nibert, 2003: 20–1)

Surely this is true of (green) criminologists too. So, offering a new direction in green criminology, a veganising of the field starts with becoming an ethical vegan.

Notes

[1] Jace Valcore et al argue that some of the problems lie with feminist theory itself, because 'a narrow and outdated notion of feminism that embraces a biologically deterministic view of sex and gender [is] wholly dismissive of intersectional and transinclusive perspectives' (2021: 689).

[2] Catherine Johnston writes that '[a]t its most basic level, animal geographers seek to make nonhumans visible in order to ensure that their material (and in some cases, emotional) needs are not unthinkingly ignored or automatically placed below our own' (2008: 646).

[3] Even in critical victimology, 'victims of environmental harms have largely been overlooked in the literature' (Hall, 2014: 131).

[4] Matthew Cole and Karen Morgan (2011) point to discourses used by the UK news media that are prejudiced against vegans.

[5] For example, at least 33 per cent of sharks (who are used in soups) meet the International Union for the Conservation of Nature Red List Criteria for being threatened with extinction (Zhou et al, 2021).

References

Adams, C.J. (1990) *The Politics of Meat*. Cambridge: Polity.

Arluke, A. and Sanders, C. (1996) *Regarding Animals*. Philadelphia: Temple University Press.

Arrigo, B.A. and Bersot, H.Y. (2016) Revolutionizing academic activism: Transpraxis, critical pedagogy, and justice for a people yet to be. *Critical Criminology*, 24(4), 549–64.

Bauman, Z. (2007) *Consuming Life*. Cambridge: Polity.

Beirne, P. (1999) For a nonspeciesist criminology: Animal abuse as an object of study. *Criminology*, 37(1), 117–48.

Beirne, P. (2009) *Confronting Animal Abuse: Law, Criminology and Human-Animal Relations*. Lanham: Rowman & Littlefield.

Beirne, P. (2011) Animal abuse and criminology: Introduction to a special issue. *Crime, Law, and Social Change*, 55(5), 349–57.

Beirne, P. (2014) Theriocide: Naming animal killing. *International Journal for Crime, Justice and Social Democracy*, 3(2), 49–66.

Beirne, P. and South, N. (2007) Introduction: Approaching green criminology. In P. Beirne, and N. South (eds) *Issues in Green Criminology*. Cullompton: Willan, pp xiii–xxii.

Beirne, P. with O'Donnell, I. and Janssen, J. (2018) *Murdering Animals: Writings on Theriocide, Homicide and Nonspeciesist Criminology*. London: Palgrave Macmillan.

Belknap, J. (2015) Activist criminology: Criminologists' responsibility to advocate for social and legal justice: Activist criminology. *Criminology*, 53(1), 1–22.

Bielefeldt, H. (2021) Moving beyond anthropocentrism? Human rights and the charge of speciesism. *Human Rights Quarterly*, 43(3), 515–37.

Bittman, M. (2019) *Vegan Before 6: Lose Weight and Restore Your Health With The Flexible Diet You Can Really Stick To*. London: Sphere.

Boss, J.A. and Boss, A.V. (1994) Paradigm shifts, scientific revolutions and the moral justification of experimentation on nonhuman animals. *Between the Species*, Summer and Fall, 119–30.

Brewer, J. and Trentmann, F. (eds) (2006) *Consuming Cultures, Global Perspectives: Historical Trajectories, Transnational Exchanges*. Oxford: Berg.

Brisman, A. and South, N. (2018) Green criminology and environmental crimes and harms. *Sociology Compass*, 13(1), 1–12.

Bryant, C.D. (1979) The zoological connection: Animal-related human behavior. *Social Forces*, 58(2), 399–421.

Castricano, J. and Simonsen, R.R. (eds) (2016) *Critical Perspectives on Veganism*. Houndmills: Palgrave Macmillan.

Chesney-Lind, M. (1973) Judicial enforcement of the female sex role. *Issues in Criminology*, 8, 51–70.

Chesney-Lind, M. (2006) Patriarchy, crime, and justice: Feminist criminology in an era of backlash. *Feminist Criminology*, 1(1), 6–26.

Cochrane, A. and Cojocaru, M.-D. (2023) Veganism as political solidarity: Beyond 'ethical veganism'. *Journal of Social Philosophy*, 54(1), 59–76

Coelho, S. (2018) Dolphin protection. *DW*, 24 May. Available at: www.dw.com/en/dolphins-gain-unprecedented-protection-in-india/a-16834519 (accessed 29 July 2022).

Cole, M. and Morgan, K. (2011) Vegaphobia: Derogatory discourses of veganism and the reproduction of speciesism UK national newspapers. *The British Journal of Sociology*, 62(1), 134–53.

Cook, K.J. (2016) Has criminology awakened from its 'androcentric slumber'? *Feminist Criminology*, 11(4), 334–53.

Copson, L. and Boukli, A (2020) Queer utopias and queer criminology. *Criminology & Criminal Justice*, 20(5), 510–22.

de Vel-Palumbo, M., Howarth, L. and Brewer, M.B. (2019) 'Once a sex offender always a sex offender'? Essentialism and attitudes towards criminal justice. *Policy Psychology Crime and Law*, 25(2), 1–40.

Deckha, M. (2009) Intersectionality and posthumanist visions of equality. *Wisconsin Journal of Law, Gender & Society*, 23(2), 249–67.

Derrida, J. (1992) Force of law: The mystical foundation of authority. In D. Cornell, M. Rosenfeld and D. Gray Carlson (eds) *Deconstruction and the Possibility of Justice*. Abingdon: Routledge, pp 3–67.

Derrida, J. (2008) *The Animal That Therefore I Am*, edited by M.-L. Mallet. New York: Fordham University Press.

Durkheim, E. (2005) The dualisim of human nature and it social conditions. *Durkheimian Studies*, 2, 35–45.

Düwell, M., Bos, G. and van Steenbergen, N. (2018) *Towards the Ethics of a Green Future: The Theory and Practice of Human Rights for Future People*. London: Routledge.

The Economist (2017) Argentina's vegan Mondays. *The Economist*, 19 October. Available at: www.economist.com/the-economist-explains/2017/10/19/argentinas-vegan-mondays (accessed 13 April 2022).

Factmonster (2017) Estimated number of animal and plant species on earth. *Factmonster*. Available at: www.factmonster.com/math-science/biology/plants-animals/estimated-number-of-animal-and-plant-species-on-earth (accessed 26 July 2022).

Faria, C. and Paez, E. (2019) It's splitsville: Why animal ethics and environmental ethics are incompatible. *The American Behavioral Scientist*, 63(8), 1047–60.

Fitzgerald, A.J., Kalof, L. and Dietz, T. (2009) Slaughterhouses and increased crime rates: An empirical analysis of the spillover from 'the jungle' into the surrounding community. *Organization & Environment*, 22(2), 158–84.

Flynn, M. (2021) Opinion: Veganising criminology. *The Vegan Society*, 12 July. Available at: www.vegansociety.com/about-us/research/research-news/expert-series-1-veganising-criminology (accessed 19 June 2022).

Flynn, M. and Hall, M. (2017) The case for a victimology of nonhuman animal harms. *Contemporary Justice Review*, 20(3), 299–318.

Fuller, S. (2006) *The New Sociological Imagination*. London: SAGE.

Ghahari, J.M. and McAdam, J.A. (2018) Combating climate change one bite at a time: Environmental sustainability of veganism (with a socio-behavioral comparison of vegans and omnivores). *Journal of Social Science*, 14(1), 1–11.

Gheihman, N. (2021) Veganism as a lifestyle movement. *Sociology Compass*, 15(5), 1–14.

Griffin, N.S. (2017) *Understanding Veganism: Biography and Identity*. Basingstoke: Palgrave Macmillan.

Hall, M. (2014) Environmental harm and environmental victims: Scoping out a 'green victimology'. *International Review of Victimology*, 20(1), 129–43.

Harris, T. (2017) The problem is not the people it's the system: The Canadian industrial complex. In D. Nibert (ed) *Animal Oppression and Capitalism*, vol 1. Santa Barbara: Praeger, pp 57–75.

Hauser, M. (2000) *Wild Minds: What Animals Really Think*. London: Allen Lane/Penguin.

Heidensohn, F. (1985) *Women and Crime*. Basingstoke: Macmillan.

Heimer, K. (2019) Inequalities and crime. *Criminology*, 57(3), 377–94.

Hill Collins, P. and Bilge, S. (2016) *Intersectionality*. Cambridge: Polity.

Holdier, A.G. (2016) Speciesist veganism: An anthropocentric argument. In J. Castricano and R.R. Simonsen (eds) *Critical Perspectives on Veganism*. London: Palgrave Macmillan, pp 41–66.

Horta, O. (2010) What is speciesism. *Journal of Agricultural & Environmental Ethics*, 23(3), 243–66.

Hovorka, A.J. (2012) Women/chickens vs. men/cattle: Insights on gender-species intersectionality. *Geoforum*, 43(4), 875–84.

Hovorka, A.J. (2019) Animal geographies III: Species relations of power. *Progress in Human Geography*, 43(4), 749–57.

Johnston, C. (2008) Beyond the clearing: Towards a dwelt animal geography. *Progress in Human Geography*, 32(5), 633–49.

Jones, R.C. (2016) Veganisms. In J. Castricano and R.R. Simonsen (eds) *Critical Perspectives on Veganism*. London: Palgrave Macmillan, pp 15–40.

Joy, M. (2011) *Why We Love Dogs, Eat Pigs, and Wear Cows: An Introduction to Carnism*. San Francisco: Conari Press.

Kalte, D. (2021) Political veganism: An empirical analysis of vegans' motives, aims, and political engagement. *Political Studies*, 69(4), 814–33.

Kopnina, H., Washington, H. Taylor, B. and Piccolo, J.J. (2018) Anthropocentrism: More than just a misunderstood problem. *Journal of Agricultural and Environmental Ethics*, 3(1), 109–27.

Kortetmäki, T. and Oksanen, M. (2020) Is there a convincing case for climate veganism? *Agriculture and Human Values*, 38(3), 729–40.

Laland, K. and Seed, A. (2021) Understanding human cognitive uniqueness. *Annual Review of Psychology*, 72(1), 689–716.

Lawson, N. (2009) *All Consuming*. London: Penguin.

Lockie, S. (2015) What is environmental sociology? *Environmental Sociology*, 1(3), 139–42.

Lynch, M.J. (2015) The classlessness state of criminology and why criminology without class is rather meaningless. *Crime, Law, and Social Change*, 63(1–2), 65–90.

Lynch, M.J. and Stretesky, P. (2016) *Exploring Green Criminology: Toward a Green Criminological Revolution*. London: Routledge.

Makkar, H.P.S. and Ankers, P. (2014) Towards sustainable animal diets: A survey-based study. *Animal Feed Science and Technology*, 198, 309–22.

McClanahan, B. (2020) Earth–world–planet: Rural ecologies of horror and dark green criminology. *Theoretical Criminology*, 24(4), 633–50.

McKeown, P. and Dunn, R.A. (2021) A 'life-style choice' or a philosophical belief? The argument for veganism and vegetarianism to be a protected philosophical belief and the position in England and Wales. *The Liverpool Law Review*, 42(2), 207–41.

Midgley, M. (2002 [1979]) *Beast and Man*. London: Routledge.

Midgley, M. (2004) *The Myths We Live By*. London: Routledge.

Nachiappan, A. (2020) Experts get their teeth into idea of vegan hate crime. *The Times*, 8 August. Available at: www.thetimes.co.uk/article/experts-get-their-teeth-into-idea-of-vegan-hate-crime-65nsf6c02 (accessed 19 July 2022).

Newburn, T. (2017) *Criminology*, 3rd edn. London: Routledge.

Nibert, D. (2002) *Animal Rights/Human Rights: Entanglement of Oppression and Liberation*. Plymouth: Rowman & Littlefield.

Nibert, D. (2003) Humans and other animals: Sociology's moral and intellectual challenge. *International Journal of Sociology and Social Policy*, 23(3), 5–25.

Nibert, D. (2013) *Animal Oppression and Human Violence: Domesecration, Capitalism, and Global Conflict*. New York: Columbia University Press.

North, M., Kothe, E., Klas, A. and Ling, M. (2021) How to define 'vegan': An exploratory study of definition preferences among omnivores, vegetarians, and vegans. *Food Quality and Preference*, 93, 104246–54.

Noske, B. (1989) *Humans and Other Animals*. London: Pluto Press.

Nurse, A. (2016) *An Introduction to Green Criminology and Environmental Justice*. London: SAGE.

Paik, L. (2017) Critical perspectives on intersectionality and criminology: Introduction. *Theoretical Criminology*, 21(1), 4–10.

Peggs, K. (2009) A hostile world for nonhuman animals: Human identification and the oppression of nonhuman animals for human good. *Sociology*, 43(1), 85–102.

Peggs, K. (2012) *Animals and Sociology*. Basingstoke: Palgrave Macmillan.

Peggs, K. (2020) Veganism. In C. Rojek and G. Ritzer (ed) *Wiley Blackwell Encyclopedia of Sociology*, 2nd edn. London: Wiley Blackwell.

Peggs, K. (2023) Eating animals: A critical criminology of the domestic. In P. Davies and M. Rowe (eds) *Criminology of the Domestic*. London: Routledge, pp 101–16

Regan, T. (2004) *Empty Cages: Facing the Challenge of Animal Rights.* Lanham: Rowman & Littlefield.

Ritvo, H. (1987) *The Animal Estate: The English and Other Creatures in the Victorian Age.* Cambridge, MA: Harvard University Press.

Robertson, A. (2018) Criminal past of vegan activist who claims cows bred for milk is 'MURDER' is exposed live on GMB as Piers reveals he was jailed for carrying a gun. *Daily Mail Online*, 27 June. Available at: www.dailymail.co.uk/news/article-5890853/Criminal-past-vegan-activist-expo sed-live-GMB.html (accessed 20 September 2022).

Rose, N. (2000) The biology of culpability: Pathological identity and crime control in a biological culture. *Theoretical Criminology*, 4(1), 5–34.

Rowlands, M. (2002) *Animals Like Us.* London: Verso.

Ryder, R.D. (1983 [1975]) *Victims of Science: The Use of Animals in Research*, revised edn. London: National Anti-Vivisection Society.

Schuster, J. (2016) The vegan and the sovereign. In J. Castricano and R.R. Simonsen (eds) *Critical Perspectives on Veganism*. London: Palgrave Macmillan, pp 203–23.

Shear, M. (1986) Media watch: Celebrating women's words. *New Directions for Women*, 15(3), 6.

Singer, P. (2019) All animals are equal. In T. Ball, R. Dagger and D.I. O'Neill (eds) *Ideals and Ideologies: A Reader*. New York: Routledge, pp 413–23.

Smart, C. (1976) *Women, Crime and Criminology: A Feminist Critique.* London: Routledge and Kegan Paul.

Sollund, R. (2011) Expressions of speciesism: The effects of keeping companion animals on animal abuse, animal trafficking and species decline. *Crime, Law, and Social Change*, 55(5), 437–51.

Sollund, R. (2021a) Green criminology: Its foundation in critical criminology and the way forward. *Howard Journal of Crime and Justice*, 60(3), 304–22.

Sollund, R. (2021b) Nonspeciesist criminology, wildlife trade, and animal victimization. In *Oxford Research Encyclopedia of Criminology*. Available at: https://oxfordre.com/criminology/view/10.1093/acrefore/978019 0264079.001.0001/acrefore-9780190264079-e-608 (accessed 30 May 2022).

Speed, C. (2006) Anthropocentrism and sustainable development: Oxymoron or symbiosis? *WIT Transactions on Ecology and the Environment*, 93, 323–32.

Strickland, E. (2008) Great apes have the right to life and liberty, Spain says. *Discover*, 28 June. Available at: www.discovermagazine.com/planet-earth/great-apes-have-the-right-to-life-and-liberty-spain-says (accessed 19 September 2022).

Sutcliffe, F.E. (1968) Introduction. In R. Descartes, *Discourse on Method and Other Writings*. Middlesex: Penguin, pp 7–23.

Taylor, N. and Fitzgerald, A. (2018) Understanding animal (ab)use: Green criminological contributions, missed opportunities and a way forward. *Theoretical Criminology*, 22(3), 402–25.

Taylor, S. (2017) *Beasts of Burden: Animal and Disability Liberation*. New York: The New Press.

Urban Dictionary (2008) Pesco vegan. *Urban Dictionary*, 17 December. Available at: www.urbandictionary.com/define.php?term=Pesco%20Vegan (accessed 13 April 2022).

Valcore, J., Fradella, H.F., Guadalupe-Diaz, X., Ball, M.J., Dejong, C., Walker, A., Wodda, A. and Worthen, M.G.F. (2021) Building an intersectional and trans-inclusive criminology: Responding to the emergence of 'gender critical' perspectives in feminist criminology. *Critical Criminology*, 29(4), 687–706.

Vegh Weis, V. and León, K.S. (2021) Critical criminology and race: Re-examining the whiteness of US criminological thought. *Howard Journal of Crime and Justice*, 60(3), 388–408.

Wunsch, N.-G. (2021) Number of vegans in Great Britain from 2014 to 2019. *Statista*, 22 June. Available at: www.statista.com/statistics/1062104/number-of-vegans-in-great-britain/ (accessed 26 October 2022).

Yilmaz, A.F. (2019) Contemporary feminist politics of veganism: Carol J. Adams' *The Sexual Politics of Meat* and alternative approaches. *Global Media Journal Canadian Edition*, 11, 23–38.

Young, J. (2011) *The Criminological Imagination*. Cambridge: Polity.

Zampa, M. (2018) How many animals are killed for food every day? *Sentient Media*. Available at: https://sentientmedia.org/how-many-animals-are-killed-for-food-every-day/#:~:text=%20%20%201%20Farmed%20fish.%20.%20More,2%20million%20geese%20are%20killed%20for...%20More%20 (accessed 11 November 2021).

Zhou, X., Booth, H., Li, M., Song, Z., MacMillan, D.C., Zhang, W., Wang, Q. and Veríssimo, D. (2021) Leveraging shark-fin consumer preferences to deliver sustainable fisheries. *Conservation Letters*, 14(6), 1–8.

4

Men and the Climate Crisis:
Why Masculinities Matter
for Green Criminology

Stephen R. Burrell

Introduction

There is much for social scientists to examine when it comes to the climate emergency. Indeed, it is urgent that more start doing so – including criminologists. Irrespective of official definitions, can there be any crime greater than anthropogenic climate change – continuing to burn fossil fuels and release greenhouse gases into the atmosphere, while being fully aware of the massive harm this is causing to human populations across the globe, and to all life on earth (White, 2018)? Criminologists can use our knowledge of the causes, impacts and prevention of harmful human behaviour – as well as what can obstruct such solutions – to help advance efforts to mitigate and adapt to climate change, and bring about more just responses to it.

Central to this is understanding how global heating is shaped by social inequalities, including gender. It is becoming increasingly widely recognised that the climate crisis is gendered. There is a growing body of multidisciplinary research demonstrating the disproportionate impacts that climate breakdown is having on women and girls, as well as LGBTQ+ communities, in countries across the globe (Pearse, 2017). This is of much relevance to green criminology, as it means that women and girls are being inordinately affected by green crimes, of which climate change is the ultimate example.

There are a number of reasons for this. One of the biggest manifestations of gender inequality is that women, girls and LGBTQ+ groups tend to have fewer socioeconomic resources, which can make it harder to deal with and recover from natural disasters aggravated by global heating (WHO, 2014). They may also not be encouraged to the same extent as men and boys to develop skills which can aid survival in extreme weather events. For instance, research suggests that in countries which have faced extreme flooding, women have often been more likely to die, in part because they have not been taught how to swim (Sultana, 2014). The fact that women are more likely to have caregiving responsibilities can make it harder to escape disasters, as it means they are often left having to take care of others in addition to themselves, such as children or older family members (Pearse, 2017). In many contexts, women are also more reliant on natural resources hampered by climate breakdown such as water, fuel and food, because they are expected to cultivate and collect them (WHO, 2014).

In addition to the direct effects of climate breakdown, its social consequences can also intensify gendered and other inequalities. One example of this is how different forms of gender-based violence are often exacerbated in the wake of disasters. We have seen during COVID-19 that in times of crisis, domestic abuse can intensify, and research suggests that the same is true after events such as wildfires (Thurston et al, 2021). Hewitt (2016) has written about how disasters can breach the privacy of the domestic sphere and bring what is typically hidden into more public view. COVID-19 led to increased public awareness about the harms of domestic abuse, even if policy makers were slow to recognise that home is not a safe space for all.

This is not to suggest that disasters somehow cause violence and abuse themselves, as this exculpates the agency and responsibility of perpetrators. It is important to recognise that the use of violence is a choice, made by humans. However, given that domestic abuse is rooted in the exertion of power and control, in disasters when one experiences a substantial loss of control in other parts of one's life, some people – typically men – may seek to deal with these insecurities by increasing their dominance in the private sphere, including over their partners and/or children. Crises often uproot dominant social norms, at least temporarily, and this can instigate wider shifts in gender relations (Le Masson et al, 2016). However, some men may respond to this by seeking to reassert the extant social order, and entrench it further. Extreme weather can also increase stress and anxiety, which can feed into violence and abuse, and there is evidence that different forms of violent crime increase in rising temperatures (Tiihonen et al, 2017).

Climate breakdown can also exacerbate forms of sexual violence and abuse. For instance, sexual exploitation and trafficking often grows in the wake of disasters, as some seek to take advantage of the chaos when many women and children become forcibly displaced or are left economically desperate (Le Masson et al, 2016). Similarly, child marriage can increase as families struggle to meet basic needs and turn to marrying off young daughters to alleviate financial difficulties (Castañeda Carney et al, 2020). Sexual violence can also be used as a weapon against women environmental activists and land defenders, including by arms of the state such as the police (Castañeda Carney et al, 2020). Violence against environmental activists appears to be increasing, with a record number murdered in 2020 (Global Witness, 2021).

Yet it is notable that men are often concealed in conversations about gender and climate catastrophe, even when they are in fact playing a significant role in the issues discussed. For instance, the vast majority of the aforementioned violence and abuse is being perpetrated by men. This chapter therefore examines some of the manifold ways in which men and boys fit into understandings around gendering climate change. It argues that this kind of gendered analysis is crucial to making sense of the *causes* of global heating and environmental destruction, as well as how to tackle and prevent them. This is particularly important from a green criminology perspective, because when we consider the crimes involved in climate change, men and masculinities are often playing a substantial part in driving them.

The masters of climate change

One obvious place to start in this respect is to examine who is presiding over the climate crisis. Who has been making the decisions that have led us to this situation, and seem to be unable to take the necessary kinds of actions to seriously address it? It has been pointed out by feminist researchers that women have been consistently under-represented in environmental decision-making. At COP-1 in 1995, women made up 18 per cent of delegations sent by parties to the convention (Kruse, 2014). While this number has risen, it was still only 37 per cent at COP-26 in 2022, while women accounted for 29 per cent of total speaking time (UNFCCC Secretariat, 2022). What this also means is that the vast majority of people in positions of power making the decisions perpetuating unsustainable carbon emissions and maintaining climate inaction have been men. For instance, in 2021 only 22 per cent of government ministers worldwide were women (IPU and UN Women, 2021), and the proportion of women environment ministers was even smaller, at 15 per cent (IUCN, 2021). Of course, politics is not the only setting where decisions are made with significant implications for the climate. In the business sector, for example, only 15 per cent of Fortune 500 chief executive officers were women in 2021 (Bucholz, 2022). What's more, while these figures are small, they actually illustrate *progress* in women's participation in business and politics over recent decades. So when considering who has been in charge, it is clear that global heating can be seen as a 'man-made' crisis. And it is also important to highlight that these decision-makers are not just any men, but typically men who are heterosexual and cisgender, white, and middle or upper class (Nagel, 2017).

Statistics about who holds power provide crude illustrations of the persistence of patriarchal gender orders. While they suggest this has a big part to play in helping us to understand the social dynamics of the climate crisis, this does not mean that simply having more women in leadership positions will solve the problem. We do not have to look far to find examples of women leaders also making decisions which wreak environmental and social harm, and perpetuate patriarchal inequalities. In fact, within highly masculinised institutions such as politics, some degree of compliance with patriarchy may be required for women to get to such positions in the first place (Connell, 2009). However, while more women and LGBTQ+ leaders will clearly not on its own bring about gender equality or climate justice, it is still an important step forward as part of addressing both these problems. There is research which suggests that women's empowerment not only has positive social impacts, but also environmental ones, and that there is a mutually beneficial relationship between different forms of equality and environmental quality (Ergas et al, 2021).

Masculinity in a changing climate

The influence of constructions of masculinity goes beyond just those in charge. It extends in considerable ways to men and boys' relationships with nature more broadly. For instance, research suggests that men tend to contribute more to carbon emissions than women (Räty and Carlsson-Kanyama, 2010; EIGE, 2012; Carlsson-Kanyama et al, 2021). In part this is because economic inequalities mean that men have more resources in the first place, and thus greater capacity to consume and impact the environment. However, it is also because there are often connections between dominant ideas of masculinity and unsustainable, extractive interactions with nature. Indeed, some of the most environmentally destructive practices are frequently associated with hegemonic masculinities. In other words, they are seen as desirable behaviours for men performing the most socially valued forms of masculinity to engage in.

One example of this is meat-eating, which is often celebrated and promoted as a deeply masculine activity, especially when it involves red meat such as steak (that is, food involving the largest animals and the biggest impacts on the environment) (Rothgerber, 2013). Meanwhile, not eating meat, and being vegan or vegetarian, can be constructed as emasculating, exemplified by the popular pejorative term 'soy boy' (Aavik and Velgan, 2021). It is of little surprise then that women tend to eat less meat than men, including for environmentally motivated reasons (EIGE, 2012).

Men are also more likely to fly and to drive than women (Graham and Metz, 2017; Balkmar, 2019), and these practices are again often marketed as highly desirable 'masculine' endeavours, associated with expressing dominance and freedom over the natural environment (Groombridge, 1998). There is much research into men and 'car cultures', for example, with vehicles providing potent masculine status symbols and demonstrations of phallic virility (Balkmar, 2019). Research by Carlsson-Kanyama et al (1999) found that men's transportation patterns contributed 53 per cent more to carbon emissions than women's, with longer distances and choice of vehicle key factors. Indeed, Daggett (2018) has written about how, in countries like the United States, many (white) men's identities are wrapped up in the consumption of fossil fuels, which have been relied upon to cheaply supply the core components of what she calls 'petro-masculinity'. Key markers of this include driving cars, living in the suburbs and being the head of the nuclear family.

This points to how many men's jobs are reliant upon fossil fuel consumption. Of course this is also the case for women, but men appear to have greater involvement in work that contributes to higher greenhouse gas emissions (Cohen, 2014). Many of the most environmentally damaging industries, such as fossil fuels, mining and weapons production, are highly masculinised. The World Petroleum Council estimates that only 22 per cent of people working in the oil and gas industry are women, for instance (Rick et al, 2017). This is another example of how masculine identities are often closely entangled with global heating. It demonstrates the need to take gender into account in efforts to bring about 'just transitions' to more sustainable economies. In industries which need to be scaled back or transformed, in addition to ensuring that good quality green replacement jobs are available, it is important to engage with men about how masculine expectations might be holding them back, and why they do not have to be the breadwinner.

Gender norms also appear to influence people's perceptions of the problem. Survey research in different countries, including the UK, indicates that women are more likely to believe climate change is happening, worry about its consequences, perceive more risks from it, express more knowledge about it (while often underestimating that knowledge compared to men), and view it as posing a threat within their lifetimes (Pearson et al, 2017; ONS, 2021a). Women also appear more likely to change their behaviour in response to climate change than men (Pearson et al, 2017; ONS, 2021a). Of course, many men are highly concerned as well, but it appears that such perceptions are more openly and widely articulated among women. This could in part be because men and boys fear that openly expressing care and concern about the environment is not a very 'manly' thing to do, and that they may face gendered disparagements as a result. Research suggests that this is a factor in why fewer men engage in 'eco-friendly' practices such as recycling (Brough et al, 2016).

Nature and 'man'

At a broader level, it is important to consider the role of patriarchal social orders in causing anthropogenic climate change, which eco-feminist theorists have powerfully illustrated (Mies and Shiva, 2014). If industrial capitalism is the engine of global heating, then it is a socioeconomic system which has been shaped considerably by masculinist logics. For instance, in how it is driven by the quest for constant growth, the domination and conquering of our surroundings, and the maximum exploitation of resources – and in the very notion of 'natural resources' itself, whose primary function is human extraction and consumption. The overlaps with colonialism here are also clear to see, demonstrating the interwoven, mutually reinforcing nature of different systems of power.

Underpinning these exploitative relations is a sense of ownership of the natural world; of (masculine) entitlement to nature and land, and limitless liberty to do as one wishes with it. This in turn is legitimised by enlightenment ideas of *man*kind being separate from and above fauna and flora. There are manifold connections between these approaches to the environment and hegemonic ideals of manhood in many Global North contexts; that 'rational' men should be dispassionate and detached from others, whether that be 'emotional' and 'vulnerable' women and children, or 'primeval' nature (Pease, 2019a). Hultman and Pulé (2019) argue that this is closely connected to 'industrial/breadwinner' constructions of masculinity, which have been a central component of modernity.

Eco-feminists have illustrated that dominant constructions of masculinity legitimise men's control of women and nature alike (MacGregor, 2009) – and that this is a key way of proving one's hegemonic status to oneself and others (Hultman and Pulé, 2019). This is not to suggest that women and nature are uniquely one and the same, or that women are somehow innately more connected to nature than men. The point is, we are *all* part of nature and cannot be separated from or seen as 'superior' to it. All humans are animals, no matter how much we may seek to ignore or conceal this reality, or remove ourselves from it in concrete megapolises. And even if these logics preserve the dominance of (a small number of) men in the short term, they ultimately serve no one's interests, because they are incommensurate with the survival of life on earth.

Climate change adaptation and mitigation

Men and masculinities are also shaping societal *responses* to the climate crisis. When considering who is perceived to hold the solutions to climate change, we are frequently encouraged to look to science and technology – both highly masculinised sectors (Nagel, 2017; Pease, 2019a). MacGregor (2009) has described this as the masculinisation of environmentalism. Of course, science and technology have vital roles to play. However, it is no doubt comforting for those in positions of power to think that few changes to how society is organised and run are required, because new technologies created by heroic (male) entrepreneurs will save us. Yet many of these technologies, such as carbon capture and storage, or low-carbon air travel, barely even exist yet, and the evidence that they will be able to be implemented at scale in an affordable and timely way is dubious (Hultman and Pulé, 2019). For instance, while electric vehicles are undoubtedly an important part of the picture, we also need to dramatically reduce the number of cars being produced, sold and driven in the first place – something which many car industry executives, and men who benefit from the socioeconomic status quo, may find difficult to acknowledge.

In this respect, Hultman and Pulé (2019) refer to another form of masculinity which has emerged in recent decades; that of 'ecomodern' masculinities. This describes men who recognise that the climate crisis is happening and that action needs to be taken, but are unwilling to question the wider social relations which have brought it about in the first place. Indeed, they seek to preserve the social order through their faith in the market and technological innovations, inability to let go of materialist addictions to growth, and prioritisation of human needs and desires above the wider ecology. Hultman and Pulé refer to the actor and former governor of California, Arnold Schwarzenegger, and tech-entrepreneur Elon Musk as two archetypal examples of ecomodern masculinities.

Meanwhile, the voices of people already being most affected by climate change, such as Indigenous communities and people living in the Global South, are frequently marginalised in high-level climate talks. This points to a striking example of epistemic injustice (Bacevic, 2021), in which the very people who have played a disproportionate role in creating the climate crisis – white men in the Global North – are presumed to be the 'authorities' on how to tackle it. Meanwhile, the knowledge, expertise and solutions held by those already possessing plentiful experience of dealing with environmental degradation, and finding more harmonious ways of living on the earth, are ignored.

It is notable too that in economic responses to the climate crisis, similar to recovery schemes in the wake of COVID-19, there is typically an emphasis on industry and male-dominated occupations. Cohen and MacGregor (2020) have discussed how this can be the case even in left-wing 'green new deal' proposals, while sectors in which women are more likely to work, such as care and education, are often ignored, even though many of these jobs are already low in carbon emissions. This is one example of why more involvement of women and marginalised groups at all levels of climate decision-making is a vital first step to ensuring that transitions to greener economies take existing social inequities into account.

Climate change denial

A major crime of the (male-dominated) fossil fuel companies has not only been their direct role in heating the planet itself, but also in seeking to cover it up and hide the harms they have caused. This has included funnelling resources for decades into pseudo-scientific research and commentaries playing down the impacts of anthropogenic climate change, or denying its existence altogether (White, 2018). It is notable that a disproportionate number of those individuals who express climate change scepticism appear to be men (Pearson et al, 2017). This was found in research in Sweden by Anshelm and Hultman (2014), for example, who argued that many such men's identities appeared to be wrapped up in a masculinity attached to industrial modernity, which they perceived to be under threat by social change triggered by the climate crisis.

Indeed, masculine attempts to defend the patriarchal status quo can be observed in some responses to environmental activists (Castañeda Camey et al, 2020). The misogynistic vitriol directed at campaigners such as Greta Thunberg illustrates this, with high-profile male politicians and commentators such as former US president Donald Trump seemingly uncomfortable with a young woman being unafraid to speak publicly in a way which challenges dominant sociopolitical narratives and unapologetically holds men in power to account (White, 2021). There are notable overlaps in membership and discourses among climate change deniers (and sometimes also eco-fascists) and anti-feminist groups, especially in virtual spaces such as the 'manosphere' (White, 2021).

It is important to recognise that in at least some countries, the influence of explicit climate change scepticism appears to be waning, as it becomes increasingly widely accepted that global heating is happening, is caused by human actions, and poses significant threats to human and more-than-human life. Instead, one of the biggest political obstacles now appears to be claims that efforts to tackle the climate crisis are excessively expensive, panic-driven and elitist, especially in the context of economic downturns and cost-of-living increases, which climate action is presented as contributing to, rather than as part of the solution to. For instance, a 'Net Zero Scrutiny Group' of UK Members of Parliament was established in 2021 (Atkins, 2022), and it is notable that 18 of the 20 publicly named members at the time of writing are men. It is also noteworthy that political scepticism about the economic affordability of climate action still often seems to rapidly descend into scepticism about climate change itself (Atkins, 2022).

However, an even bigger danger may be that of the 'business as usual' perspective, in which actors such as corporations may recognise that climate change is happening, but seem unwilling or incapable of changing how they operate in order to seriously address it, seemingly waiting for some as yet unknown magical solutions to appear in the future. This approach also represents a denial of reality, and appears alarmingly influential across large swathes of the business sector (Wright and Nyberg, 2017).

Given that these logics appear to be somewhat hegemonic in the private sector, particularly in the large multinational corporations whose actions have the biggest impact, an important avenue of research lies in analysing the men who dominate the senior levels of business organisations, and how gender influences the decisions and legitimisations they make. For instance, how in their views on the environment might senior managers and business owners be influenced by masculine notions of competitive individualism, risk-taking, self-reliance and the prioritisation of profit above all? Connell and Wood's (2005) exploration of 'transnational business masculinities' provides a valuable example of this. The prospect of environmental regulations and leaving vast amounts of fossil-fuelled profits in the ground may seem like a personal affront to the freedoms of some such men.

Violence against the environment

In a range of ways, then, it appears that men and masculinities are playing a particularly significant role in contributing to the climate crisis. Indeed, it could be argued that global heating, as with other forms of environmental harm and 'ecocide', has a number of connections with different forms of men's violence (Pease, 2019a). Not in the sense that phenomena such as violence against women and violence against the planet are one and the same, because they are not, and as already discussed, it is important to avoid essentialist equations of femininity with nature. But in the sense that men's violence – be it against women, other men, or oneself (Kaufman, 1987) – is so often rooted in attempts to accomplish the power and control which society expects men to possess, and proving to oneself and to others that one is indeed a 'real man'.

Frequently this expression and attainment of masculine dominance and entitlement seems to be a key factor in environmental violence too – whether it is the direct violence of actions such as burning down a forest or releasing toxic chemicals into the ocean, or the structural violence (Galtung, 1990) of continuing to burn fossil fuels while the planet warms. Nixon's (2011) concept of 'slow violence' is highly pertinent in capturing the harm caused by anthropogenic climate change, which is gradual, incremental, often invisible, but incurring irrecoverable damage to people and planet, typically in communities with less power or influence.

It could therefore be argued that these different kinds of environmental violence, as with other forms of violence, should be understood as substantially shaped by patriarchal power relations and by masculinity. Of course, we all hold some degree of complicity in them – especially those of us living in the Global North. One of the issues with structural violence is the difficulty pinpointing and identifying specific 'perpetrators'. But some are playing a much more direct and active role than others, and those people tend to be men. Here too, though, it is important to recognise that men are not a homogeneous group – some have much more power than others, and they tend to be white, older, middle- or upper-class, heterosexual, cisgender and able-bodied. Nonetheless, gender does appear to be a crucial factor when we are seeking to understand who are the major 'climate criminals'.

When considering environmental violence, one important – and highly masculinised – institution to acknowledge is the military. As well as causing substantial harm to human populations, war also brings about huge destruction to the environment. Yet this is rarely recognised, and often ignored in the masculinist logics – such as the need to 'show strength', not 'back down', and exert one's rightful dominance over others – which are often used to justify war and its 'collateral damage'. More broadly, militaries and the 'treadmill of destruction' of expanding militarism involve massive consumption of fossil fuels and other resources, yet the resulting emissions are rarely discussed or considered as part of efforts to decarbonise (Jorgenson and Clark, 2016). Militaries are therefore one clear – yet largely invisible – example of how masculine violence against humans and the planet overlap.

Caring for the planet

While hegemonic constructions of masculinity are often associated with domination, exploitation and violence towards the environment, it is important to recognise that men and masculinities are incredibly diverse, and many do engage in more caring relationships with nature. There are numerous men involved in the environmental movement, for example, which in lots of ways challenges patriarchal relations (MacGregor, 2009; Pease, 2019b). For instance, it increasingly appears to be led by women, LGBTQ+ people and young people, seeks to organise in horizontal and democratic ways, and frequently advocates non-violence (Togami and Staggenborg, 2022). That said, it does seem that men are still often over-represented at senior levels of environmental organising, even if less commonly found at the grassroots (Buckingham, 2017).

Some men are also practising care for nature more directly, for example through involvement in nature conservation and restoration work. Men engaging in activities such as farming, gardening and allotments also spend much time close to nature, and it is possible to carry out these activities in ways which are beneficial for humans and the wider ecosystem alike. However, this is often not how they are practised currently, and even activities such as conservation may sometimes be motivated by the perceived need to exert human control over nature as much as living in harmony within it. Similarly, it cannot be assumed that environmental activism is always a bastion of gender equality; it is entirely possible for unequal power relations and harmful gender norms to be replicated within environmentalist spaces too (Buckingham, 2017). For instance, in the wake of the #MeToo movement a number of senior leaders within the US-based non-profit organisation the Nature Conservancy resigned following allegations of sexual harassment within the organisation, which was rebuked for failing to take the complaints seriously (Rhode, 2019).

Nonetheless, it does seem that environmental activism often involves deviating to some extent from hegemonic expectations of masculinity. This was something Connell (1990) identified in research with men in the environmental movement over 30 years ago, finding that they were separating themselves from hegemonic masculinity, engaging in a remaking of the masculinised self, and shifting towards collective politics. The challenges involved in this remaking of masculinity may help to explain why more men do not become involved, and why those that do are disproportionately white and middle class, because men who are less privileged often face greater repercussions for defying masculine norms. However, this does suggest that the environmental movement offers valuable lessons for how to shift rigid and restrictive constructions of gender. For instance, in research with men involved in activism to end violence against women, participants felt that environmentalist men were a group more likely to speak out against gender-based violence, and thus important to build alliances with (Westmarland et al, 2021).

Another issue to consider is that men are often 'victims' of climate breakdown as well. Even if the impacts tend to be particularly severe on women, many men also suffer and die as a result of extreme weather events and disasters. For instance, in wildfires men are typically over-represented among the dead because firefighting is so male-dominated (Enarson and Pease, 2016). One highly vulnerable yet hidden group is incarcerated people, who are predominantly men, and frequently face inhospitable conditions in prisons during heatwaves, while already having a disproportionate amount of physical and mental health comorbidities (Skarha et al, 2020).

In addition to the harm caused to physical health, disasters can cause considerable trauma, loss and longer-term mental health problems. Many men may struggle to deal with such after-effects because they have not been encouraged to develop the skills to reflect on and communicate their emotions or seek help when struggling. They may be encouraged instead to 'bottle up' these difficult feelings, or turn to maladaptive coping strategies. There is evidence that rates of substance abuse and suicide increase among men in the wake of disasters, for instance, as well as the aforementioned exacerbation of domestic abuse (Enarson and Pease, 2016).

This points to wider difficulties men may experience in adapting to a changing climate. Not only are they more likely to be invested in the status quo, men may find it harder to accept that humans are inherently vulnerable in relation to nature, and dependent on it for our survival. This in turn may make it difficult to deal with the fear, stress and anxiety associated with unfolding climate breakdown, and adjust to the changes to one's way of life that it demands of us.

Paths forward

All of this demonstrates the urgent need to engage more with men and boys, from an early age and across the life-course, about climate change and environmental harm. Of course, this should be the case across the board, but there are particular issues which need to be taken into account when working with men and boys, which require a gender-sensitive – and indeed gender-transformative – approach. Work of this kind could be seen as a vital form of crime prevention, in terms of stopping environmental violence from happening in the first place – and there are clear overlaps here with existing violence prevention work. Research demonstrates that in-depth gender-transformative work with men and boys around opening up rigid, restrictive ideas of masculinity can bring about meaningful changes in attitudes and behaviours towards preventing violence and abuse (Flood, 2019). Work of this kind could also play a role in shifting men's relationships with the environment in healthier directions. Hultman and Pulé (2019) describe this as fostering 'ecological masculinities', although it could be argued that there is a need to move away from specific sets of expectations around masculinity altogether, rather than simply constructing more positive forms.

In addition to education about the climate crisis and other environmental issues, a cornerstone of this work could be becoming more mindful about the impacts of our actions, both on other people and the planet. It could involve helping men and boys to develop more deep-rooted connections with nature, and an ethic of care for the environment (Pease, 2019b) – while encouraging collective critical reflection about wider gender norms and expectations and how these may discourage such practices. Research suggests that engaging in caregiving such as active fatherhood can have transformative impacts on men, and help them to become more rooted in and connected to wider society (Barker et al, 2021). It is quite possible that becoming actively involved in caring for the environment – which can involve a wide range of different activities – could have similar effects, and help foster a deeper sense of connectedness to the earth and to other living things, as well as to society.

Work of this kind could also therefore have significant positive impacts on men and boys themselves. There is a wealth of research demonstrating the benefits for physical and mental health of spending time in green spaces, caring for the environment and fostering a deeper sense of connectedness with nature (Martin et al, 2020). An emphasis on being independent, invulnerable and detached from others frequently has detrimental consequences for the wellbeing of men and boys themselves – so becoming more connected to the environment can provide a powerful antidote to that. However, the ultimate goal of such work should also be to stimulate more awareness and action regarding climate change mitigation and adaptation – which is clearly also in men and boys' own interests, given the considerable damage it will have, and is already having, on people's health and wellbeing.

Conclusion

When addressing the role of men and boys in the climate crisis, it is also important not to become focused solely on individual responsibilities or solutions. This is a systemic problem which requires major structural change if it is going to be seriously tackled. Individual men and boys changing their mindsets and behaviours in relation to the environment is not sufficient to reduce global heating on its own. This also applies to constructions of masculinity, which are not simply reproduced by and among individuals, but embedded within and reinforced by the institutions and structures of society. This demonstrates the importance of social movements pushing for deeper and more systemic social change – and more men and boys becoming actively involved in them. Of course, each of us as individuals have important contributions to make to such movements, as well as engaging with the people around us and pushing for transformations in the organisations and communities we are part of. Furthermore, calling for dramatic action to tackle climate change surely requires us to embody this in our own practices, too. Given that men continue to hold most positions of power – and thus many of the levers of change – within society, this is particularly important to reflect on for men and boys.

This also underlines that some men do have a lot more power and influence, and thus responsibility for the climate crisis, than others. The purpose of this chapter is not to blame individual men for structurally embedded inequalities, but some individuals do need to be held to account for the massive harm their decisions are causing to people and planet, and there are important questions about what meaningful climate 'justice' would look like in this regard (White, 2018). Currently, the wealthiest individuals – who are overwhelmingly men – are far less likely to feel the worst effects of climate change than others, despite their primary role in causing it. So while this is not a problem of individuals, we must all take responsibility for our own interactions with the environment, commensurate with the impact they have.

It is also important to recognise that it is not just men who are contributing to climate change – we all are, to varying degrees. There are ways in which dominant constructions of femininity, what Connell (2009) describes as 'emphasised femininity', serve to legitimise fossil fuel consumption and environmental destruction. For instance, the pressures exerted on women and girls to live up to narrow beauty standards by the cosmetics and fashion industries has significant environmental ramifications (Scott, 2010). The point is that patriarchal gender orders as a whole – which are defined by male domination but which most of us help to reproduce to some extent – are fundamental in shaping human societies which are unsustainable with the welfare of the planet.

There is nothing 'natural' about this social order, and if men are contributing more to global heating, it is not because they are somehow 'innately' less nurturing of nature than women. Indeed, dominant social norms and expectations can suppress the caring behaviours and environmental connectedness that many boys express as they grow up, and gradually learn that 'real men' do not openly express care and concern for other living things. It is also important to avoid painting simplistic or essentialist dichotomies of men as 'bad' and women as 'good', 'passive victims'. There are numerous examples of people of all genders demonstrating an ethic of care for the environment and resistance to 'business as usual' – and vice versa, of complicity in a status quo which they as individuals may derive benefits from. However, there are also clear social patterns and inequalities which can be observed, which are shaped in no small part by constructions of gender.

Gender must therefore be central to the analysis and solutions that green criminologists – and criminologists more broadly – put forward. Men and masculinities are a crucial component of this, given their sizeable yet often hidden role in causing the climate crisis and a range of other environmental harms (Groombridge, 1998). However, there are also risks associated with such a focus; that in the process, men can be re-centred even more than is already the case within criminology, or that they come to be seen as the main 'victims' of masculinity (Pease, 2019a). A relational approach is therefore vital, recognising that men and masculinities are never operating in isolation. They are always affected by and impacting on others, both human and nonhuman life, and shaped significantly by other intersecting systems of power.

We are at a pivotal moment when it comes to social relations with the environment; there is still time to avert the worst effects of climate change, but only dramatic transformations to how society is organised can achieve this. There was hope that the COVID-19 pandemic would help to instigate such shifts, given that as a zoonotic virus it demonstrated the unsustainability of human interactions with nature, with the threat of more regular pandemics to come. Generally, however, there has been a dash by the forces of capital to return to the status quo and the ceaseless pursuit of profit and growth, while inequality, war and division has deepened in many contexts. Nonetheless, COVID-19 did expose that a different world is possible, with governments forced to act on an unprecedented scale in the name of public health, and communities coming together to support and care for each other. It also led to changes in people's relationships with nature, with spending time in green space one of the only things people were allowed to do outside of their homes during lockdowns (ONS, 2021b), and the slowdown of day-to-day life caused many to pause and reflect. All of this shows that it is quite possible for societies to engage in the kind of radical action needed to tackle the climate crisis, and cultivate more interconnected relationships with the environment that we depend on. However, this kind of change will only happen if more of us, including men and boys, start advocating for it.

References

Aavik, K. and Velgan, M. (2021) Vegan men's food and health practices: A recipe for a more health-conscious masculinity? *American Journal of Men's Health*, 15(5), 1–14.

Anshelm, J. and Hultman, M. (2014) A green fatwā? Climate change as a threat to the masculinity of industrial modernity. *NORMA: International Journal for Masculinity Studies*, 9(2), 84–96.

Atkins, E. (2022) 'Bigger than Brexit': Exploring right-wing populism and net-zero policies in the United Kingdom. *Energy Research and Social Science*, 90, 102681, 1–5.

Bacevic, J. (2021) Epistemic injustice and epistemic positioning: Towards an intersectional political economy. *Current Sociology*. doi: 10.1177/00113921211057609

Balkmar, D. (2019) Men on the move: Masculinities, (auto) mobility and car cultures. In L. Gottzén, U. Mellström and T. Shefer (eds) *Routledge International Handbook of Masculinity Studies*. Abingdon: Routledge, pp 351–9.

Barker, G., Garg, A., Heilman, B., van der Gaag, N. and Mehaffey, R. (2021) *State of the World's Fathers: Structural Solutions to Achieve Equality in Care Work*. Washington, DC: Promundo-US.

Brough, A.R., Wilkie, J.E., Ma, J., Isaac, M.S. and Gal, D. (2016) Is eco-friendly unmanly? The green-feminine stereotype and its effect on sustainable consumption. *Journal of Consumer Research*, 43(4), 567–82.

Bucholz, K. (2022) How has the number of female CEOs in Fortune 500 companies changed over the last 20 years? *World Economic Forum*. Available at: www.weforum.org/agenda/2022/03/ceos-fortune-500-companies-female (accessed 10 July 2022).

Buckingham, S. (2017) Gender and climate change politics. In S. MacGregor (ed) *Routledge Handbook of Gender and Environment*. Abingdon: Routledge, pp 384–97.

Carlsson-Kanyama, A., Linden, A.-L. and Thelander, A. (1999) Gender differences in environmental impacts from patterns of transportation: A case study from Sweden. *Society and Natural Resources*, 12(4), 355–69.

Carlsson-Kanyama, A., Nässén, J. and Benders, R. (2021) Shifting expenditure on food, holidays, and furnishings could lower greenhouse gas emissions by almost 40%. *Journal of Industrial Ecology*, 25(6), 1602–16.

Castañeda Camey, I., Sabater, L., Owren, C. and Boyer, A.E. (2020) *Gender-based Violence and Environment Linkages: The Violence of Inequality*. Gland: IUCN.

Cohen, M. (2014) Gendered emissions: Counting greenhouse gas emissions by gender and why it matters. *Alternate Routes: A Journal of Critical Social Research*, 25, 55–80.

Cohen, M. and MacGregor, S. (2020) *Towards a Feminist Green New Deal for the UK: A Paper for the WBG [Women's Budget Group] Commission on a Gender-equal Economy*. London: WBG and WEN.

Connell, R.W. (1990) A whole new world: Remaking masculinity in the context of the environmental movement. *Gender and Society*, 4(4), 452–78.

Connell, R.W. (2009) *Gender: In World Perspective*. Cambridge: Polity.

Connell, R.W. and Wood, J. (2005) Globalization and business masculinities. *Men and Masculinities*, 7(4), 347–64.

Daggett, C. (2018) Petro-masculinity: Fossil fuels and authoritarian desire. *Millennium: Journal of International Studies*, 47(1), 25–44.

EIGE (European Institute for Gender Equality) (2012) *Review of the Implementation in the EU of Area K of the Beijing Platform for Action: Women and the Environment – Gender Equality and Climate Change*. Luxembourg: EIGE.

Enarson, E. and Pease, B. (2016) The gendered terrain of disaster. In E. Enarson and B. Pease (eds) *Men, Masculinities and Disaster*. Abingdon: Routledge, pp 3–20.

Ergas, C., Greiner, P.T., McGee, J.A. and Clement, M.T. (2021) Does gender climate influence climate change? The multidimensionality of gender equality and its countervailing effects on the carbon intensity of well-being. *Sustainability*, 13(7), 3956, 1–23.

Flood, M. (2019) *Engaging Men and Boys in Violence Prevention*. Basingstoke: Palgrave Macmillan.

Galtung, J. (1990) Cultural violence. *Journal of Peace Research*, 27(3), 291–305.

Global Witness (2021) *Last Line of Defence: The Industries Causing the Climate Crisis and Attacks against Land and Environmental Defenders*. London: Global Witness.

Graham, A. and Metz, D. (2017) Limits to air travel growth: The case of infrequent flyers. *Journal of Air Transport Management*, 62, 109–20.

Groombridge, N. (1998) Masculinities and crimes against the environment. *Theoretical Criminology*, 2(2), 249–67.

Hewitt, K. (2016) Foreword. In E. Enarson and B. Pease (eds) *Men, Masculinities and Disaster*. Abingdon: Routledge, pp xvii–xxi.

Hultman, M. and Pulé, P.M. (2019) *Ecological Masculinities: Theoretical Foundations and Practical Guidance*. Abingdon: Routledge.

IPU (Inter-Parliamentary Union) and UN Women (2021) *Women in Politics 2021 Map*. Available at: www.unwomen.org/en/digital-library/publicati ons/2021/03/women-in-politics-map-2021 (accessed 10 July 2022).

IUCN (International Union for Conservation of Nature) (2021) *New Data Reveals Slow Progress in Achieving Gender Equality in Environmental Decision Making*. Available at: www.iucn.org/news/gender/202103/new-data-reve als-slow-progress-achieving-gender-equality-environmental-decision-mak ing (accessed 10 July 2022).

Jorgenson, A.K. and Clark, B. (2016) The temporal stability and developmental differences in the environmental impacts of militarism: The treadmill of destruction and consumption-based carbon emissions. *Sustainability Science*, 11(3), 505–14.

Kaufman, M. (1987) The construction of masculinity and the triad of men's violence. In M. Kaufman (ed) *Beyond Patriarchy: Essays by Men on Pleasure, Power, and Change*. Oxford: Oxford University Press, pp 1–29.

Kruse, J. (2014) Women's representation in the UN climate change negotiations: A quantitative analysis of state delegations, 1995–2011. *International Environmental Agreements: Politics, Law and Economics*, 14(4), 349–70.

Le Masson, V., Lim, S., Budimir, M. and Selih Podboj, J. (2016) *Disasters and Violence against Women and Girls: Can Disasters Shake Social Norms and Power Relations?* London: Overseas Development Institute.

MacGregor, S. (2009) A stranger silence still: The need for feminist social research on climate change. *The Sociological Review*, 57(2_suppl), 124–40.

Martin, L., White, M.P., Hunt, A., Richardson, M., Pahl, S. and Burt, J. (2020) Nature contact, nature connectedness and associations with health, wellbeing and pro-environmental behaviours. *Journal of Environmental Psychology*, 68, 101389, 1–12.

Mies, M. and Shiva, V. (2014) *Ecofeminism*, 2nd edn. London: Zed Books.

Nagel, J. (2017) The continuing significance of masculinity. *Ethnic and Racial Studies*, 40(9), 1450–9.

Nixon, R. (2011) *Slow Violence and the Environmentalism of the Poor*. London: Harvard University Press.

ONS (Office for National Statistics) (2021a) *Data on Public Attitudes to the Environment and the Impact of Climate Change, Great Britain*. Available at: www.ons.gov.uk/peoplepopulationandcommunity/wellbeing/datasets/ dataonpublicattitudestotheenvironmentandtheimpactofclimatechangegreat britain (accessed 10 July 2022).

ONS (Office for National Statistics) (2021b) *How Has Lockdown Changed Our Relationship with Nature?* Available at: www.ons.gov.uk/economy/ environmentalaccounts/articles/howhaslockdownchangedourrelationshi pwithnature/2021-04-26 (accessed 10 July 2022).

Pearse, R. (2017) Gender and climate change. *Wiley Interdisciplinary Reviews: Climate Change*, 8(2), e451.

Pearson, A.R., Ballew, M.T., Naiman, S. and Schuldt, J.P. (2017) Climate change communication in relation to race, class, and gender. In *Oxford Research Encyclopedia of Climate Science*. Oxford: Oxford University Press, pp 1–38.

Pease, B. (2019a) *Facing Patriarchy: From a Violent Gender Order to a Culture of Peace*. London: Zed Books.

Pease, B. (2019b) Recreating men's relationship with nature: Toward a profeminist environmentalism. *Men and Masculinities*, 22(1), 113–23.

Räty, R. and Carlsson-Kanyama, A. (2010) Energy consumption by gender in some European countries. *Energy Policy*, 38(1), 646–9.

Rhode, D.L. (2019) #MeToo: Why now? What next? *Duke Law Journal*, 69(2), 377–428.

Rick, K., Martén, I. and Von Lonski, U. (2017) *Untapped Reserves: Promoting Gender Balance in Oil and Gas*. Boston, MA: World Petroleum Council and Boston Consulting Group.

Rothgerber, H. (2013) Real men don't eat (vegetable) quiche: Masculinity and the justification of meat consumption. *Psychology of Men and Masculinity*, 14(4), 363–75.

Scott, B.A. (2010) Babes and the woods: Women's objectification and the feminine beauty ideal as ecological hazards. *Ecopsychology*, 2(3), 147–58.

Skarha, J., Peterson, M., Rich, J.D. and Dosa, D. (2020) An overlooked crisis: Extreme temperature exposures in incarceration settings. *American Journal of Public Health*, 110(S1), S41–S42.

Sultana, F. (2014) Gendering climate change: Geographical insights. *The Professional Geographer*, 66(3), 372–81.

Thurston, A.M., Stöckl, H. and Ranganathan, M. (2021) Natural hazards, disasters and violence against women and girls: A global mixed-methods systematic review. *BMJ Global Health*, 6, e004377.

Tiihonen, J., Halonen, P., Tiihonen, L., Kautiainen, H., Storvik, M. and Callaway, J. (2017) The association of ambient temperature and violent crime. *Scientific Reports*, 7(1), 1–7.

Togami, C. and Staggenborg, S. (2022) Gender and environmental movements. In M. Grasso and M. Giugni (eds) *Routledge Handbook of Environmental Movements*. Abingdon: Routledge, pp 419–33.

UNFCCC Secretariat (2022) *Gender Composition and Progress on Implementation: Report by the Secretariat*. United Nations FCCC/CP/2022/3.

Westmarland, N., Almqvist, A.-L., Egeberg Holmgren, L., Ruxton, S., Burrell, S.R. and Delgado Valbuena, C. (2021) *Men's Activism to End Violence Against Women: Voices from Spain, Sweden and the UK*. Bristol: Policy Press.

White, M. (2021) Greta Thunberg is 'giving a face' to climate activism: Confronting anti-feminist, anti-environmentalist, and ableist memes. *Australian Feminist Studies*, 36(110), 396–413.

White, R. (2018) *Climate Change Criminology*. Bristol: Policy Press.

WHO (World Health Organization) (2014) *Gender, Climate Change and Health*. Geneva: WHO.

Wright, C. and Nyberg, D. (2017) An inconvenient truth: How organizations translate climate change into business as usual. *Academy of Management Journal*, 60(5), 1633–61.

5

Reconceptualising Gendered Dimensions of Illegal Wildlife Trade in Sub-Saharan Africa through Legal, Policy and Programmatic Means

Helen U. Agu, Josiah C. Ogbuka and Meredith L. Gore

Introduction: Overview of illegal wildlife trade

Illegal wildlife trade (IWT) refers to the unlawful, unregulated and unsustainable catch, trafficking, utilisation, acquisition and extermination of animals or plants in violation of local and international laws, conventions and treaties (Kurland et al, 2017). These crimes constitute considerable business at local, national, regional and global scales, which maximally threaten natural balance, state and individual economic stability, human health and livelihood, as well as protection of natural resources and criminal jurisprudence. According to Elliott (2007), IWT is worth 25 per cent of the global wildlife trade, which significantly poses threat to biodiversity conservation and management. It has also been reported that IWT is worth an estimated US$20 billion annually, without including the unlawful timber trade and unlawful fishing (Barber-Meyer, 2010; Wilson-Wilde, 2010), while Fison (2011) and GFI (2011) reported estimated ranges of US$7.8–10 billion and US$10–20 billion respectively each year. The WWF (2012) observed that the combination of IWT, illegal forest trade, and fishing IWT constitute the fourth largest illegal business after human trafficking, drugs and adulterated goods and products. To provide more insight into the species-specific effect of IWT, the Annual Progress Assessment (2015) reported the depletion of approximately 20 per cent of the African elephant population to 400,000 in the past decades, while 20 wild rhinos were eliminated in one year alone through rustling, according to conservationists' estimates.

Recently, emerging issues from research have suggested the existence of gender dimensions of wildlife crime, although limited or few empirical studies have emphasised the claim. In contrast, most authors have conceptualised wildlife crime from conservationists' perspectives (Kareiva and Marvier, 2012), without considering the relevance of crime scientists in examining and determining the multivariate complexities in wildlife crime (cf Moreto, 2015), as well as the gender dimensions of these unlawful acts. Generally, the gender concept entails the sociocultural customs or rules relating to what are appropriate for men and women in any society. These rules or customs about men and women inherently determine the demand and supply of these products, which also underpin the sociocultural values associated with their purported natural attributes or properties (FAO, 2016). In essence, it is more or less a perspective based on stereotyped customary definition or relativist conditioning of male and female roles and responsibilities, as well as their rights and privileges in any society. It has been suggested that some of the precursors that influence IWT and other related crimes are gender-induced or gender-related (Elliott, 2007). For example, in some countries, it is generally accepted that feeding on some components of wild animals will automatically transmit their innate attributes to humans, such attributes as the ferocity of tigers and the sexual vigour of rhinoceros for men, as well as the meekness of deer for women (McElwee, 2012; Milliken and Shaw, 2012). In addition, while men are typically involved in illegally traded animal-dependent health products, gendered investigations involving some animal-dependent health products utilised for remedying health cases, such as bile from bear, remain understudied. Rather non-gendered trafficking of animal-dependent health products has been reported to attract equal usage by both men and women (Drury, 2011).

This chapter aims to advance discourse about gendered dimensions of IWT in sub-Saharan Africa in the context of the existing regional and national wildlife legal frameworks, policies and programmes to explore the potentials of gender-driven wildlife management interventions in the region and member states. The chapter outlines gendered perceptions of and motivations for wildlife trafficking in sub-Saharan Africa, key attributes of the Sub-Saharan Africa wildlife legal ecosystem in the context of gender and IWT, and towards reinventing laws, policies and programmes for combating wildlife trafficking in sub-Saharan Africa. Enhanced discourse can help contribute to more sustainable wildlife management and conservation in the region and among member states because gender impacts the effectiveness of interventions in environmental management (Cirelli and Morgera, 2010). Deepening understanding about gendered differentiations towards strengthening the regional and national wildlife legal instruments, policies, regulations and programmes in sub-Saharan Africa can also influence constituent countries. The insights gained from this contribution can also assist other wildlife-dependent regions and nations in streamlining their wildlife legal frameworks, regulations, policies and programmes to reflect gendered differentiations such as gender-based perceptions of and motivations to engage in IWT and help minimise IWT globally, regionally, nationally or locally.

The gendered dimensions of the illegal wildlife trade need not be ignored

IWT, the illegal capture and/or trade of protected species and goods produced from them, poses multiple direct and indirect threats to species and their ecosystems (Wyatt, 2013). IWT is considered a serious crime in part because of its global spread and magnitude (UNODC, 2020). Efforts to combat IWT commonly focus on identifying and analysing IWT drivers, with the prevailing narrative that the key driver is the symbiosis of poverty and need with individuals having substantial financial resources and demand (Nellemann et al, 2016). This commonly acknowledged view is reflected in the majority of activities to combat IWT, focusing on two-fold economic survival challenges of poverty and demand across geographical scales; such activities mostly ignore that IWT has developed and evolved in a world divided by gender (Seager, 2020). Considering the gendered differentiation and inequalities of IWT is both prudent *and* prescient. The United Nations affirmed this need when it adopted a resolution charging member states to mainstream gender in IWT in part to help achieve more inclusive and productive activities (A/RES/75/311) (UNGA, 2021). Broadly, the resolution supported gender-receptive IWT projects to generate multiple wins for gender equality, human rights and conservation. Implementing non-gender-receptive measures to combating IWT produces bias across activities and procedures. Anthem (2018) noted that ignoring the gendered dimensions of IWT was akin to tackling the problem with one hand tied behind our back.

It is well known that men and women's perceptions and interactions with their environment, biodiversity and natural resources can differ in many ways; for example, their reliance on natural resources, resource use, conservation or resource management priorities, and knowledge about the state of environmental resources may differ (UNEP, 2016; Westerman, 2017). Because views about nature and forms of interactions between nature and its resources can be gender-differentiated, it is appropriate to consider the differences in the context of efforts to combat IWT. Specifically, as new legal frameworks, policies and programmes are being conceived to combat serious trends or motivations to engage in the trade, spotlighting gendered dimensions of IWT across these interventions can be considered a necessity (Martino, 2008). Gender-based differentiations, involving perceptions of and motivations for IWT, are rarely studied or represented in sub-Saharan countries, even though many countries have developed national and/or regional wildlife legal frameworks, policies and programmes or increased penalties for wildlife-related rule breaking (for example, Angola, Kenya, Malawi, Nigeria, South Africa) (Cirelli and Morgera, 2010; Price, 2017).

The basis for introducing new, existing and further developing discussions and assessments exploring gendered motivations associated with IWT rests upon the need to refocus efforts to combat IWT. The consequences and advantages of mainstreaming gender into efforts to combat IWT are also gender-differentiated. Men and women encounter different challenges (for example, protecting the rights to traditional land ownership, especially in traditional African communities has been characterised by gendered advantage to men as against women, which commonly results in limits to women's decision-making roles in resource use and management [Deere et al, 2012; Kieran et al, 2015; Doss et al, 2017; Chigbu et al, 2019]). Gender-differentiated motivations relating to IWT can be driven by diverse gender-based codes of conduct, demands, perception and generalisations, as well as economic, social, political and cultural characteristics associated with being a woman or a man (Seager, 2021). For instance, one investigation conducted adjacent to Zambia's Kafue National Park showed 56 per cent of men and 39 per cent of women firmly consented to the importance of elephants in the environment, while many of the men and women sampled in the same survey rejected the killing of elephants that constitute threats to the safety of families and crops (GRI/IFAW, 2017). This example typifies a gendered perception and motivation towards the conservation and protection of elephants, involving men and women around the park, indicating attitudinal dominance of men over women in favour of conservation of the elephant population.

Gendered perceptions of and motivations for wildlife trafficking in sub-Saharan Africa

Although there have been some considerable efforts at a global scale to alter gendered ethics affecting wildlife trafficking, such as MenEngage Alliance, in collaboration with allied bodies such as the Indian-based Forum to Engage Men, Kenyan-based MenEngage and the Caribbean Men Action Network, gendered dimensions of IWT appeared not to attract the attention of these groups or other wildlife researchers (Seager, 2021). Gender-based codes of conduct, capacities and generalisations about masculinity and femininity, as well as economic, social, political and cultural characteristics associated with being a woman or a man, all influence perceptions and incentives underpinning wildlife trafficking or IWT at the local, national, regional and trans-boundary levels, including in the sub-Saharan African region (Seager, 2021). Almost all resource-dependent cultures around the world, for example, have men who hunt and women who process the food. Gendered codes of conduct and roles, on the other hand, are defined and manifest on a local level, which can also contribute to shaping gendered orientation towards wildlife use, protection and management Arlbrandt et al, 2021). Example include the predominance of men and poor representation of women in most global and local organisations that participate actively in policies relating to IWT, including governments, research institutions, non-governmental organisations (NGOs), academia, and so on (GRI/IFAW, 2017). On the other hand, gender-based codes of conduct and roles can also incentivise unusual/extreme gender differentiation on IWT. For example, the pressure on men to demonstrate manly attributes may be increasing as illegal wildlife harvesting and anti-illegal wildlife harvesting operations in some African regions become more sophisticated and dangerous (Henson et al, 2016). This can be illustrated by the catastrophic increase in unlawful logging of scarce timber species in Madagascar, possibly because of gender-motivated pressure manifest in the massive entry of casual male workers into the small communities in the area, thereby generating more sophisticated and destructive impacts on wildlife (Seager, 2021). Men who do not embrace these ethics may be at risk of falling short of being adequate economic providers, as well as breaking the cultural pressure driving men to exhibit masculine traits and attraction to the opposite sex, without which they are regarded as feminine and publicly intolerable. In the same way, women can raise the stakes on manliness by challenging men to provide for their families and criticising them if they do not do so to their satisfaction. Seager (2021) reported that according to several persons interviewed, men often feel cajoled to participate in illegal wildlife harvesting to fulfil their gender-norm role as family breadwinners. Furthermore, being a skilled hunter fits many men's perceptions of how to propagate masculinity, suggesting that this perception or pattern that considers the act of manliness as a product of experience in hunting drives (some) IWT. On the other hand, Iori (2020) observed that the drivers and patterns determining wildlife crime among women also relate to gender-based attractions to IWT. Such drivers included, among others, meeting the demand for resources (for example, medications, money and food), protection (for example, socioeconomic, cultural marginalisation), exploitation (for example, sex-for-food, sex-for-accommodation), job crisis in a legitimate economy (for example, high rate of unemployment especially in legal markets).

Investigations into gendered perceptions of and motivations for IWT in sub-Saharan Africa remain sparse. However, several gendered case studies in the region exemplify the gender-based patterns and motivations. Mmassy and

Røskaft (2013) and Lowassa et al (2012) showed a few of the ways women make undue demands on men to hunt wildlife during group interactions in western Serengeti, Tanzania. The first example indicated how women's verbal expressions and behaviour to their husbands or men can incentivise IWT. For instance, the price of meat from commercial meat sellers, which may sometimes be costly, makes certain women perceive and express confidence in the ability of their husbands to provide better alternatives through IWT rather than legal means. In fact, some wives may taunt their husbands, pressurising them to assert their manliness and authority through illegal hunting, even threatening to deny food if men do not provide wild meat (Lowassa et al, 2012). The second example indicated men's involvement in IWT as a motivation to satisfy women's appreciation for money. This pattern reflects a widespread or generalised scenario, not just in the Serengeti, but in other parts of sub-Saharan Africa. It is common knowledge in many contexts that a man without money is despised. To satisfy monetary/material demands from wives and discourage them having extramarital relationships with other men, some husbands reluctantly resort to IWT to save face. The trend promotes a 'get rich quick attitude' and seemingly discourages legitimate and legal traditional food-production activities among men, such as farming. As a result, some men venture into IWT to express their love and care and provide financially for their women or wives. In contrast to the aforementioned feminine-motivated sentiments or attitudes that promote IWT among men, men's ambition to attain greatness, fame, wealth, prestige and revered social status, especially in traditional African communities such as the southeast region of Nigeria, has also been found to lead to the killing and consumption of certain wild meats, for example, lion or tiger. Often, this is generally acknowledged as a great exploit, which sometimes earns the person a chieftaincy title or special recognition in the community, such as 'Ogbuagu' (killer of tiger) or 'Otagburuagu' (the one that tore tiger to death). In sum, four overarching factors have been generally identified as drivers of IWT in this regard, namely:

1. a lack of alternative revenue sources for men to provide for women who depend on them;
2. gender norms encouraging women to support and perpetuate illegal activities;
3. a reluctance to wait for farming harvest period, where the yield is often viewed as seasonal and unpredictable (Seager, 2021); and
4. men's ambition to attain greatness, fame, wealth, prestige and revered social status.

Lunstrum and Givá (2020) illustrated, through a field survey in Mozambique, how well-known illegal adult men hunters motivated and pressured boys to participate in IWT. For example, some boys participate in rhino trafficking as a reaction to economic pressure and ridicule perpetrated by well-known illegal wildlife hunters and occasionally by family members. Evidence from a regional leader showed adult men may castigate boys as being weaklings and underperformers, attributing the perceived weakness and fear to their non-participation in IWT. Sometimes, wives and relatives of elder men cajole youths, which ultimately motivates them to identify with an IWT network. The outcome of Lunstrum and Givá's (2020) research suggested a gendered differentiation in perception and motivation of IWT among men only, thereby generating further questions into intergenerational motivation to IWT (Seager, 2021). Such IWT driven by gendered differentiation in perception and motivation demonstrated by men only is known to influence the legal and illegal felling of some forest species, with the associated impacts on the communities and the populace. For example, after 2009, illegal logging of rare wood in Madagascar accelerated dramatically. Unlike IWT, commercial-scale illegal logging usually involves a massive influx of transient working men, who set up medium- to long-term camp sites in the forests or boost the population of small, isolated villages and towns. The lawful harvest in the Masoala Park and portions of Makira Park was accompanied by the entrance of thousands of migrants from all over Madagascar into the municipalities of Ambohitralanana and Ampanavoana in particular between October 2013 and January 2014. Between June and September of 2014, the same scenario occurred again. The transitory movements of men for illegal logging increased insecurity and criminal behaviour directed at local people, such as increased amphetamine and/or alcohol consumption among boys, rape of girls, and an increase in prostitution (Ratsimbazafy et al, 2016). In the Ambohitralanana community, roughly one-third of the girls over 14 was engaged in sex-for-money during the peak season of timber exploitation (Ratsimbazafy et al, 2016). More so, some scholars have estimated that between 40 per cent and 50 per cent of women over the age of 12 dropped out of school to pursue sex-for-money and other related activities in many other communities (Ratsimbazafy et al, 2016).

On the other hand, within Kafue National Park, periodic illegal wildlife hunting was motivated by sexual activities involving unmarried women, widows and men illegally hunting, who take advantage of the economically disadvantaged and vulnerable women (Seager, 2021). The illegal hunters rent houses for disadvantaged and vulnerable women, which is usually paid for either by cash or in the form of illegally hunted meat, near their hunting camps in order to engage in casual sex with them (personal communication with Game Rangers International representative). The concealment of this sex-for-accommodation by these underprivileged women assisted in shielding them from public disgrace, and most importantly, encourages illegal hunters to continue with their illegal business. Another gender dimension to IWT that results in disadvantaging the women includes gender-driven cultural norms, which restricts women from fishing in boats in most regions. The restriction of women from boat fishing can contribute to the conventional behaviour that accords men greater access to economic power than women. For example, this scenario is characterised by narrow livelihood options for women, often as a result of climate change, and illegal wildlife hunting-driven ecosystem instability, and sexual exploitation to produce 'fish-for-sex' schemes, which are now common in both inland and marine fishing systems in East Africa (UNEP, 2016). One observer in Lake Victoria noted:

> Fishermen have the power to choose which woman will unload their catch, and in exchange, they give the women a share of the catch. How much fish is fair payment is up to the fisher's decision. Fish can be used as payment for women's processing enterprises, which dry and fry fish for resale as a finished product. (Kosome et al, 2020)

In fact, women must pay the fishers (equivalent to US$4) for the right to work for them, as well as have sex with them, to get work unloading fish and to receive a portion of the catch for their processing companies. Research among Zambian fish workers revealed that 31 per cent of women fish sellers had an institutionalised fish-for-sex connection (Béné et al, 2007). To break this gendered illegal wildlife hunting-based anti-social relationship, women's efforts to establish their own fishing programmes to avoid fish-for-sex exploitation have met with success, uncertainty and, in some cases, additional bribery demands from men in authority positions (Smith, 2017). As a result, the bribery demands from men put financial pressure on women, which has the potential to promote the fish-for-sex relationship due to financially disadvantaged condition of local women, who find it difficult to satisfy these demands. Similar scenarios have been reported at Sinohydro Karuma project in Uganda, where a surge in casual labour has been observed, suggesting a general undetectable synergy between sex and wildlife trafficking (Hongjie, 2019; Watchdog, 2020). The Karuma building project has seen an influx of approximately 6,000 young men construction workers. The presence of this labour force in this area (and possibly elsewhere) creates local demand markets for wildlife items, particularly pangolin, for food or traditional medicine. Women's participation in IWT for traditional medicine is then inextricably tied to a parallel sex work which may or may not be done with their consent. Many widows, school drop-outs, separated or divorced women, and school-age girls have been drawn to the Karuma building project from as far as Nebbi, Gulu, Oyam, Wakiso, Kampala and Mbarara, in order to satisfy men's sexual desires (Seager, 2021).

Another dimension of gender-motivated attitudes promoting IWT can be connected to reluctance in uncovering the crime. The attitude involving reluctance to uncover IWT has been demonstrated among men and women in different contexts. For example, in South Africa, women from the poorest households were found to be less willing to report than men, which the authors concluded might be connected to fear of threats to life and diverse security sensitivities (Sundstrom et al, 2019). In Zambia, on the other hand, there were significant overall gender differences in willingness to report to the extent that 60 per cent of women and 47 per cent of men said they were afraid to report IWT activities to authorities (GRI/IFAW, 2017). The reasons for this may include a high propensity to IWT activities due to the illicit income it generates for economically disadvantaged local women, and fear of being attacked for either gender. The higher proportion of women unwilling to report IWT in Zambia suggests that the risk perception and likely consequences of reporting such crime to the authorities is higher among women than men.

From the standpoint of wildlife consumption patterns, a study in northern Uganda revealed gendered differentiation between men and women regarding the pattern and motivation for eating wild meat. For instance, the cooks, mostly women, reported cherishing the taste of domestic meats (for example, chicken, goat), whereas the men hunters loved the taste of wild meat (Seager, 2021). The study suggested men's preference for wild meat may play a role in wild meat consumption, as four of the top five preferred foods by hunters were wild animals rather than domestic options. However, this finding is not supported by the cooks' preferences, which prefer domestic meat options and believe domestic meat options are more nutritious to wild meat, suggesting that men family members may have more influence over household food choices (Dell et al, 2020). Men's preference for wild meat may also correlate with the dominance of men in cases involving seized wildlife harvested illegally from 2015 to 2020 in Tanzania, Uganda, Kenya and Ethiopia, where 38 per cent of seized wild meat illegally traded was linked to men suspects compared to 15 per cent linked to women suspects (Seager, 2020). This finding also elaborated the dominance of men in illegal wildlife activities, amounting to 96 per cent in East Africa (Seager, 2020).

Key attributes of the sub-Saharan Africa wildlife legal ecosystem in the context of gender and illegal wildlife trade

Wildlife regulation and IWT have been long implemented through international legal frameworks, such as Convention on Biological Diversity (Rio de Janeiro, 1992), Convention concerning the Protection of the World Cultural and Natural Heritage (WHC, Paris, 1972), Convention on the Conservation of Migratory Species of Wild Animals (Bonn, 1979), United

Nations Convention Against Corruption (New York, 2003) and United Nations Convention against Transnational Organised Crime (New York, 2003). Essentially, these international legal instruments are pivotal to drawing up and reviewing functional regional and country-specific legal frameworks enhancing responsible wildlife management practices (Birnie and Boyle, 2002; Morgera and Wingard, 2009).

On a global spectrum, the Convention on International Trade in Endangered Species of Wild Fauna and Flora CITES can be regarded as the umbrella legal framework with provisions relevant to other regional, national and local legal frameworks for wildlife trade, use, management and conservation, which supports protection of wildlife species drifting towards total disappearance (Appendix I) because of overexploitation, illegal hunting and IWT. The second category of species regulated under CITES include vulnerable species, whose exploitation beyond the permissible conservation and protection limits may translate to potential extinction (Appendix II), while the third category is referred to as rare species, which require strict protection due to scarceness and patchy nature of their population (Appendix III) (Cirelli and Morgera, 2010). At least 19 sub-Saharan countries have ratified CITES (Price, 2017).

Within the sub-Saharan Africa region, regional wildlife-centred legal instruments include the Lusaka Agreement on Co-operative Enforcement Operations Directed at Illegal Trade in Wild Fauna and Flora (LATF) (Lusaka, 1994), with the Republics of Kenya, Congo, Uganda, Tanzania, Liberia, Zambia and the Kingdom of Lesotho, and signatories including South Africa, Ethiopia and Kingdom of Swaziland; other African nations have the right of assent to the instrument. The main objective of the LATF is to promote collaboration among parties to mitigate and in due course eradicate wildlife crime in Africa. The Arusha Declaration on Regional Conservation and Combating Wildlife/Environmental Crime (November 2014) calls for an inclusive record of actions towards enhancing across border cooperation on tackling wildlife/environment crime in order to encourage wise use. Among the signatories to the Arusha Declaration include the East African Community and South African Development Community (SADC) countries. Within the SADC, the Protocol on Wildlife Conservation and Law Enforcement to the SADC Treaty (Cirelli and Morgera, 2010) was enacted to set up inside the guiding principle of national legal instruments of parties a harmonised approach to responsible and sustainable use of wildlife products and to promote successful compliance to the regulations administering the resources (Selier et al, 2016). The Protocol also motivates parties to standardise legal instruments governing wildlife, put in place wildlife management activities, and develop region-based documentation for the state of wildlife and its management. The East African Community EAC enacted article 116 of the EAC Treaty (endorsed in 1999, amended in 2006), which mandates parties to set up joint and well-articulated policy for responsible and sustainable wildlife use and recreational locations across the six countries in the region. The article called on parties to put together policies that will assist in conserving inside and outside protected areas, as well as organise initiatives to control and monitor trespassing and illegal wildlife hunting.

A considerable number of sub-Saharan Africa nations have established national legal frameworks on wildlife use, management and protection (Table 5.1). It is useful to consider case studies of wildlife national legal frameworks of some sub-Saharan Africa countries in the context of gendered differentiations in wildlife use, protection and management, especially in relation to gender-based motivations to IWT in the region.

Reinventing laws, policies and programmes for combating wildlife trafficking in sub-Saharan Africa

The non-inclusion of gender differentiations and gendered motivations in wildlife legal and/or regulatory frameworks of the sub-Saharan African member nations have created a fundamental challenge demanding critical attention and interventions aimed at reinventing regional and national legal, policies and programmes for tackling IWT. The main objective of this section is, therefore, to provide some realistic and inclusive strategies aimed at mainstreaming gendered perspectives, motivations, codes of conduct, needs, capacities, generalisations, and so on, in developing/designing legal frameworks, policies and programmes that can assist in addressing IWT, which can contribute to sustainable wildlife management in sub-Saharan Africa. The place or relevance of mainstreaming gendered differentiations and gender-motivated activities in sustainable wildlife management is recently and progressively more recognised or identified (Agu and Gore, 2020), but there remains a considerable knowledge gap in the areas of legal provisions, policies and programmes (UNODC, 2020), aimed at combating illegal wildlife activities in sub-Saharan Africa. Even an interregional collaboration between Africa and Asia-Pacific nations known as the Africa Asia-Pacific Symposium on Strengthening Legal Frameworks to combat Wildlife Crime did not consider or include gender issues in the proposed amendments to the wildlife legal frameworks of nations that make up the two continents (Africa-Asia Pacific Symposium on Strengthening Legal Frameworks to Combat Wildlife Crime, 2017). The symposium had in attendance representatives of various wildlife institutions from 11 sub-Saharan Africa countries, including Nigeria, and 11 from the Asia-Pacific countries, and legislators from Lao People's Democratic Republic, the Philippines, Thailand and the United Republic of Tanzania.

Table 5.1: Summary of country-specific assessments of the state of gender and illegal wildlife trade in national wildlife legal and regulatory frameworks of some member states of the sub-Saharan African region

Serial number	Country	Wildlife legal and/or regulatory framework(s)	Gendered focus on IWT	Reference(s)
1	Angola	Decree No. 40.040 implemented by the Hunting Regulation of 1957 (as Revised in 1972), Forest, Wildlife and Protected Areas Law (2006)	Unrepresented	Cirelli and Morgera (2010)
2	Botswana	Wildlife Conservation and National Parks Act of 1992 (Principal legislation): Wildlife Conservation and National Parks (Lions) (Killing Restriction) Order (22nd April, 2005); Wildlife Conservation and National Parks (Cheetahs) (Killing Suspension) Order (22nd April, 2005); Wildlife Conservation and National Parks (Hunting and Licensing) Regulations (Section 92) (10th August, 2001); Wildlife Conservation (Possession and Ownership of Elephant Tusks or Ivory) Regulations (Section 92) (15th March, 1999); National Parks and Game Reserves Regulations (Under Section 92) (2000)	Unrepresented	Cirelli and Morgera (2010); Mauck (2013); EIA (2016); Price (2017)
3	Democratic Republic of Congo	Hunting Law of 1982	Unrepresented	Cirelli and Morgera (2010)
4	Kingdom of Lesotho	Historical Monuments, Relics, Fauna and Flora Act of 1967; Forestry Act of 1998 and Environmental Act of 2001	Unrepresented	Cirelli and Morgera (2010)
5	Kenya	Wildlife (Conservation and Management) Act of 1976; Prevention of Organised Crime Act; The Proceeds of Crime and Anti-Money Laundering Act; The Evidence Act; The Criminal Procedure Code; Environmental Management and Coordination Act; and The EAC Customs Act.	Unrepresented	Cirelli and Morgera (2010); Weru (2016); Price (2017)
6	Madagascar	Law No. 90–033 (principal legislation): Ordinance No. 60–126 regulates wildlife capture, fishing and safeguarding of wildlife. Law No. 2001-005 provides code for Protected Area Management in relation to wildlife	Unrepresented	Cirelli and Morgera (2010)

(continued)

Table 5.1: Summary of country-specific assessments of the state of gender and illegal wildlife trade in national wildlife legal and regulatory frameworks of some member states of the sub-Saharan African region (continued)

Serial number	Country	Wildlife legal and/or regulatory framework(s)	Gendered focus on IWT	Reference(s)
7	Malawi	National Parks and Wildlife Act of 2004; National Parks and Wildlife (Protected Areas) Regulations, 1994	Unrepresented	Cirelli and Morgera (2010)
8	Mauritius	Wildlife and National Parks Act of 1993; For the purpose of implementation, National Parks and Reserves Regulations, 1996, the Wildlife Regulations, 1998, and the Wildlife (Amendment of Schedule) Regulations, 2004 constitute the guiding rules that assist in enforcing the Act	Unrepresented	Cirelli and Morgera (2010)
9	Mozambique	Forest and Wildlife Law no. 10/99 of 1999, and the enforcement instrument includes the Forests and Wildlife Regulations (Decree no. 12/200)	Unrepresented	Cirelli and Morgera (2010)
10	Namibia	Nature Conservation Ordinance of 1975 (principal legislation)	Unrepresented	Cirelli and Morgera (2010)
11	Nigeria	The Endangered Species (Control of International Trade and Traffic) Act; Nigerian National Park Service Act, 2006; National Environmental Standards and Regulations Enforcement Agency (Establishment) Act (NESREA) 2007; National Environmental (Protection of Endangered Species in International Trade) Regulations, 2011	Unrepresented	
12	Seychelles	Wild Animals and Birds Protection Act of 1961 (main legislation); National Parks and Nature Conservancy Act of 1969; (Wild Animals (Giant Land Tortoises) Protection Regulations, Wild Animals (Seychelles Pond Turtle) Protection Regulations, and Wild Animals (Turtles) Protection Regulations)	Unrepresented	Cirelli and Morgera (2010)

Table 5.1: Summary of country-specific assessments of the state of gender and illegal wildlife trade in national wildlife legal and regulatory frameworks of some member states of the sub-Saharan African region (continued)

Serial number	Country	Wildlife legal and/or regulatory framework(s)	Gendered focus on IWT	Reference(s)
13	South Africa	National Environmental Management Biodiversity Act of 2004; National Environmental Management Protected Areas Act of 2003; National Parks Act of 1976; Forest Act of 1998	Unrepresented	Cirelli and Morgera (2010); DLA Piper (2015); Price (2017)
15	Zimbabwe	Parks and Wildlife Act (Cap 20:14) (principal legislation); Parks and Wildlife (Amendment) Regulations	Unrepresented	Cirelli and Morgera (2010); Chibememe et al (2014); Price (2017)

Reconceptualising gendered motivations in wildlife legal frameworks

Effective assessment of wildlife legal frameworks should reflect diverse factors, which include gender issues, among others, but gendered investigations have yet to attract such attention in many nations, including sub-Saharan Africa. In other words, gender concerns are sparse in wildlife legal frameworks, and this situation can be significantly challenging, especially in wildlife use in traditionally and culturally organised settings like sub-Saharan Africa, where women are typically marginalised in making decisions or accessing certain rights. What seems like a gendered consideration can be found in Zambia where the law was specific on membership of wildlife advisory agencies in a manner that reflects 'equitable gender participation' (Cirelli and Morgera, 2010). This Zambian wildlife legal provision for gender equity in representation for wildlife advisory agencies implies a gender provision limited to administrative roles (advisory), while opportunities for a more comprehensive, broader, holistic and multipurpose gender mainstreaming approach to the entire wildlife value chain is envisaged in the proposed reimagining process of gender in wildlife legal frameworks in sub-Saharan Africa. UNODC (2020) and Roman (2020) corroborated women's normative rules of ethics, which include being more trusted in the communities, especially given the current decline in public trust on the police exhibiting high standards in interpersonal relationships, involvement in calming men enforcement officers in times of stress and work pressure, demonstrating high positive influence, transparency and fairness, as well as more likelihood of using less force in crime situations. The inclusion of more women in the wildlife law enforcement workforce may reduce the escalation and militarisation of wildlife enforcement activities (Strobl, 2019), coupled with the ability to search, investigate and prosecute women wildlife crime suspects, which men are culturally restricted from doing.

The wildlife national legal frameworks for sub-Saharan Africa nations can be amended to incorporate gendered differentiations and gender-based codes of conduct, capacities and generalisations, as well as the use of the International Consortium on Combating Wildlife Crime (ICCWC) Indicator Framework to assess and monitor their effectiveness, especially in relation to the response to IWT. The data and indicators for assessing the effectiveness of the laws and performance of law enforcement agencies/agents should reflect diverse demands and interests based on sex, sexual characteristics and gender identity.

Reconceptualising wildlife legal frameworks in sub-Saharan Africa can contribute to meeting Sustainable Development Goal 5 by incorporating gender initiatives that can assist in eradicating all forms of gender inequality, especially against women and girls, as well as protecting their rights to access, use, enforcement, protection and management of wildlife in sub-Saharan Africa as a region and its component nations. In addition, reconceptualising wildlife legal frameworks in sub-Saharan Africa can also assist in delivering Sustainable Development Goal 16's targets by promoting and strengthening institutional wildlife governance that will enhance an effective and gender-inclusive criminal justice system for adjudicating wildlife trafficking in the region and its component nations. Conversely, gender-blind wildlife legal frameworks at regional or national levels can hinder lawful socioeconomic activities, global peace and security, human rights, gender equality and gender-based empowerment. A typical example of gendered wildlife legal practice is through a law enforcement initiative exclusively for women in wildlife anti-trafficking ranger forces, namely, the Black Mamba Anti-Poaching Unit in South Africa, Team Lioness in Kenya and the Akashinga in Zimbabwe (Seager, 2021). Since the Black Mambas were established, the rate of IWT has decreased 76 per cent (UNODC, 2020). Another way to institutionalise gender in wildlife legal framework, especially in law enforcement positioning to reflect gender equality, is through legislation. Recent developments have shown that a quota-based mandatory representation of women in government establishments and corporations has started gaining prominence. For instance, evidence has shown that governments that involve a considerable proportion of women produce more gender-equality legislations (Wängnerud and Sundell, 2012; Alexander, 2012) and environmental-friendly legislations (UNEP, 2016).

Innovating wildlife policies and programmes for combating illegal wildlife trade

Considering the emerging evidences of gendered dynamics linked to IWT in this chapter, it becomes pertinent to innovate/reframe existing and future wildlife policies and programmes around the rubric of gender. Some of the innovative approaches to retool wildlife policies and programmes to address diverse intricacies of gender are enumerated here.

Formulating wildlife gender policies and programmes around the four-pillar and Actors, Drivers, Impacts and Responses frameworks

The four-pillar conceptual model developed by the Trade Records Analysis of Flora and Fauna in Commerce and World Wildlife Fund for Nature in 2012 as well as the Actors, Drivers, Impacts and Responses framework, represent patterns that can be analysed in the context of gender-based motivations to IWT and combating wildlife crime, which can inform gender interventions and outcomes. The model includes 'stop the poaching, stop the trafficking, stop the buying and international policy' as step-by-step actions that can inform policies and programmes aimed at combating gender dynamics of IWT. These two frameworks are interrelated and complementary in the way that each of the four pillars, namely, poaching, trafficking, buying and policy, correspond with actors, drivers, impacts and responses. Actors entail the gender-differentiated roles men and women play as defaulters, defenders, enforcement agents, informants, motivators, facilitators and consumers in IWT. Drivers include pressures, motivations and situations that boost IWT such as gender inequality and sexual exploitation. Impacts imply the gender-differentiated effects of IWT on men and women (for example, men–dominated enforcement operations will impact the effectiveness of the activity). Responses reflect gendered propositions entrenched in enforcement, policies and programmatic interventions to IWT, coupled with how these interventions combat or intensify IWT.

Redesigning and designing strategies

Incorporating gender-differentiated attitudes, demands, capacities, interests, generalisations, and so on, in relation to IWT can be originated in redesigning the existing and designing new/future wildlife policies and programmes in order to access the gendered opportunities for inclusive, standardised and sustainable wildlife management in sub-Saharan Africa. Generally, gendered wildlife policies and programmes designs need to align consistency with the UN commitment on gender equal opportunities and women empowerment, as well as the priorities of the Convention on the Elimination of All forms of Discrimination against Women (UNODC, 2020). Innovating wildlife policies and programmes for combating IWT can begin with design-based approach (for example, results framework, activities and policy/programme performance indices) catalysed by country-specific or region-based (sub-Saharan Africa) situational analysis of gender-based perceptions, motivations, roles, needs and participations. Gender-motivated situational analysis can assist in identifying and developing key issues for engendering needed awareness and potential interventions (Muhammad et al, 2017). Characterising and evaluating these gender-based needs and priorities or problems can include designing procedures that consider sex-segregated data gathering and analysis processes. For instance, gender-segregated data are missing in most of the international, regional, national and local laws, policies and programmes on wildlife, especially in relation to combating IWT. As a result, it has been challenging to deal with gendered codes of conduct, capacities, demands, generalisations, motivations and responses relating to IWT.

Application of the International Consortium on Combating Wildlife Crime's Wildlife and Forest Crime Analytic Toolkit

The ICCWC facilitates wildlife and forest crime enforcement cooperation involving CITES, the United Nations Office on Drugs and Crimes, Interpol, the World Bank and World Customs Organization, as well as other international agencies like the United Nations Environmental Programme and some nations. The ICCWC constitutes a key innovative opportunity that can be useful in scaling up gender-based design for wildlife policies and programmes to combat IWT, that is, the ICCWC's Wildlife and Forest Crime Analytic Toolkit. The ICCWC Analytic Toolkit is a technical instrument that assists nations in conducting national assessment of the major concerns concerning wildlife and forest crimes (IWT) and to investigate national-based deterrent and criminal justice actions. It also assists wildlife authorities to conduct all-encompassing investigation of the potential ways and techniques of protecting wildlife and forest resources or products. Sub-Saharan countries such as Angola have commenced the process of requesting the Toolkit, while the Democratic Republic of the Congo, Kenya and Senegal are at the planning phase, and implementation has been completed in Botswana, Gabon and Tanzania (2017. The Toolkit can be mainstreamed into the sub-Saharan Africa region and country-based wildlife policies and programmes to enhance opportunities to achieve a widespread wildlife national assessment outcome.

Empowering updates of diverse alternatives to illegal wildlife trade based livelihood opportunities

Wildlife policies and programmes can be reconstructed through the conception of variety of alternative livelihood opportunities. Such opportunities can minimise women's socioeconomic pressure on men, their reliance on wildlife stealing by men, and broaden the range of alternative non-IWT livelihood activities for men and women around wildlife territories. In Kafue National Park, Zambia, Game Rangers International offers many non-IWT livelihood programmes, which safeguards women from IWT-motivated exploitations such as sexual abuse and economic marginalisation. It was reported that within four months of setting up a bakery in the park area, the income of participating women had doubled (Seager, 2021). Involvement in non-IWT-related livelihood opportunities can further reduce or discourage gendered demands for wildlife and forest products. Forces of demand and supply can promote or discourage IWT, and this implies that high demand for wildlife encourages an increase in the supply chain, which in turn increases the flow of profit to the culprits. Effectively, a relatively new development in IWT consumption has given effect to the rapidly growing study on 'demand reduction', which has been recognised as an effective policy and programmatic strategy to mitigating IWT impacts (Kennaugh, 2015; Burgess, 2016). In the sub-Saharan Africa region and member states, opportunities for policies and programmes targeted towards wildlife demand reduction could be prioritised through the involvement of civil society organisations and mass media in developing culturally fitting means of communication to the appropriate audience and revisiting such communications or awareness programmes from time to time.

Engaging non-governmental organisations and youth groups

The involvement of NGOs and youth groups (boys and girls) in wildlife conservation and protection activities has recently been found to be useful in disincentivising gender differentiated elements that influence IWT in some locations. For instance, Conserving Natural Capital and Enhancing Collaborative Management of Transboundary Resources, an initiative established by a consortium of NGOs involved in IWT issues in East Africa, has as its main goal enhancing the interface with conservation and dedication to wildlife among diverse constituent groups, with a special focus on women and youths (IUCN, 2019). NGOs involved in campaigns against IWT have also engaged the use of radio programmes to reach women to communicate the importance of behavioural changes to handling wildlife species and products. The Game Rangers International awareness programme in Zambia has acknowledged the use of radio as an effective means of reaching the communities on wildlife issues (GRI/IFAW, 2017). Notably, out of 517 people sampled in the Zambian community, 348 depended on radio news and programmes for information relating to elephants (GRI/IFAW, 2017). On the other hand, the World Wildlife Fund for Nature assists women in Madagascar through women's 'radio listener groups' to partake in conservation programmes and hold interactive talks with other women in the neighbourhood to share with them that they have learnt from the radio programmes.

Conclusion

This chapter provides a conceptual foundation to link gender and IWT to the wider conceptual and action frameworks involving gender and the environment, which has become known in the previous two decades (since 2000). Although mainstreaming gender-based perceptions and motivations into IWT interventions is not a self-sufficient solution approach, it can be considered as a contributor or predictor to achieving sustainability by exploring a wide-ranging consideration of challenges, strategies, policies, legal and programmatic actions not offered in other wildlife-related studies. It helps offer opportunities for wildlife-based organisations to reshape and realign their policies, strategies and programmes to reflect gendered diversities in wildlife use, protection and management. De-emphasising gendered perceptions of and motivations for IWT at international, regional, national and local scales, especially in sub-Saharan Africa, creates gaps in the knowledge and understanding of IWT activities, courses of action and prospects of legal, policy and programmatic interventions that can contribute to redefining the orientation and commonalities changing the extremities in the existing gendered codes of conduct, attitudes, motivations, demands, and so on that may incentivise IWT. To help support the narrative regarding lack of gendered dynamics, especially gender-based motivations in anti-IWT activities and initiatives, this chapter has considered some of the legal, policies and programmatic possibilities that can contribute to discouraging the conventional orientations and radicalised wildlife trafficking trends in the sub-Saharan African region and its member states. Wildlife organisations can prioritise conservation training and education structured around gendered differentiations, which can also enhance effective legal, policy and programme priorities for sustainable wildlife conservation and combating IWT. The training can be targeted mainly at the under-represented or unrepresented gender to assist in balancing the diversity and inclusion gaps, as well as enhancing wildlife personnel capacities on managing gender and IWT. Sub-Saharan African countries can retool their wildlife legal frameworks, policies and programmes in alignment with gender dynamics in IWT, especially gender-based motivations in wildlife crime.

References

Africa-Asia Pacific Symposium on Strengthening Legal Frameworks to Combat Wildlife Crime (2017). United Nations Inter-Agency Task Force on Illicit Trade in Wildlife and Forest Products Symposium Report. Bangkok, 4–5 July.

Agu, H.U. and Gore, M.L. (2020) Women in wildlife trafficking in Africa: A synthesis of literature. *Global Ecology and Conservation*, 23, 1–12.

Alexander, A.C. (2012) Change in women's descriptive representation and the belief in women's ability to govern: A virtuous cycle. *Politics & Gender*, 8(4), 437–64.

Anthem, H. (2018) Gender blindness: Are we tackling wildlife crime with one hand tied behind our back? *Fauna & Flora International*. Available at: www.fauna-flora.org/news/gender-blindness-tackling-wildlife-crime-one-hand-tied-behind-back (accessed 13 April 2022).

Annual Progress Assessment (2015) US National Strategy for Combating Wildlife Trafficking, pp 1–32.

Arlbrandt, L., Simomeau, N., Parry-Jones, R. and Leger, T. (2021) A gendered approach to the illegal wildlife trade could engender an anti-trafficking revolution (commentary). *Mongabay*. Available at: https://news.mongabay.com/2021/10/a-gendered-approach-on-the-illegal-wildlife-trade-could-engender-an-anti-trafficking-revolution-co (accessed 13 April 2022).

Barber-Meyer, S.M. (2010) Dealing with the clandestine nature of wildlife trade market surveys. *Conservation Biology*, 24(4), 918–23.

Béné, C., Steel, E. Kambala Luadia, B. and Gordon, A. (2007 Fish as the 'bank in the water': Evidence from chronic-poor communities in Congo. *Food Policy*, 34(1), 108–18.

Birnie, P. and Boyle, A. (2002) *International Law and the Environment*. Oxford: Oxford University Press.

Burgess, G. (2016) Powers of persuasion. *TRAFFIC Bulletin*, 28(2). Available at: www.traffic.org/site/assets/files/3385/powers-of-persuasion.pdf (accessed 5 May 2022).

Burgess, G. and Zain, S. (2018) *Reducing Demand for Illegal Wildlife Products*. Harare: Defra/TRAFFIC.

Chibememe, G., Dhliwayo, M., Edson, G., Mtisi, S., Never, M. and Laiza, K.O. (2014) *Review of National Laws & Policies that Support or Undermine Indigenous Peoples and Local Communities: Zimbabwe*. A Report for Natural Justice. Available at: http://naturaljustice.org/wpcontent/uploads/2015/09/Zimbabwe-Legal-Review.pdf (accessed 5 May 2022).

Chigbu, U., Paradza, G. and Dachaga, W. (2019) Differentiations in women's land tenure experiences: Implications for women's land access and tenure security in sub-Saharan Africa. *Land*, 8(22), 1–22.

Cirelli, M.T. and Morgera, E. (2010) *Wildlife Law in the Southern African Development Community*. Budapest: FAO and CIC.

Deere, C.D., Alvarado, G.E. and Twyman, J. (2012) Gender inequality in asset ownership in Latin America: Female owners vs household heads. *Development and Change*, 43(2), 505–30.

Dell, B.M., Souza, M.J. and Willcox, A.S. (2020) Attitudes, practices, and zoonoses awareness of community members involved in the bushmeat trade near Murchison Falls National Park, northern Uganda. *PLOS ONE*, 15(9), e0239599.

DLA Piper (2015) *Empty Threat 2015: Does the Law Combat Illegal Wildlife Trade?* Review commissioned by The Royal Foundation on behalf of the United for Wildlife partnership. Available at: www.dlapiper.com/~/media/Files/News/2015/05/IllegalWildlifeTradeReport2015.pdf (accessed 26 March 2022).

Doss, C., Kieran, C. and Kilic, T. (2017) *Measuring Ownership, Control, and Use of Assets*. Policy research working paper no. WPS 8146. Washington, DC: World Bank Group.

Drury, R. (2011) Hungry for success: Urban consumer demand for wild animal products in Vietnam. *Conservation and Society*, 9(3), 247–57.

EIA (2016) *Time for Action: End the Criminality and Corruption Fuelling Wildlife Crime*. Environmental Investigation Agency. Available at: https://eia-international.org/wp-content/uploads/EIA-Time-for-Action-FINAL-1.pdf (accessed 2 April 2022).

Elliott, L. (2007) Transnational environmental crime in the Asia Pacific: An un(der)securitized security problem? *The Pacific Review*, 20(4), 499–522.

FAO (2016) *Sustainable Wildlife Management and Gender*. The Collaborative Partnership on Sustainable Wildlife Management (CPW) Fact Sheet. FAO, pp 1–6.

Fison, M. (2011) The £6bn trade in animal smuggling. *The Independent.* Available at: www.independent.co.uk/environment/nature/the-1636bn-trade-in-animal smuggling-2233608.html (accessed 22 March 2022).

GFI (2011) *Transnational Crime in the Developing World.* Washington, DC: Global Financial Integrity. Available at: www.gfintegrity.org/stor age/gfip/documents/reports/transcrime/gfi_transnational_crime_web. pdf (accessed 13 March 2022).

GRI/IFAW (2017) *Strategic Communications for Poaching and Orphan Reporting.* Drive the Poaching of Endangered Animals.

Henson, D.W., Malpas, R.C. and D'Udine, F.A.C. (2016) *Wildlife Law Enforcement in Sub-Saharan African Protected Areas: A Review of Best Practices.* Occasional Paper of the IUCN Species Survival Commission No. 58. Cambridge and Gland: IUCN.

Hongjie, L. (2019) China calls on citizens in Africa to stop wildlife trafficking. *China Daily*, 26 March. Available at: www.chinadaily.com.cn/a/201903/ 26/WS5c99e6e6a3104842260b2ac1.html (accessed 15 April 2022).

ICCWC (2017 *Implementation of the ICCWC Wildlife and Forest Crime Analytic Toolkit Summary of Progress.* CITES Secretariat and UNODC, updated 30 May 2017. Available at: https://cites.org/sites/default/files/ eng/prog/iccwc/ICCWC_Toolkit_implementation_table_rev30May17-web.pdf (accessed 21 March 2022).

Iori, G. (2020) *LCA Webinar: Women, Wildlife Crime and Conservation.* Available at: www.youtube.com/watch?v=NgPzDjuhrQc&feature=youtu. be (accessed 21 March 2022).

IUCN (2019) 2019 IUCN Annual Report.

Kareiva, P. and Marvier, M. (2012) What is conservation science? *BioScience*, 62(11), 962–9.

Kennaugh, A. (2015) *Rhino Rage: What Is Driving Illegal Consumer Demand for Rhino Horn.* New York: NRDC. Available at: www.rhinoresourcecen ter.com/index. php?s=1&act=pdfviewer&id=1483702532&folder=148 (accessed 8 May 2022).

Kieran, C., Sproule, K., Doss, C., Quisumbing, A. and Kim, S.M. (2015) Examining gender inequalities in land rights indicators in Asia. *Agricultural Economics*, 46(S1), 119–38.

Kosome, V., Davis, R. and Silver, M. (2020) Life was improving for 'no sex for fish.' Then came the flood. A publication of WUSF Public Media, Tampa, Florida, USA by Viola Kosome, Rebecca Davis, and Marc Silver on November 1, 2020

Kurland, J., Pires, S.F., McFann, F.C. and Moreto, W.D. (2017) Wildlife crime: A conceptual integration, literature review, and methodological critique. *Crime Science*, 6(4), 1–16.

Lowassa, A., Tadie, D. and Fischer, A. (2012) On the role of women in bushmeat hunting: Insights from Tanzania and Ethiopia. *Journal of Rural Studies*, 28(4), 622–30.

Lunstrum, E. and Givá, N. (2020) What drives commercial poaching? From poverty to economic inequality. *Biological Conservation*, 245, 1–14.

Margulies, J.D., Wong, R. and Duffy, R. (2019) The imaginary 'Asian Super Consumer': A critique of demand reduction campaigns for the illegal wildlife trade. *Geoforum*, 107, 216–19.

Martino, D. (2008) Gender and urban perceptions of nature and protected areas in Bañados del Este biosphere reserve. *Environmental Management*, 41(5), 654–62.

Mauck, G. (2013) *Wildlife Legislation in Sub-Saharan Africa: Criminal Offences*. Research project commissioned by the Conservation Action Trust (CAT). Available at: https://conservationaction.co.za/wpcontent/uploads/2013/11/Wildlife-Legislation-in-SS-Africa-Nov-13.pdf (accessed 2 May 2022).

McElwee, P. (2012) The gender dimensions of the illegal trade in wildlife: Local and global connections in Vietnam. In M.L. Cruz-Torrez and P. McElwee (eds) *Gender and Sustainability Lessons from Asia and Latin America*. Tucson: University of Arizona Press, pp 71–93.

Milliken, T. and Shaw, J. (2012) *The South Africa—Viet Nam Rhino Horn Trade Nexus: A Deadly Combination of Institutional Lapses, Corrupt Wildlife Industry Professionals and Asian Crime Syndicates*. Johannesburg: TRAFFIC.

Mmassy, E.C. and Røskaft, E. (2013) Knowledge of birds of conservation interest among the people living close to protected areas in Serengeti, Northern Tanzania. *International Journal of Biodiversity Science, Ecosystem Services & Management*, 9(2), 114–22.

Moreto, W.D. (2015) Introducing intelligence-led conservation: Bridging crime and conservation science. *Crime Science*, 4(15), 1–15

Morgera, E., and Wingard, J. (2009) Principles for developing sustainable wildlife management laws. *FAO Legal Paper Online*, 75. Available at: www.fao.org/legal/prs-ol/lpo75.pdf (accessed 22 March 2022).

Muhammad, F.A., Elias, M., Lamers, H., Omard, S., Brooke, P. and Hussin, M.H. (2017) Participatory research to elicit gender differentiated knowledge of fruit trees. *Biodiversity International*, Sarawak, Malaysia. Available at: www.bioversityinternational.org/fileadmin/user_upload/Participatory_Muhammad.pdf (accessed 22 March 2022).

Nellemann, C., Henriksen, R., Raxter, P., Ash, N. and Mrema, E. (eds) (2016) *The Environmental Crime Crisis: Threats to Sustainable Development from Illegal Exploitation and Trade in Wildlife and Forest Resources*. A UNEP Rapid Response Assessment. Nairobi and Arendal: United Nations Environment Programme and GRID-Arendal.

Price, R.A. (2017) *National and Regional Legal Frameworks to Control the Illegal Wildlife Trade in Sub Saharan Africa.* K4D Helpdesk Report. Brighton: Institute of Development Studies. Available at: www.cbd.int/financial/monterreytradetech/unep-illegaltrade.pdf (accessed 12 May 2022).

Ratsimbazafy, C., Newton, D.J. and Ringuet, S. (2016) *Timber Island: The Rosewood and Ebony Trade of Madagascar.* Cambridge: TRAFFIC.

Roman, I. (2020) Women in policing: The numbers fall far short of the need. *Police Chief Online*, 22 April. Available at: www.policechiefmagazine.org/women-in-policing (accessed 11 May 2022).

Seager, J. (2020) *Gender Learning Review of 'Leading the Change', Madagascar and Namibia.* WWF. Madagascar and Namibia.

Seager, J. (2021) *Gender and Illegal Wildlife Trade: Overlooked and Underestimated.* Gland: WWF.

Selier, S.A.J., Slotow, R., Blackmore, A. and Trouwborst, A. (2016) The legal challenges of transboundary wildlife management at the population level: The case of a trilateral elephant population in Southern Africa. *Journal of International Wildlife Law & Policy*, 19(2), 1–37.

Smith, H. (2017) Small fish, big problems: Gender based violence in Lake Victoria's fisheries. Available at: humanrights.fhi.duke.edu/small-fish-big-problems-gender-based-violence-in-lake-victorias-fisheries (accessed on May 26, 2022).

Strobl, S. (2019) *Towards a Women-Oriented Approach to Countering Wildlife, Forest and Fisheries Crime (WFFC).* University of Wisconsin-Platteville, USA.

Sundstrom, A., Gore, M.L., Linell, A., Ntuli, H. and Sjostedt, M. (2019) Gender differences in poaching attitudes: Insights from communities in Mozambique, South Africa, and Zimbabwe living near the great Limpopo. *Conservation Letters*, May, 1–8.

UNEP (2016) *Global Gender and Environment Outlook.* Nairobi: UN Environment.

UNGA (2021) Resolution adopted by the General Assembly on 23 July 2021. A/RES/75/311.

UNODC (2020) *The Time Is Now: Addressing the Gender Dimensions of Corruption.* Vienna: UNODC. Available at: www.unodc.org/documents/corruption/Publications/2020/THE_TIME_IS_NOW_2020_12_08.pdf (accessed 27 May 2022).

Wängnerud, L. and Sundell, A. (2012). Do politics matter? Women in Swedish local elected assemblies 1970–2010 and gender equality in outcomes. *European Political Science Review*, 4(1), 97–120.

Watchdog, A. (2020) World Pangolins Day, 15 February. Available at: www.albertinewatchdog.org/2020/02/15/world-pangolins-day-let-us-protect-endangered-threatened-andvulnerable-african-pangolins-in-uganda; personal communication with interviewee (accessed 19 March 2022).

Weru, S. (2016) Wildlife protection and trafficking assessment in Kenya: Drivers and trends of transnational wildlife crime in Kenya and its role as a transit point for trafficked species in East Africa. *TRAFFIC*. Available at: www.trafficj.org/publication/16_Wildlife_Protection_and_Trafficking_Assessment_Kenya.pdf (accessed 13 May 2022).

Westerman, K. (2017) Interview in *Why Gender Matters in Conservation*. Mongabay. Available at: news.mongabay.com/2017/05/qa-with-a-champion-of-the-gendered-approach-to-conservation (accessed on April 25, 2022).

Wilson-Wilde, L. (2010) Wildlife crime: A global problem. *Forensic Science, Medicine and Pathology*, 6(3), 221–2.

WWF (2012) *Fighting Illicit Wildlife Trafficking: A Consultation with Governments*. WWF Report. World Wildlife Fund International. Available at: http://d2ouvy59p0dg6k.cloudfront.net/downloads/wwffightingillicitwildlifetraffickinglr1.pdf (accessed 15 April 2022).

Wyatt, T. (2013) *Wildlife Trafficking: A Deconstruction of the Crime, the Victims, and the Offenders*. Basingstoke: Palgrave Macmillan.

6

The Attitudes of People with Different Gender Identities and Different Perceptions of Gender Roles towards Nonhuman Animals and Their Welfare

Aphra Hope-Forest, Ekaterina Gladkova and Tanya Wyatt

Introduction

The UK claims to have a long tradition of animal welfare, including the creation of the world's first animal welfare charity, the Society for the Prevention of Cruelty to Animals, which was founded in 1824 and which became the Royal Society for the Prevention of Cruelty to Animals (RSPCA, 2022). The UK also claims its legislation goes beyond European Union (EU) requirements (World Animal Protection, 2020). Yet, 6.4 billion nonhuman animals are killed annually to support the UK food supply (Animal Clock UK, 2020) and more than 3.4 million animals were used for the first time in experiments in 2018 (Animal Aid, 2018). Jail terms for animal cruelty in England and Wales are the lowest in Europe (Bell, 2017). In order to decrease and ultimately eliminate nonhuman animal suffering, there needs to be better understanding of attitudes towards nonhuman animals. While previous studies have explored attitudes towards nonhuman animals using a range of variables, those that have studied gender as a variable have used the traditional binary definition of gender, 'men' and 'women' (Wells and Hepper, 1997; Evans et al, 1998; Woodward and Bauer, 2007; Lee et al, 2010; Tesform and Birch, 2013; Kendall et al, 2016; Byrd et al,

2017; Knight et al, 2017). These studies found that women are more likely to be concerned with nonhuman animal suffering (Wells and Hepper, 1997; Evans et al, 1998).

However, gender is not a biological construct or even a distinct category. Butler (1990: 31) defines gender identity as 'a performative accomplishment compelled by social sanction and taboo. … Gender is … an identity instituted through a repetition of acts'. While she recognises the limits to how much agency individuals performing gender can have (Butler, 2004), the idea of gender as a performative accomplishment redirects the attention from internal individual matters towards interactional and institutional realms (West and Zimmerman, 1987). Thus, gender identity becomes a matter of both bodies and cultures (Fausto-Sterling, 2019). The recognition of non-binary (NB) gender identity is relatively recent, with Australia offering a third gender 'X' option on all passports in 2011 (*The Guardian*, 2011), and the state of Oregon in the United States recognising non-binary gender identity in 2016 (Shupe, 2016). NB gender identity is not currently legally recognised in the UK, although plans to recognise NB gender in Scotland were made in 2017 (Brooks, 2017).

Very few studies have looked at relationships between people of diverse genders and animal companions (Riggs et al, 2018; Taylor et al, 2018a,b), thus echoing Grubbs' (2012) idea that queer thought and animal studies exist in isolation from each other, and such investigations focused primarily on the role of animal companions in the context of the family and domestic violence. The field of critical animal studies has encouraged cross-fertilisation between gender and nonhuman animals by asking how 'gender' becomes performed whenever nonhuman animals are observed (Birke, 2002) and looks at the issue through the lens of veganism (Simonsen, 2012). Yet, the question of attitudes towards nonhuman animals of people with different gender identities has thus far remained unexplored. We suggest that gender identity is likely to contribute to people's attitudes towards nonhuman animals as previous studies have found when looking at gender as a binary.

Furthermore, it is important to further explore the possibility that it is not only gender identity that contributes to a person's attitude towards nonhuman animals, but the person's perception of gender roles. In this regard, socialisation regarding what has been traditionally viewed as binary gender roles − women are nurturing and family-oriented, men are the breadwinners − may affect attitudes towards nonhuman animals. This is also presumably relevant for people who have modern or non-traditional perceptions of gender roles. A very small amount of research (Wyckoff, 2014; Allcorn and Ogletree, 2018) has been conducted to explore people's perceptions of gender roles rather than gender and the accompanying attitudes towards nonhuman animals. The lack of research on perceptions of gender roles overlooks the complex interplay of historic–cultural socialisation regarding attitudes that are categorised as 'masculine' or 'feminine' and the resulting relationship with nonhuman animals.

Our chapter moves beyond binary gender investigations to an inclusive exploration of the range of gender identities and people's perceptions of nonhuman animals and their welfare. Our aim here is to subvert 'hegemonic understandings of speciesism through a queer framework' (Grubbs, 2012: 4). In addition, we also explore perceptions of gender roles and the way they inform attitudes towards nonhuman animals. Here, we aim to advance an understanding of the commonality of oppressions (Best et al, 2007). We investigate the attitudes of people self-identifying as binary or NB towards nonhuman animals. We also investigate the attitudes of people's own views of their gender roles (see the 'Methods and data' section). To illustrate our thinking on these questions, we consider the case study of attitudes towards status dogs typically used in dog fighting. The case study method has featured in the methodological toolkit of critical animal studies before (Thomas, 2013). We, thus, aim to explore the knowledge and/or acceptance of cruelty to companion animals by people with different gender identities and with different perceptions of gender roles. The two research questions we sought to answer are: What are the attitudes towards nonhuman animals of people with different gender identities and with different perceptions of gender roles? What are the attitudes towards status dogs of people with different gender identities and with different perceptions of gender roles? Our chapter recognises the existing critique of nonhuman animal welfarism in critical animal studies (Pedersen and Stanescu, 2014). Yet, our chapter is aligned with Sztybel's (2007) pragmatist approach to animal rights that blurs the opposition between welfare and rights; welfare policies are thus promoted without endorsing the speciesist belief that humans are superior to animals.

We first will provide background information from previous studies' findings on the pattern of attitudes towards nonhuman animals of 'men' and 'women', including the research on attitudes towards status dogs. We then will outline the methodological approach we used to investigate attitudes towards nonhuman animals and status dogs specifically, across the range of gender identities and those with differing perceptions of gender roles. This will be followed by our findings and their analysis and will conclude with suggestions for further study.

The gender patterning of attitudes to nonhuman animals

As mentioned, most previous studies of attitudes towards nonhuman animals took a binary approach to gender by defining gender as either 'man' or 'woman', thus excluding individuals who identify as a NB gender. Such research reduced representativeness and overlooked the attitudes of an entire social group. Research has shown that there are fairly clear patterns relating to attitudes towards nonhuman animals in relation to 'man' and 'woman'. The discussion that follows unpacks such attitudes and the possible explanations behind them. It also identifies and critically evaluates the existing research that links different gender identities and perceptions of gender roles with attitudes towards nonhuman animals. The discussion also recognises the limitations of existing research.

Kendall et al (2016) found that gender affected attitudes, but the most significant effect on attitude to animal welfare was childhood experience. In relation to gender, women were more concerned with animal wellbeing than men were, which, according to Kendall et al (2016) is due to socialisation into gender roles, which encourage women to be more nurturing than men. Women are socialised into more nurturing roles, which increases sensitivity to animal suffering, while men are socialised into more utilitarian roles, focusing on practical uses of animals (Herzog et al, 1991). However, it is unknown if gender differences in attitudes still apply in contexts where binary gender is rejected and/or where traditional gender roles are rejected. Similarly, Woodward and Bauer (2007) found women were more attached to their companion animals than men were, and women viewed their companion animals' behaviour more positively. However, attachment to the companion animal was also positively correlated with time spent with him/her (Woodward and Bauer, 2007). This means, as was also found by Kendall et al (2016), that gender is not the only factor affecting attitudes, and a man who spends more time with his companion animal would be more attached to him/her than a woman who spends less time with her companion animal (Woodward and Bauer, 2007).

In addition, Lee et al (2010) found that women reported more agreeableness, general sympathy, sympathy for animal suffering and sympathy for a bait dog (used to train fighting dogs) viewed during their study's film, than the male participants. However, the study reveals a complex picture: agreeableness mediated the relationship between gender and sympathy, which means gender does not directly affect sympathy, but rather affects agreeableness, which in turn affects sympathy (Lee et al, 2010). Other studies have shown women were less likely to support the use of animals as entertainment (for example, in circuses or horse/dog racing) than men, and that they were more concerned for farm animal welfare than men (although for some uses there were no significant differences between attitudes of men and women [Byrd et al, 2017]). Tesform and Birch (2013) showed that men were more likely to obtain a companion animal as part of their self-image than women were, and men were more likely than women to consider the breed of the dog as an important factor while choosing the dog. This might partly explain why it is men who predominantly engage in dog fighting and in having a status dog, as these are ways of constructing a masculine identity and status (Evans et al, 1998; Hughes et al, 2011; Smith, 2011; Harding, 2013). This resonates with Butler's (1990: 172) idea that acts specific to each gender produce 'a false stabilization of gender in the interests of the heterosexual construction'.

Women show less support for uses of animals than men, but men's support for some uses is still limited (Knight et al, 2017). For women and men, uses that led to the injury or death of the animal were regarded more negatively than other uses (Wells and Hepper, 1997). Belief that animals have minds is more common in women than men (Herzog and Galvin, 1997), and was found to be the strongest predictor of attitudes towards animal use (Knight et al, 2017). Furthermore, Peek et al (1997) argue that women and animals occupy positions in the social hierarchy inferior to men, so women are more sensitive than men to animal suffering that stems from animals' similar inferior status. If this is the case, this raises the possibility that NB gender identities would likely also be more sensitive to nonhuman animal suffering as NB gender identities also occupy inferior positions in the social hierarchy to men. However, in line with Kendall et al (2016), and Peek et al's (1997) argument, previous experiences with animals might be a better predictor of attitudes to animals than gender.

As mentioned, few studies take a NB approach to examining attitudes towards nonhuman animals. Riggs et al's (2018) research does so through the case of companion animals, linking the role of the latter in lives of the people of diverse genders with experiences of discrimination and abuse among this population. They conclude that animal companions may facilitate social support and reduce psychological distress. Taylor et al (2018a) use the same sample of participants for exploring the link between human- and nonhuman-animal-directed violence. Interestingly, they suggest that animals often serve as a positive affirmation of an individual's gender identity. However, Riggs et al (2018) and Taylor et al (2018b), while discussing the role of nonhuman animal companions in human lives in the context of domestic violence, do not explore attitudes towards nonhuman animals in depth. In addition, it is important to recognise that the idea of companion animal ownership is often debated (Hillsburg, 2010), particularly in light of 'pets' having the status of property (Francione, 2004).

Studies that explore the link between gender roles and attitudes towards nonhuman animals approach the subject from the perspective of the 'linked oppression thesis', drawing a link between sexism and speciesism. Allcorn and Ogletree (2018: 465) analyse attitudes towards animals through the lens of meat eating, concluding that 'traditional gender role attitudes positively correlated with pro meat-eating attitudes/justifications while non-traditional, gender transcendent attitudes correlated negatively with pro meat-eating attitudes'. This idea resonates with Adams's (2010) and Cudworth's (2010) theses that ways of eating are employed to maintain gender boundaries and meat consumption becomes a way of performing one's masculinity.

Overall, studies that adopt a binary approach to gender display a documented pattern that women are more sympathetic to the suffering and uses of nonhuman animals than men are. Nonetheless, this pattern is complicated by the fact that exposure to or experience with nonhuman animals as well as belief in nonhuman animal minds also shape one's attitudes towards nonhuman animals. Status (especially attempts to display masculine status) is another aspect of the complex relationship between gender identity and nonhuman animals, particularly in relation to companion animals like status dogs. Status dogs are a particular category of dogs used by their owners to convey an image of toughness or aggression and are closely linked to dog fighting. This is different to other types of dogs that may convey status like designer dogs and rare breeds (that is, a teacup chihuahua or lilac-coloured French bulldog). Therefore, studies that adopt a NB approach may further benefit from exploring the attitudes towards nonhuman animals in more detail. Moreover, previous studies also assume that gender roles, in addition to gender identities, are also likely to contribute to attitudes towards nonhuman animals and this link needs to be further explored. We now detail our methods for investigating attitudes towards nonhuman animals across the range of gender identities and those with differing perceptions of gender roles.

Methods and data

In order to answer our two research questions, we undertook a literature review and online survey. The project received full ethical clearance from Northumbria University's Department of Social Sciences ethics committee before data collection began. A literature review was carried out into the differing attitudes towards nonhuman animals between genders as well as differing attitudes of people with different perceptions of gender roles in order to inform the direction the rest of the study should take. Only peer-reviewed sources written in English and published since 1990 were included. Searches were made via Cardiff University Library Search, Northumbria University Library Search and Science Direct. Key words included 'status dogs' and 'dog fighting' as well as combinations of 'gender', 'non-human animal' and 'attitudes'/'perceptions'.

An online survey, created using Bristol Online Surveys, was used to solicit attitudinal and perception data. The survey opened on 8 February 2018 and closed on 28 February 2018 and was open to UK residents aged 18 or over. A URL to the online survey was distributed via flyers at a Scottish Society for the Prevention of Cruelty to Animals conference, the Facebook groups for Cardiff University Biosciences Society and the Wildlife and Conservation Society, and in two online forums.[1] Data from the survey were analysed in R (R Core Team, 2017). The survey collected demographic information such as marital status, qualifications, ethnicity and age. It asked participants to define their gender identity from 14 pre-defined options or in their own words. The range of gender identities included were adapted from those used in a survey of NB people by the Scottish Trans Alliance (Valentine, 2016). Respondents were free to choose as many definitions as they wished, and/or to provide their own definition in an 'other' category. A transgender identity is neither indicative nor mutually exclusive of a NB gender identity. Prior to analysis, any answers indicating non-cisgender gender identities (Table 6.1) were collapsed into one category (see the 'Findings' section).

The online survey drew on queer methodologies to assess participants' own definitions of their gender identity. Queer methodologies challenge traditional theories of sexuality and gender by focusing on the experiences of minority groups who do not fit into traditional gender/sex categories (Browne and Nash, 2010). Survey participants were therefore encouraged to define their gender identity however they chose. This is important as it enabled the study to focus on the experiences of gender minorities, rather

Table 6.1: The pre-defined gender identities and their definitions, as used in the survey

Gender identity	Definition
Agender, non-gendered	Does not identify as having a gender identity
Androgyne	Has a gender identity which comprises masculinity and femininity, not necessarily in equal amounts
Bigender	Experiences two gender identities, which are not necessarily male and female, and which can be simultaneous or fluctuating
Demigirl	Partially (not fully) identifies as a woman, regardless of assigned gender at birth
Demiboy	Partially (not fully) identifies as a man, regardless of assigned gender at birth
Cisgender man	Assigned male at birth and currently identifies as man
Cisgender woman	Assigned female at birth and currently identifies as woman
Don't define	
Genderfluid	Does not identify as having a fixed gender
Neutrois	Identifies as a neutral gender
Transgender man	Not assigned male at birth, but currently identifies as man
Transgender woman	Not assigned female at birth, but currently identifies as woman
Third gender	Used in some societies to describe a recognised gender other than man or woman
Unsure or questioning	

than the binary of men and women that underpins so much existing research on gender.

The survey received 324 responses. The survey sample included a higher percentage of people with NB gender identities than expected, but this difference was not significant. Statistics on NB gender identity were obtained from Practical Androgyny (2014). A response to the question on gender identity was provided by 319 respondents and 14.2% per cent (n=45) used one or more term to describe a NB or transgender gender identity, while 86.7 per cent (n=274) described their gender identity as cisgender. As stated in the methods and data section, 33 participants (10.3 per cent) were removed from the sample, as they did not answer all of the questions. Thus, the sample size was 286 comprising 40 cisgender men, 208 cisgender women and 38 non-cisgender individuals. The non-cisgender participants identified as 25 unique gender identities (see Table 6.2). As indicated, for analysis we combined all non-cisgenders into a single category. However, we believe that each gender identity is likely to have different, unique perspectives about companion animals. Further studies with appropriate sample sizes of each gender identity and qualitative methods should be undertaken to further unpack these important dimensions.

The results of the 2011 UK census results (Office for National Statistics, 2014) were used to obtain expected statistics for age, marital status, ethnicity, qualifications and student status. Statistics on having a companion animal were obtained from the Pet Food Manufacturer's Association (2017).

The survey asked whether companion animals – 'pets' – were ever owned, owned during childhood and/or whether they were owned currently. Participants were then asked about attitudes to animals, coded such that a higher score (out of 109) indicated a more favourable attitude to animal welfare. Question 1 required the respondents to indicate the extent to which they are in agreement/disagreement with eight statements (see Appendix). For example, 'humans have too little respect for the lives of animals' with responses on a Likert scale from strongly disagree (1) to strongly agree (5). Question 2 asked about participation in six non-injurious recreational or educational uses of animals (see Appendix). For example, visiting zoos, petting zoos, safari parks or aquariums were assessed using the following Likert scale response options: never (1); less than once a month (2); once a month (3); more than once a month, but not every week (4); once per week (5); more than once a week, but not every day (6); and at least once daily (7). Participants were also asked (in questions 4 and 5) about the extent to which they find status dogs (as detailed earlier) and dog fighting acceptable, with the following response options: entirely unacceptable (5); mostly unacceptable (4); unsure (3); mostly acceptable (2); and entirely acceptable (1). These questions were adapted from Wells and Hepper (1997), Byrd et al (2017), Knight et al (2017) and Kendall et al (2016).

Table 6.2: Non-cisgender identities

Gender identity	n
Agender	7
Androgyne	5
Bigender	2
Demiboy	1
Demigirl	1
Don't define	2
Genderfluid	2
Genderqueer/nonbinary	1
Neutrois	2
Androgyne, Unsure/questioning	1
Transgender woman	2
Transgender man	3
Unsure/questioning	3
Agender, genderfluid	1
Agender, Genderfluid, Unsure/questioning	1
Agender, Don't define, Genderfluid, Unsure/questioning	1
Agender, Demiboy, Don't define, Neutrois	1
Androgyne, Demigirl, Don't define, Genderfluid	1
Androgyne, Cisgender woman	2
Androgyne, Transgender man	1
Bigender, Cisgender man, Demiboy	1
Cisgender woman, Demigirl	1
Cisgender man, Unsure/questioning	1
Cisgender woman, Unsure/questioning	1
Neutrois, Genderqueer/nonbinary	1

In relation to status dogs, participants could not be asked if they had ever been involved in dog fighting for ethical reasons. Instead, participation in lawful hunting of animals, and watching animal racing on television or in real life, were assessed in Question 3 with possible answers ranging from never (7) to at least once daily (1). Participants were also asked if they were aware of status dogs and dog fighting prior to the survey, but these answers were not coded and did not contribute to the calculation of animal score (out of 109). There were six 'unsure' answers to this question and two missing answers. Eight answers in total were therefore removed for this question. Participants who did not answer one or more of the questions had their data removed from the analysis. In many cases the same participant did not answer more than one question.

Regarding perceptions of gender roles and participants' attitudes towards nonhuman animals, we asked survey respondents to describe their views of gender roles using the following options: 'Completely modern', 'Mostly modern but some traditional', 'Equally traditional and modern', 'Mostly traditional but some modern' and 'Completely traditional'. This is a novel question not adapted from pre-existing research. It provides participants with an existing definition of gender roles to ensure a similar understanding among all respondents. We then correlated this to the animal score mentioned earlier (see Appendix). The specific questions as well as definitions of traditional and modern views of gender roles can be seen in the Appendix.

Findings

Gender identities

In order to assess attitudes associated with gender identities towards nonhuman animals, we explored current caretaking of companion animals (see Table 6.3). Past companion animal caretaking was also explored (though we do not report on this here). For all gender identities (cisgender woman, cisgender man and non-cisgender), a higher proportion of people have a companion animal than do not have a companion animal.

Following on from the exploration of companion animals and gender identity, we then investigated gender identity and animal score, for the purpose of unpacking general attitudes towards nonhuman animals rather than companion animals only.

Animal score is out of 109 and had a non-normal distribution. The median scores for our three categories were 80 for cisgender women, 79.5 for cisgender men and 78 for non-cisgender individuals. These differences were not statistically significant (Kruskal-Wallis test, chi-squared=1.4369, df=2, p-value=0.4875).

In order to delve deeper into attitudes of people with different gender identities towards nonhuman animal welfare, we explored status dogs and dog fighting as a specific example of cruelty to companion animals.

Table 6.3: Gender identity and ownership of a companion animal

Currently has a companion animal	Cisgender woman	Cisgender man	Non-cisgender
No	42	15	16
Yes	166	25	22

Table 6.4: Gender identity and awareness of status dogs/dog fighting

Aware of status dogs/dog fighting	Cisgender men	Cisgender women	Non-cisgender
No	0	4	1
Yes	40	204	37

Table 6.5: View of gender roles and pet ownership

View of gender roles	Current pet ownership	
	Yes	No
Completely modern	81	32
Mostly modern	83	31
Equal	30	5
Mostly traditional	18	4
Completely traditional	1	1

Interestingly, the median score for acceptance of status dogs was the same (4) for all three categories. This was statistically significant (Kruskal–Wallis test, chi-squared=16.728, df=2, p-value=0.0002331). Table 6.4 details the percentage of individuals of each gender identity who report being aware or unaware of the existence of status dogs and dog fighting prior to the survey taking place. A vast majority of participants are aware of this form of cruelty. These differences were not statistically significant using both a Pearson's Chi-squared test (x-squared=0.92137, df=2, p-value=0.6309) and a Fisher's Exact Test for Count Data (p-value=0.6024), which was run due to low (<4) expected frequencies.

Gender roles

The same series of questions on the survey were correlated to gender roles as opposed to gender identity in order to investigate if attitudes towards nonhuman animals can be understood through a connection to socialisation of perceived characteristics of the traditional views of binary gender (see Appendix). Again, this was broken down into companion animals, nonhuman animals (through the animal score), and status dogs and dog fighting. Table 6.5 shows the number of people with companion animals compared to people without companion animals in relation to the five possible views of gender roles – all modern, mostly modern, equal, mostly traditional, and all traditional. As mentioned, these percentages are above the expected portions of the population who are thought to have

Table 6.6: View of gender roles and awareness of status dogs

Aware of status dogs/dog fighting	Completely traditional	Mostly traditional	Equal	Mostly modern	Completely modern
No	0	0	1	3	1
Yes	2	22	34	111	112

companion animals. These differences in companion animal ownership were not statistically significant (Pearson's Chi-squared test X-squared=4.2109, df=4, p-value=0.3782. This was then followed up with a Fisher's exact test due to some low expected frequencies: Fisher's Exact Test p-value=0.3237 [still not significant]).

Respondents who reported their views of gender roles as completely traditional had the lowest median animal score (less favourable attitude towards nonhuman animal welfare – 55). Respondents who reported their views as completely modern had the highest score (more favourable attitude – 82). This was followed by mostly traditional as second highest (80.5), then mostly modern (79) and equal (78). Respondents with mostly modern views of gender roles had the largest range of animal scores. These differences were not statistically significant (Kruskal-Wallis, chi-squared= 6.1159, df=4, p-value=0.1907).

As with gender identities and acceptance of status dogs, the median score for all five categories of gender roles was the same (5). In this case, however, this was not statistically significant (Kruskal-Wallis chi-squared=1.3268, df=4, p-value=0.8568). Table 6.6 shows the awareness, or lack thereof, of status dogs in relation to gender roles. Again, as with gender identity, a vast majority of participants are aware of this form of animal cruelty. Originally, we ran a Pearson's Chi-squared test (X-squared=1.6857, df=4, p-value= 0.7933), but some expected frequencies were low (<4), so we also ran a Fisher's Exact Test for Count Data (p-value=0.7105). In both cases, there was no statistical significance.

Analysis

In regard to over-representation in the sample, two in particular have been noticed in other studies – women participants and participants with companion animals (Wyatt et al, 2017). In the case of this research, more women than men (cis in this instance) chose to take part. Moreover, studies, in general, attract self-selecting participants, who are interested in the area of research. In this research, a significant percentage of participants with companion animals chose to take part. These sample dynamics might skew the results, encouraging the assumption that participants are more concerned about nonhuman animal welfare, when this may not be the case or at least not to the extent found in our survey.

Unpacking gender identity and attitudes towards nonhuman animals

The initial assumption made in this research was that non-cisgender identities would have higher animal scores than cisgender identities. First, a non-cisgender identity may indicate a progressive and open-minded approach to one sphere of life, that is, gender, and thus may be linked to a similar approach to other spheres of life, such as a progressive attitude of greater concern for animal welfare (George et al, 2016). In addition to that, we suggested that those identifying as NB, similarly to cisgender females, occupy positions in the social hierarchy inferior to cisgender males. Based on this, we assumed that NB gender identities might empathise more with nonhuman animals than binary identities, particularly cisgender male identities, and thus have higher animal scores. To some degree, this was true. Non-cisgender individuals had higher animal scores than cisgender men, indicating more favourable attitudes towards animal welfare. Non-cisgender identities did not, however, have more favourable attitudes towards animal welfare than participants identifying as cisgender female did. Further studies would need to be done to uncover the origins of the differences in attitudes towards animal welfare to understand why non-cisgender individuals sit between cisgender females and males.

Attitudes towards status dogs showed no variation. All three categories had the same median score related to status dogs and there was an extremely high number of participants aware of status dogs and dog fighting. We speculate that the lack of variation might be due to the extreme nature of status dogs and dog fighting in relation to animal welfare. Furthermore, as already mentioned, a majority of people in the study had companion animals, which is not the case with the wider public in the UK. This might also indicate our sample was more favourable in general to animal welfare and would thus be unaccepting of status dogs and dog fighting.

Gender roles and nonhuman animals

The initial assumption made in this research was that one's perception of gender roles might affect their view of nonhuman animal welfare. More specifically, we hypothesised that gender roles and attitudes towards animals are connected: dismissal of binary gender roles results in both men and women being more sympathetic towards suffering of nonhuman animals. This assumption is based on the previous conclusion that individuals who regard women as different from or subordinate to men are also more likely to view animals as inferior to humans (Allcorn and Ogletree, 2018), and thus have little regard for animal welfare. This assumption did not prove to be true. While those with all traditional views did have the lowest animal score and those with all modern the highest animal score, these findings were not significant. All five perspectives on gender roles – 'Completely modern', 'Mostly modern but some traditional', 'Equally traditional and modern', 'Mostly traditional but some modern' and 'Completely traditional' – scored the same median score of four, out of a possible range of 0–4, in regard to acceptance of status dogs. Our study, then, does not support that women and nonhuman animals should be viewed as a continuum (Allcorn and Ogletree, 2018), or that gender equality progress is closely intertwined with the social perception of nonhuman animals. We did expect that the lowest animal score would be people with all traditional views, which presuppose a male-dominated society where both women and nonhuman animals are identified as nature and men are geared towards domination of nature (Wyckoff, 2014), but, again, this was not significant.

Limitations of this study

There are several limitations to our study. First, since the survey used a convenience sample, it was not representative of the wider population. There are significant differences between participants' demographics and the general population's demographics for every category apart from gender, ethnicity and marital status, particularly age (a portion of the participants in the sample are younger than the general population's age).

Second, a key problem with the survey in relation to gender identity was the relatively small sample size of each NB gender identity, which in some cases only contained two individuals. This reduced the representation of these groups in the sample. As such, NB genders were aggregated into one category, which lessens the specificity to distinct gender identities. Care should be taken when addressing the ways in which NB gender identity is defined, because the life experiences associated with each identity are likely to be very different, and it is important to avoid invalidating any individual's sense of identity.

Although not statistically significant, there were more NB gender identifying participants than the estimated UK national percentage (0.4 per cent nationally, 14.2 per cent in the study). This perhaps may indicate that the estimate in the overall population is too low or that possibly census questions might be failing to capture the diversity of identifiers of gender. Furthermore, we would not have expected to see a significant over-representation of people with NB gender identities in the sample. The URL of the survey and its introductory text do not give any indication that the survey included a range of NB options for gender identity, and the survey was not distributed in places expected to have a higher proportion of NB individuals. There were therefore no obvious intrinsic qualities of the survey which could make it attract an over-representative number of NB respondents. There is a chance the survey was spread between people with NB gender identities via word of mouth, effectively resulting in a snowball sample. However, there is no way to determine whether this was the case, and if it was, it did not significantly affect the representativeness of the sample.

However, this does not mean the sample was not biased. Survey respondents were demographically not representative of the wider population, and the survey was advertised in places likely to attract predominantly younger individuals, who were more likely to be in or have completed higher education and were more likely to already be interested in nonhuman animal welfare. This bias means that while the study is an initial investigation into how attitudes towards nonhuman animal welfare varies with gender identity, it cannot claim to offer findings which are generalisable to wider society.

Quantitative statistical analyses such as those used in this study are also inherently limited because they do not provide insight into respondents' unique experiences. Quantitative methods alone are not enough to explain the reasons behind the attitudes found in the survey and do not truly reflect the range of subjective attitudes likely to be held by respondents. However, the collection of quantitative data means the sample size can be larger than if purely qualitative methods were used.

Some respondents to the survey did not appear to understand the question about gender identity. For example, one respondent answered 'straight', which is describing a heterosexual sexual orientation rather than gender identity. This shows a more comprehensive definition of gender identity, with emphasis on its difference to sexual orientation, is needed when investigating this topic. More than one respondent simply answered 'woman' or 'female'. It is possible these answers were given due to confusion over the range of terms available, and these respondents may have meant 'cisgender female'. However, there is no way to determine this for sure, as the respondent may have deliberately chosen the term 'female' rather than 'cisgender female' to describe their gender identity. Furthermore, another respondent argued that 'there's only two genders, male and female'. This view is common, though misleading: even before NB gender identities became more accepted, it was recognised that intersex individuals have a range of genetic and physical characteristics, which do not fit the gender binary (Hine and Martin, 2016).

Furthermore, there are limitations around the questions used to assess attitudes to animals, and gender roles. It is important to acknowledge some of the animal attitudes questions were leading, however, these questions were adapted from Knight et al (2017), and care was taken to ensure that not all questions were leading. For example, some questions asked participants to rate the frequency with which they engage with nonhuman animals for recreational or educational purposes, which were not emotive questions but elicited a factual response. Using a range of questions such as this does increase the chance of capturing participants' real experiences and views, but care should still be taken when applying the findings to wider society. In addition, as there was no prior research into attitudes to nonhuman animals among different gender identities, the gender roles question was just one possible way of assessing participants' attitudes and there is scope for improving the ways in which this is addressed in future research. Future questions should use inclusive language and could also draw on existing measures such as the Bem Sex Role Inventory (Bem, 1974), which measures an equal number of masculine, feminine and gender-neutral traits.

Conclusion

Our study has advanced current knowledge of human-animal and gender studies by investigating attitudes towards nonhuman animals among those with NB gender identities. The study has also confirmed the findings from previous research and showed that cisgender women have more favourable attitudes to nonhuman animals than cisgender men. It also demonstrated that non-cisgender identities did not have more favourable attitudes towards nonhuman animals than participants identifying as cisgender women.

Moreover, the study has explored the attitudes towards nonhuman animals among those with different perceptions of gender roles. While not statistically significant, those with what we have termed 'Completely modern' views of gender roles tend to have a more favourable attitude towards nonhuman animal welfare, but overall, the study does not indicate that gender equality progress is closely intertwined with the social perception of nonhuman animals. It is difficult to draw concrete conclusions about the views held by both cisgender and non-cisgender individuals as well as those with different views of gender roles, suggesting that there is scope for future research on this topic. Future studies could address relative levels of interest in animal welfare issues and gender issues, to determine if an increase in one is associated with a decrease in the other and explore the reasons behind it. Future studies could also help explore the variation in attitudes of those with different perceptions of gender roles, qualitatively scrutinising the views that shape survey responses.

This study (and the others we propose) has hopefully contributed some understanding to the complex issue of gender identity, gender roles, and the relationship to nonhuman animals and their welfare. We believe that through deconstructing attitudes we can uncover aspects of the intersecting struggles for LGBTQ and nonhuman animal liberation and thus develop strategies to resist bias and stereotypical hierarchies in all forms. Yet, we also recognise that theoretical knowledge alone is not sufficient for such a resistance; as Best et al (2007) suggest, knowledge deepens in practice, in and through political struggle and social movements. Intersectional alliances are crucial for the mission of the abolition of nonhuman animal oppression (Nocella II, 2010; Loadenthal, 2012).

Note

[1]   www.thestudentroom.co.uk and www.reddit.com/r/SampleSize

Appendix

Animal score was calculated from the following sets of questions, and was out of a total of 109:

1. Please indicate the extent to which you agree or disagree with the following statements:
   - Humans have too little respect for the lives of animals.
   - There should be more regulations about the welfare of farmed animals.
   - Animal agriculture raises serious ethical questions about animal welfare.
   - Animals should not be used to test products such as cosmetics, soap and cleaners.
   - Hunting animals for sport is not an acceptable recreational activity.
   - The consequences of abusing animals should be as severe as the consequences of abusing children.
   - Animals should not be used to produce any food products, even those which do not result in the death of the animal.
   - Animals should not be used to produce clothing or decorative items (for example, leather or fur clothing, shoes or accessories, or fur cushions or furnishings).

These questions were coded so a response of 'strongly agree' was equal to 5 and a response of 'strongly disagree' was equal to 1. (Intermediate options were slightly agree, neither agree nor disagree, slightly disagree, strongly disagree.)

2. The following are examples of using animals for recreational or educational purposes. Please indicate the frequency with which you engage in each activity:
   - Watching or reading non-fiction about animals (for example, watching nature documentaries, reading educational books about animals).
   - Watching or reading fiction about animals (for example, films or books that feature animals or have animals as characters in them).
   - Visiting zoos, petting zoos, safari parks or aquariums.
   - Playing with pets (which belong to me or someone in my household).

- Playing with pets (which do not belong to me or someone in my household).
- Volunteering with animals in any capacity, for example, in a rescue centre, as a dog walker, and so on.

These questions were coded so 'never' was equal to 1 and 'at least once daily' was equal to 7. (Intermediate options were less than once a month; once a month; more than once a month, but not every week; once per week; more than once a week, but not every day.)

3. The following are examples of using animals for recreational purposes. Please indicate the frequency with which you engage in each activity:
   - Lawfully hunting animals (for example, shooting certain species of birds during the shooting season).
   - Watching animals racing (for example, dog or horse racing) on TV or in real life.

These questions were coded so 'never' was equal to 7 and 'at least once daily' was equal to 1. (Intermediate options were the same as for Question 2.)

4. A 'status dog' is a dog that is used by its owners to convey an image of toughness or aggression. Owners of these dogs often think that their dog will make the owner themselves appear tough or powerful. From this definition and any prior knowledge you may have, what do you think of the use of dogs as status dogs?
5. Dog fighting in this context is when humans force two dogs to fight. It is illegal in the UK and can result in horrific injuries to the dogs and the dogs involved may die during the fight. Losing dogs are sometimes killed or beaten by their owners. Dog fighting is also associated with other serious crimes such as owning illegal weapons and using illegal drugs. However, many of the people involved in dog fighting report that it is a very close-knit and loyal community with a strong sense of honour, and many people think that it is in a dog's nature to fight and therefore they should be allowed and encouraged to do so because ultimately the dog will be happier for it. From this definition and any prior knowledge you may have, what do you think of the use of dogs in dog fighting?

Questions 4 and 5 had the response options 'entirely acceptable' (scored 1), 'mostly acceptable', 'neither acceptable nor unacceptable', 'mostly unacceptable' and 'entirely unacceptable' (scored 5).

The survey asked the following question in relation to gender roles: 'Gender roles are defined as the expected attitudes and behaviour of each sex. Traditional views of gender roles often describe women as better suited to nurturing, traditionally feminine tasks such as childcare and homemaking, while men are seen as better suited to traditionally masculine roles such as working full-time and doing DIY tasks. Modern views of gender roles often recognise that there are no differences between men and women in how naturally well-suited they are to different tasks. Most women now work full-time and more men are responsible for childcare or homemaking. Based on these definitions and any prior knowledge you may have, how would you describe your own views of gender roles?'

Response options included:

- My views are completely traditional.
- My views are mostly traditional, but some are more modern.
- My views are equally traditional and modern.
- My views are mostly modern, but some are traditional.
- My views are completely modern.

References

Adams, C. (2010) *The Sexual Politics of Meat: A Feminist-Vegetarian Critical Theory*. New York: Continuum.

Allcorn, A. and Ogletree, S. (2018) Linked oppression: Connecting animal and gender attitudes. *Feminism & Psychology*, 28(4), 457–69.

Animal Aid (2018) *Animal experiments – scandalous annual statistics*. Available at: www.animalaid.org.uk/animal-experiments-scandalous-annual-statistics/ (accessed 15 March 2021).

Animal Clock UK (2020) *2020 UK Animal Kill Clock*. Available at: https://animalclock.org/uk/ (accessed 24 April 2021).

Bell, B. (2017) Are we tough enough on animal cruelty? *BBC*. Available at: www.bbc.co.uk/news/uk-england-39042626#:~:text=Between%202013%20and%202015%20more,in%20Germany%20it%20is%20three (accessed 2 May 2021).

Bem, S.L. (1974) The measurement of psychological androgyny. *Journal of Consulting and Clinical Psychology*, 42, 155–62.

Best, S., Nocella, A.J., Kahn, R., Gigliotti, C. and Kemmerer, L. (2007) Introducing critical animal studies. *Journal for Critical Animal Studies*, 5(1), 4–5.

Birke, L. (2002) Intimate familiarities? Feminism and human-animal studies. *Society and Animals*, 10(4), 429–36.

Brooks, L. (2017) Legal recognition for non-binary people planned in Scotland. *The Guardian*. Available at: www.theguardian.com/society/2017/nov/09/legal-recognition-for-non-binary-people-planned-in-scotland (accessed 2 February 2021).

Browne, K. and Nash, C.J. (2010) *Queer Methods and Methodologies: Intersecting Queer Theories and Social Science Research*. Burlington, VT: Ashgate.

Butler, J. (1990) *Gender Trouble: Feminism and the Subversion of Identity*. Abingdon: Routledge.

Butler, J. (2004) *Undoing Gender*. Abingdon: Routledge.

Byrd, E., Widmar, N. and Fulton, J. (2017) Of fur, feather, and fin: Human's use and concern for non-human species. *Animals (Basel)*, 7(3), 1–18.

Cudworth, E. (2010) 'The recipe for love'? Continuities and changes in the sexual politics of meat. *The Journal for Critical Animal Studies*, 8(4), 78–99.

Evans, R., Gauthier, D. and Forsyth, C. (1998) Dogfighting: Symbolic expression and validation of masculinity. *Sex Roles*, 39(11–12), 825–38.

Fausto-Sterling, A. (2019) Gender/sex, sexual orientation, and identity are in the body: How did they get there? *The Journal of Sex Research*, 56(4–5), 529–55.

Francione, G. (2004) *Animals, Property and the Law*. Philadelphia: Temple University Press.

George, K., Slagle, K., Wilson, R., Moeller, S. and Bruskotter, J. (2016) Changes in attitudes toward animals in the United States from 1978 to 2014. *Biological Conservation*, 201, 237–42.

Grubbs, J. (2012) Guest editorial. *Journal for Critical Animal Studies*, 10(3), 4–6.

The Guardian (2011) Australian passports to have third gender option. *The Guardian*. Available at: www.theguardian.com/world/2011/sep/15/austral ian-passports-third-gender-option (accessed 25 January 2021).

Harding, S. (2013) 'Bling with bite' – the rise of status and weapon dogs. *Veterinary Record*, 173, 261–3.

Herzog, H. and Galvin, S. (1997) Common sense and the mental lives of animals: An empirical approach. In R. Mitchel (ed) *Anthropomorphism, Anecdotes and Animals*. Albany: State University of New York Press, pp 237–53.

Herzog, H., Betchart, N. and Pittman, R. (1991) Gender, sex role orientation, and attitudes towards animals. *Anthrozoös*, 4, 184–91.

Hillsburg, H. (2010) My pet needs philosophy: Ambiguity, capabilities and the welfare of domestic dogs. *Journal for Critical Animal Studies*, 8(1/2), 33–46.

Hine, R. and Martin, E. (2016) *A Dictionary of Biology*, 7th edn. Oxford: Oxford University Press.

Hughes, G., Maher, J. and Lawson, C. (2011) *Status Dogs, Young People and Criminalisation: Towards a Preventative Strategy*. Research Project Report, Cardiff Centre for Crime, Law and Justice, Cardiff University.

Kendall, H., Lobao, L. and Sharp, J. (2016) Public concern with animal well-being: Place, social structural location, and individual experience. *Rural Sociology*, 71(3), 399–428.

Knight, S., Vrij, A., Cherryman, J. and Nunkoosing, K. (2017).Attitudes towards animal use and belief in animal mind. *Anthrozoos*, 17, 43–62.

Lee, S., Gibbons, J. and Short, S. (2010) Sympathetic reactions to the bait dog in a film of dog fighting: The influence of personality and gender. *Society & Animals*, 18(2), 107–25.

Loadenthal, M. (2012) Operation splash back! Queering animal liberation through the contributions of neo-insurrectionist queers. *Journal for Critical Animal Studies*, 10(3), 81–110.

Nocella II, A. (2010) Abolition a multi-tactical movement strategy. *Journal for Critical Animal Studies*, 8(1/2), 176–83.

Office for National Statistics (2014) *2011 Census*. London: Office for National Statistics. Available at: www.ons.gov.uk/census/2011census (accessed 10 January 2018).

Pedersen, H. and Stanescu, V. (2014) Conclusion: Future directions for critical animal studies. In N. Taylor and R. Twine (eds) *The Rise of Critical Animal Studies: From the Margins to the Centre*. London: Routledge, pp 262–76.

Peek, C., Dunham, C. and Dietz, B. (1997) Gender, relational role orientation, and affinity for animal rights. *Sex Roles*, 37, 905–20.

Pet Food Manufacturer's Association (2017) *Pet Population 2017*. Available at: www.pfma.org.uk/pet-population-2017 (accessed 15 January 2018).

Practical Androgyny (2014) How many people in the United Kingdom are nonbinary? *Practical Androgyny*. Available at: https://practicalandrogyny. com/2014/12/16/how-many-people-in-the-uk-are-nonbinary/ (accessed 6 March 2018).

R Core Team (2017) *R: A Language and Environment for Statistical Computing. R Foundation for Statistical Computing*. Available at: www.R-project.org/ (accessed 6 August 2018).

Riggs, D.W., Taylor, N., Signal, T., Fraser, H. and Donovan, C. (2018) People of diverse genders and/or sexualities and their animal companions: Experiences of family violence in a binational sample. *Journal of Family Issues*, 39(18), 4226–47.

RSPCA (2022) The origins of the RSPCA. *RSPCA*. Available at: www. rspca.org.uk/whatwedo/whoweare/history (accessed 26 August 2022).

Shupe, J. (2016) I am the first official genderless person in the United States. *The Guardian*, 16 June. Available at: www.theguardian.com/commentisfree/ 2016/jun/16/i-am-first-official-genderless-person-united-states (accessed 22 November 2020).

Simonsen, R. (2012) A queer vegan manifesto. *Journal for Critical Animal Studies*, 10(3), 51–80.

Smith, R. (2011) Investigating financial aspects of dog-fighting in the UK. *Journal of Financial Crime*, 4, 336–47.

Sztybel, D. (2007) Animal rights law: Fundamentalism versus pragmatism. *Animal Liberation Philosophy and Policy Journal*, 5(1), 2–54.

Taylor, J., Zalewska, A., Gates, J. and Millon, G. (2018) An exploration of the lived experiences of non-binary individuals who have presented at a gender identity clinic in the United Kingdom. *International Journal of Transgenderism*, 20(2–3), 196–204.

Taylor, N., Riggs, D.W., Donovan, C., Signal, T. and Fraser, H. (2018) People of diverse genders and/or sexualities caring for and protecting animal companions in the context of domestic violence. *Violence Against Women*. https://doi.org/10.1177/1077801218809942

Tesform, G. and Birch, N. (2013) Does definition of self predict adopter dog breed choice? *International Review on Public and Nonprofit Marketing*, 10(2), 103–27.

Thomas, J. (2013) Unpatients: The structural violence of animals in medical education. *Journal for Critical Animal Studies*, 11(1), 46–62.

Valentine, V. (2016) *Non-binary People's Experiences in the UK*. Available at: www.scottishtrans.org/wp-content/uploads/2016/11/Non-binary-report.pdf (accessed 25 May 2018).

Wells, D. and Hepper, P. (1997) Pet ownership and adults' views on the use of animals. *Society & Animals*, 5, 45–63.

West, C. and Zimmerman, D. (1987) Doing gender. *Gender & Society*, 1(2), 125–51.

Woodward, L. and Bauer, A. (2007) People and their pets: A relational perspective on interpersonal complementarity and attachment in companion animal owners. *Society & Animals*, 15, 169–89.

World Animal Protection (2020) *Animal Protection Index*. Available at: https://api.worldanimalprotection.org/ (accessed 13 October 2020).

Wyatt, T., Maher, J. and Biddle, P. (2017) *Scoping Research on the Illegal Importation and Farming of Puppies*. Edinburgh: Scottish Government and Department of Environment, Food, and Rural Affairs.

Wyckoff, J. (2014) Linking sexism and speciesism. *Hypatia*, 29(4), 721–37.

PART II

Gendered Impacts and Victimisation

Queering Green Criminology: The Impacts of Zoonotic Diseases on the LGBTQ Community

Laurence Pedroni and Benja Kromash

Introduction

As the effects of climate change continue to grow and become more evident across the globe, it becomes increasingly important to recognise how the effects of the climate disaster affect different groups. While green criminology has created expansive literature examining the differing effects of climate change across different geographies, racial groups and gender, it has so far under-examined the effects on queer people. This under-examination ignores how queer people disproportionately experience climate change and disasters (Goldsmith et al, 2021). This chapter explores the potential and meaning of queering the field of green criminology. Given the lack of connection between queerness and green criminology (in theory or practice) we ask the open-ended question of how should we queer green criminology? The field of green criminology is important for understanding the connections between different forms of marginalisation, disasters and green crimes, but we argue that this inquiry is incomplete without a queer lens.

A queer feminist green criminology should draw on a plethora of fields and methods to strive to understand the experiences of queer peoples in relation to the environment and social harm. The impacts of environmental disasters exacerbate inequality and existing forms of social marginalisation. We draw on the literatures of green criminology, eco-feminism and necropolitics to consider what a queer green criminology could look like. We use zoonotic disease pandemics to model analyses of examples of disasters that uniquely impact queer peoples to support our claims that the state's disregard for queer peoples, especially trans women of colour, constitutes a green crime. To fully understand these green crimes, we argue that green criminology needs a queer lens. The intersectional impacts of environmental disasters are complex and it is imperative that the research is done with intentionality and care. However, up to this point, a gap in the field of green criminology is the lack of attention to the vicimisation of queer peoples.

To model queer feminist green criminology our analyses take a mixed methods approach that relies on statistics, policy analysis and the voices of queer peoples. For example, the narrative by Janet Mock, a trans woman of colour, supports the claims that how queer peoples experience the world and disasters is not only unique but needs to be considered in research and policy.

In her memoir, Mock (2014) states:

As long as trans women are seen as less desirable, illegitimate, devalued women, then men will continue to frame their attraction to us as secret, shameful, and stigmatized, limiting their sexual interactions with trans women to porn and prostitution. And if a trans woman believes that the only way she can share intimate space with a man is through secret hookups or transactions, she will be led to engage in risky sexual behaviors that make her more vulnerable to criminalization, disease, and violence; she will be led to coddle a man who takes out his frustrations about his sexuality on her with his fists; she will be led to question whether she's worthy enough to protect herself with a condom when a man tells her he loves her; she will be led to believe that she is not worthy of being seen and must remain hidden. (Mock, 2014: 207)

Perspectives such as Mock's should inform green criminological research and also policy recommendations that wish to address socioenvironmental harms and injustices. Our chapter seeks to model how a queer feminist green criminology can be applied and argues that the connections between the marginalisation of queer identity and criminal exploitation of the environment need further attention in this field of research.

Statement of positionality

When discussing the effects of HIV and COVID-19 and their unequal effects on the queer community, it is important to note that we are speaking from a US-based context. This is not to diminish the effects of these zoonotic diseases in other contexts; instead, we wish to recognise that we are speaking from our positions and histories influenced by living in the United States, but also recognising that our research is in a more specific context and therefore has lower generalisability to outside the United States. We note the specific context of our research largely given the unique histories of 'race', class and gender that are located in the United States but also based on the unique histories of zoonotic diseases in other places such as Haiti (Farmer, 2004). While we note the context of our research, we maintain that a queer green criminology can and should be applied in wide-ranging contexts. To further expand our positionality, we wish to share that we both identify as white and queer, since these are layers to our identity that influence how we understand and navigate the world around us as citizens and researchers.

Disasters

Climate change marks the most extreme threat to the environment (Barry and Hoyne, 2021). Examples of the environmental degradation brought on from climate change include – but are not limited to – the rise in sea level (Griggs, 2021), extreme temperatures (Barreca, 2012), changes in precipitation patterns (Konapala et al, 2020) and extensive biodiversity loss (Habibullah et al, 2022). According to White (2022), the field of green criminology emphasises the connection between crimes by both the state and corporations, and climate change. This is largely because the perpetrators of climate change are largely governments and corporations who firmly grasp immense power, while resisting social change.

Challenges for environmental law enforcement are 'evident in regards to events and trends associated with climate change, such as natural disasters that are increasing in intensity and frequency, or the social dislocations stemming from climate induced migration' (Bergin and Allen 2008; White, 2011: 455). White and Heckenburg (2014) argue that the enactment of harm is never far from the surface in the wake of catastrophic events, and these, in turn, are influenced by patterns of global warming.

Further research from the field of green criminology research shows that women's and nonhuman animals' exposure to environmental harms and disasters, at the hands of corporations and the state, is an important source of green victimisation (Lynch, 2018; Bromwich, 2022). However, the exclusion of queer people from the green criminological research has limited our understanding of what victimisation can look like after a green crime; the exclusion of an entire portion of the population, a historically marginalised one at that, renders the field of green criminology not only incomplete but in danger of reinforcing such marginalisation. Consistent with Beirne's (2021) argument that green criminologists are *well positioned* to argue against the victimisation of animals in the wildlife trade, green criminologists are also well positioned to draw connections between the rise in disasters and their harmful effects on queer populations across the globe. We maintain that the criminological study of disasters, from pandemics to climate change, can benefit greatly from queer feminist green criminology.

Within the fields of green criminology and disaster studies, there is no singular agreed-upon definition of disaster. We draw on these fields to understand disasters as occurring when the routines of collective societal groups experience major disruptions and unplanned courses of action are required to survive (Quarantelli, 2000). Such crises are also associated with a threat to the core values of a community's life-sustaining systems. Such values include 'safety and security, welfare and health, integrity and rule of law, which become shaky or even meaningless as a result of (looming) violence, destruction, damage or other forms of adversity' (Perry, 2007: 24). Through this understanding of disaster and crisis, we posit that natural disasters evoke an intense sense of crisis in which the safety and security of one's self or loved ones are deeply threatened. Similar to the project of green criminology, our discussion requires us to ask implicit ontological questions about what a disaster is. While popular imagination of disasters includes tsunamis, earthquakes and wildfires, a disaster can also take the form of a virus, which is naked to the human eye, but which still creates massive social upheaval, leaving millions dead in its wake.

Beirne (2021) draws on the Centers for Disease Control and Prevention to establish a working definition of zoonotic diseases as the transmission from infected nonhuman animal to human through contact with saliva, blood, urine, faeces or flesh. Such spread of pathogens occurs via contact with contaminated surfaces. Examples of zoonotic diseases include Lyme disease, monkeypox, Zika virus, HIV and COVID-19. We identify the effects of these viruses as multi-scalar disasters that we understand as disasters happening on multiple levels of society. Traditionally, we think of disasters as similar to hurricanes that leave millions of people without electricity or running water. But they can also include events that decimate smaller, more localised communities. In the case of zoonotic diseases, these disasters are simultaneously taking place on multiple scales: from hiding within a person's blood, all the way to the intercontinental transmission to millions of people. The impacts of such zoonotic diseases are worsened by climate change (Bartlow et al, 2019), a deadly disaster in itself. Lawler et al (2021) argue that climate change's impacts include biodiversity loss, decreased ecosystem health and the further spread of zoonotic diseases. They conclude that responses to pandemics, especially the COVID-19 pandemic, 'must include actions aimed at safeguarding biodiversity and ecosystems, in order to avoid future emergence of zoonoses and prevent their wide-ranging effects on human health, economies, and society' (Lawler et al, 2021: e840). Our own research identifies the state's response to these disasters as green crimes. This echoes the work of Beirne when he states that: 'It is no small irony that some of the very governments who are now responding with alacrity to scientific advice about the human havoc wrought by COVID-19 continue to reject warnings about the emergency of anthropogenic climate change' (Beirne, 2021: 620).

In a similar vein, the outbreak of the monkeypox virus in the summer of 2022 is partly driven by climate change because historical data support that monkeypox most prevalently spreads during times of increased rainfall, flooding and deforestation, all of which drive animals hosting the virus into human residences (Alakunle and Okeke, 2022). Cultural commentator and author, Jonathan Van Ness, stated:

> Watching the government's botched response to monkeypox has been surreal, and in many ways, I believe it's been fueled by homophobia and transphobia. When an outbreak affects mainly men who have sex with men, some portion of our elected legislators will have no incentive to act. They think it will not touch their constituents, which is obviously messed up because people's lives are at stake, and there are queer people in all 50 states. (Van Ness, 2022: para 2)

While there are studies showing higher rates of monkeypox in queer communities, there are concerns about sampling bias that perpetuates homophobia (Singla et al, 2022), and also a risk of further stigmatising a historically marginalised community (Treisman, 2022). This is a quintessential example of how an environmental disaster, climate change, allows for the spread of a zoonotic disease, while exacerbating the social inequality faced by the LGBT community.

The effects of these disasters, both zoonotic diseases and climate change, are further amplified through the effects of a neoliberal political economy which reifies consumerism and hyper-individuality, environmental degradation, and makes and remakes stratification, which is at odds with community care and equitable public health. The approach of analysing the disaster of zoonotic diseases is an example of how disasters can be studied to understand their unique impacts on queer peoples and nature. Such an approach is in tandem with other lenses, such as 'race', gender and class. Understanding the intersectional impacts of disasters, such as zoonotic diseases, builds on Gaard's (2015) point that marginalisation is exacerbated by disasters, such as diseases, and that a queer perspective is essential to this understanding. We claim that the field of green criminology must highlight queer experiences to more fully understand disasters.

The research on the victimisation of queer peoples is well established in the literature, with some of the newest data showing that LGBT people are nearly four times as likely to be the victims of violent crime when compared to their straight counterparts (Flores et al, nd). However, these data also tend to focus on the effects and consequences of interpersonal violence (see Decker et al [2018] for a review of this literature). We are more focused on the connections between victimisation and environmental disasters, namely HIV and COVID-19. Zoonotic diseases like these draw their connections to structural problems such as the climate disaster worsened through late-stage capitalism and then filtered through harmful lenses of homophobia and transphobia, which further differentially affect queer people's daily lives.

Review of literature

This section acts as a model of analysis of what a queer feminist green criminology can look like in practice by drawing connections between the consequences of environmental degradation and the marginalisation of queer people. The green crimes being considered here are the state's response (and lack thereof) to the disasters of zoonotic diseases HIV and COVID-19. We specifically consider the impacts in the areas of housing and employment, incarceration and isolation. These are examples of areas in which queer peoples have been uniquely victimised by zoonotic disease disasters, and there are certainly many other avenues to be explored.

Employment and housing discrimination

Queer peoples are among the marginalised populations that experience higher rates of both employment discrimination and houselessness. The ways those are exacerbated in times of disaster, particularly zoonotic disease pandemics, shed light on how queer peoples are victimised by such situations. The lack of response by the state, specifically government healthcare agencies, to specifically support queer peoples constitutes these zoonotic diseases pandemics as green crimes that further call for a queering of green criminology.

In research on the many forms of stigma experienced by living with HIV, Brent (2016) finds that the stigma against those who are HIV positive results in extensive job loss or not being hired at all. This finding was reinforced through research that was conducted in nine countries, including the United States, which shows that this impact of being HIV positive is nearly universal (GNP and ILO, 2012). Maulsby et al (2020) found that people who are unemployed are more likely to receive a later HIV diagnosis and also receive delayed care, such as necessary antiretroviral therapy treatments. Research from France concludes that this dynamic relationship between HIV status and employment does not seem to be improving; Annequin et al (2015) found that despite medical innovations and improvements in access, employment discrimination against those living with HIV has not changed over time. Similarly, in the COVID-19 pandemic, queer peoples have been furloughed or laid off at higher rates than their heterosexual, cisgender counterparts (Wilson et al, 2021). That same research project also sheds light on the intersectional way the COVID-19 pandemic has had financial impacts when they report that queer people of colour are more than twice as likely to struggle to pay for household goods. The intersections of 'race', class, gender and sexual orientation are manifesting in vastly different experiences that are being exacerbated by this current crisis.

Trans women of colour experience particularly high rates of houselessness and employment discrimination (Clements-Nolle et al, 2001; Davis and Wertz, 2009; Wilson et al, 2021). As explained in our introductory quote from Mock (2014), trans women of colour are incredibly vulnerable to these crimes, and their victimisation needs to be centred in criminological research. It is important for the field of green criminology to understand how the intersections of gender and 'race' create unique experiences when navigating housing and employment discrimination. Trans women of colour are not afforded the same privileges, in consequence, they 'may turn to street-based work in order to survive. As a result, many are more likely to experience victimization, whether from the general population or from the police' (Edelman, 2014: 183). This unstable housing is linked to zoonotic diseases because while many people experienced evictions during COVID-19, in both the HIV and COVID-19 pandemics, queer people experienced higher rates of houselessness (HALSA, nd; Wilson et al, 2021). Additionally, queer people of colour are about 2.5 times as likely to rent than their straight white counterparts (Wilson et al, 2021). This acts as a positive feedback loop that further puts queer peoples at risk since houseless peoples are more likely to be exposed to COVID-19 (HRI, nd) and HIV (Spinelli et al, 2019). Queer people, particularly queer people of colour, experience forms of social marginalisation that result in worsening material conditions such as the examples that we highlight in this chapter that include higher rates of houselessness and unemployment putting them more at risk for exposure to zoonotic diseases. These worsening conditions also place queer peoples in positions where they are more likely to experience the effects of the climate crisis (Goldsmith et al, 2022). While we are focusing on the effects of zoonotic diseases, queer people are at a greater risk for experiencing the effects of the climate crisis than their heterosexual counterparts.

The higher rates of houselessness queer people experience increases their exposure to zoonotic diseases, but it also makes them more vulnerable in the face of a more familiar green crime: climate change. A review of 15 review articles by Bezgrebelna et al on the topics of housing and climate change drew important links between the two. One of the findings revealed that there is an increased risk of homelessness for people of a lower socioeconomic status and those who are vulnerably housed because of energy insecurity and hazards from climate change. A second major finding is that houseless people are most exposed to climatic events, including temperature extremes and natural disasters. Lastly, the physical and mental health of houseless people are more impacted by climate change (2021). Gibson (2019) found that individuals experiencing houselessness are far more impacted by disasters due to factors such as exposure to the elements, lack of resources and services, as well as disenfranchisement, and stigma associated with homelessness, all while experiencing greater occurrences of environmental injustice. Jarrell et al (2013) use the treadmill of production model to explain how capitalism is the primary culprit of accelerating ecological disorganisation. Through a limitless growth mindset, capitalists utilise technological innovations that reduce costs by saving on labour expenses and extracting from nature. The capitalist structure maintained by the state is the same that allows queer people to be discriminated in the housing and employment sectors, while furthering climate change.

When discussing the effects of housing and employment discrimination against queer peoples, and especially against queer people living with HIV, a necrocapitalist lens illuminates how the 'status of living dead' often conferred upon people living with HIV marks them as being extraneous and expendable to the greater machinations of capitalism and the neoliberal political economy endemic to the United States. In this, 'AIDS-related deaths are exceptionalized and normalised vis-à-vis dual processes of racialization and the spatiotemporalization of one's proximity to communities in crises' (Cheng et al, 2020). Expanding this notion of 'living dead' to include proximity to a positive serostatus[1] would include queer people, who have traditionally been targeted as a whole under the erroneous logic that HIV is a 'gay disease'.

The eco-feminist perspective on houselessness has been explored, but there is room for more applications and research. In their seminal text, *Ecofeminism*, Mies and Shiva (2014) explain that houselessness is created through ecological destruction of the 'home' and the removal of people from their homes. Environmental degradation is the destruction of the home through spiritual and ecological perspectives.

Houselessness and unemployment for queer peoples experiencing disasters are critical avenues, not only because they are necessary lifelines, but are also important for feeling a sense of belonging within society. Savage and Lopez (2020) find that the discrimination against queer peoples in these pandemics impacts their financial savings and stability. Chambers et al (2014) found that housing discrimination experienced by people living with HIV contributes to them being socially shunned and denigrated. This illustrates how housing is a critical social forum, and the state's failure to take that into consideration furthers the exclusion of queer peoples.

Incarceration

The prison-industrial complex is a site of both social harms, but also environmental harms, both within and without the prison walls that warrant further discussion, especially given the historical ways that the prison has served to uphold various hegemonic power systems within the United States and abroad. People within the prison walls are at risk for a variety of victimisations, but it is the connections between the prison and environmental victimisations that are of most interest here. Nearly one-third of state and federal prisons in the United States are within three miles of a federal superfund site[2] (with 100 being within one mile; Bernd et al, 2017). People in prison are often forced to drink contaminated water, breathe in contaminants, and find ways to survive with diseases caused by this exposure ranging from Legionnaires' disease, valley fever and Hodgkin's lymphoma. Climate change can also worsen the living conditions of people in prison with rising temperatures. John Wesley Ford, who was incarcerated at the Wallace Pack Unit in Texas, wrote that as temperatures soar, prisoners are unable to touch the metal walls or beds in their cells, and instead have to wet their sheets and cement floors in an attempt to mitigate 100-degree Fahrenheit temperatures. The response from the Texas Department of Criminal Justice was to recommend prisoners drink more water that contained elevated arsenic levels well over those permitted by the Environmental Protection Agency (Bernd et al, 2017).

The conditions in prisons further harm those already occupying stratified positions both within and without the prison walls. Pellow (2018) explains that queer people in prison are more than six times likely to be sexually assaulted than heterosexual inmates. Pellow further explores mass incarceration not only as an attack on marginalised peoples but also on the environment. There is immense environmental degradation stemming from mass incarceration, and Pellow states that '[p]risons are an environmental justice issue because they are part of the built environment, a space where marginalised bodies are made even more vulnerable; prisons are also spaces where we live, work, learn, play, and pray' (Pellow, 2018: 79–80). The intersectional impacts of zoonotic diseases are woven into the state's response through the criminalisation and victimisation of people trapped within the prison walls. Queer virologist, Osmundson (2022), reminds us that:

> In response to the HIV pandemic, some American states made sex illegal for HIV-positive people without certain types of protections, like condoms or legal disclosure of the virus. These laws didn't stop HIV; they just put people with HIV in jail. This is a deadly human adaptation to a virus. That HIV first impacted queer people, immigrants from Haiti, and injection drug users drove these horrific responses. Racism and homophobia made so many feel like HIV was not their virus, which meant it could be locked up and kept away. (Osmundson, 2022: 311)

This connects to the popular abolitionist critique of prisons being equated to 'buildings full of cages' designed to 'disappear human beings' (Davis, 2000;

Critical Resistance, 2022). In this, the prison exists as a space to not only disappear racial minorities but also to draw similar lines between acceptable forms of queerness and homonormativity, to disappear queer people who are diseased and violate social norms of heteronormativity.

Similar to houselessness and employment discrimination, incarceration is an issue that has intersectional impacts that disproportionately affect marginalised communities. It is not enough to examine the prison as a 'total institution' as Goffman (1990) did but to further a critique that recognises these spaces as fundamentally immoral spaces designed to further a white supremacist racial capitalism built on colonial histories. Various theorists have examined the prison over the years with varying approaches. Foucault (1977) saw the prison as a total institution that ranged beyond the four walls of the prison, instead infiltrating every aspect of our everyday lives through the (internalised) panopticon. Wacquant (2000) drew connections between chattel slavery and the prison as the newest form of social control designed to extract labour while maintaining the colour line. Alexander and West (2012) draw deep connections between the prison-industrial complex and waves of racial control, and mark it as 'the new Jim Crow.' Friedman (2021) examined the prison through a necrocapitalist lens to illuminate the shadow carceral state. While the connections between race and the prison-industrial complex are thoroughly studied, and should be ongoing, there is less research on how these total institutions also act as sites of oppression and marginalisation for LGBT peoples (Robinson, 2020). This gap in the research needs to be more fully explored since LGBT peoples are approximately three times more likely to experience incarceration than their heterosexual or cisgender counterparts (Meyer et al, 2017). The statistics are more daunting at the intersections of queerness with class, immigrant status and 'race', and transgender people especially are more likely to be targeted by police and the carceral system. Additionally, LGBT youth are disproportionately represented in the carceral system (Irvine, 2010). However, the research is failing to identify how the policing of queerness is essential to the overall practices of both policing and criminalisation (Robinson, 2020). There are many avenues to explore the ways mass incarceration, especially the practices in the United States, is victimising the LGBT community and nature.

Gil et al (2021) explain that the over-representation of queer peoples in prisons acts as a positive feedback loop for further harming them because of the increased exposure to diseases such as COVID-19. They demonstrate that queer peoples have high risk of COVID-19 exposure because of the nature of mandatory prison labour. They identify how incarcerated people have been forced to dig graves for and bury the victims of both the HIV/AIDS and COVID-19 pandemics in the nation's largest mass grave (Hart Island, New York). In response to COVID-19, incarcerated people have functioned as front-line workers when making preventative items, including hand sanitiser, all for exploitatively low pay. Ironically, prisons cannot use these products to mitigate transmission risks, as products with alcohol content are banned within carceral facilities (Gil et al, 2021).

It is not our project to provide a comprehensive overview of the critiques of the prison-industrial complex, and how it both fails in its stated goal of rehabilitation and the harm it creates in both the individuals caged within, their families outside the prison, and the communities the prisons exist within. This literature is simply too expansive to do it any justice. However, this brief overview serves to draw connections between a queer critique of the prison and the connections to environmental degradation.

Eco-feminism provides a useful tool where scholars and activists can draw connections to the environmental and human cost associated with prisons that are justified by many of the same practices that we see in other avenues of environmental degradation. Morse and Orias (2020) draw connections to eco-feminism built on challenging capitalist and kyriarchal[3] subordination that is constitutive instead of additive. In this, they argue against a politics of disposability that prisons are built on. Prisons are built on the notion that people, particularly those whose existence is built on a precarity marked by 'race', gender, dis/ability and poverty, are disposable (Stanley et al, 2012). Morse and Orias (2020) argue that this disposability within the prison is part of the same logics that have been used to justify policies and practices that turn the Earth into an object that can be used and disposed of in the name of profit. The connections between the micro and macro-logics of disposability help showcase the 'complex spatial and temporal causes, consequences, and possible resolutions' when discussing the climate crisis and its causes (Pellow, 2018).

Eco-feminism built on logics of 'coexistence without dominants and dominated, complementation and never exploitation' allows for a critique of prisons as hierarchical total institutions but also provides space for scholars and activists to imagine a world beyond. It is not enough to critique these systems. It is important to build alternatives. Davis (2003) reminds us that instead of trying to achieve an expansive single alternative, we should seek to radically transform society through smaller alternatives. Ferreira, using principles of eco-feminism, asks us to view the 'transgressive woman [as] not a "bad criminal", but a debtor – a person responsible for their actions that assumes the duty of reparation towards society' (2020: 248). This already challenges the traditional notion of the 'criminal' as someone who needs to be punished before they are allowed to attempt rehabilitation. In their utopian vision, Ferreira argues for ecovillages based in logics of democracy and communalism as a space where individuals can *start* to build a society where prisons are no longer necessary.

It needs to be the project of green criminologists to understand how queer peoples and nature are victims of the prison industrial complex. This research can form the basis of that eco-feminist alternative to a harmful carceral system that perpetuates the intergenerational trauma, so that both marginalised communities and nature are no longer viewed through a politics of disposability.

Isolation

While COVID-19 made headlines for the series of 'unprecedented' changes to both public health and social life, it is important to also recognise that many of these changes were actually the result of the worsening of pre-existing inequalities and enforced marginalisations across both global and domestic society. These inequalities are at least partially the result of a capitalist political economy that not only creates and maintains social marginalisation, it also is the system that justifies the degradation and exploitation of the environment. Such exploitation has set the stage for worsening environmental disasters but also the rise of zoonotic diseases like COVID-19 and HIV/AIDS (and certainly many others). While most work in the field of green criminology focuses on primary victimisation against the environment and the secondary victimisation felt by both animals (human and nonhuman) and fauna, this section will focus on a form of victimisation that is more hidden. Isolation is a more hidden form of victimisation because it does not present in more spectacular ways and is also part of our daily lives (particularly in the Global North) but is the end-result of capitalism. As Jarrell et al (2013) remind us, green crimes can present themselves in ways that are persistent and long-term (sometimes spanning decades or centuries) but also include a wide range of consequences that are not always evident right away.

One of the main challenges involved with large-scale lockdowns was the isolation associated with asking individuals to remain at home as much as possible and avoid contact with anyone they were not cohabitating, isolating them from their communities. While feelings of anxiety and lack of social support were widespread, queer people experienced higher rates of anxiety and depression, and lower rates of social support, when compared to their cisgender, heterosexual counterparts (Moore et al, 2021). These higher rates of depression and anxiety are not limited to queer people living during the COVID-19 era but are also common in people living with HIV, who experience higher rates of both current and lifetime depression (medians of 24 per cent and 42 per cent, respectively) and higher rates of cognitive impairment (Rubin and Maki, 2019). Considering the fact that queer men are 22 times more likely to become HIV positive, this is concerning given the additional burdens of living in a heteronormative society and minority stress (Tomar et al, 2021). People living with HIV also tend to have more fragile social networks based on chosen families (High et al, 2012). This was only worsened during the early stages of the COVID-19 pandemic, where queer elders were grappling with both the collective traumas of COVID-19 but also with the traumatic memories of the HIV/AIDS pandemic. The way both HIV and COVID-19 impact queerness through isolation is also explored by Osmundson (2022), who states that 'HIV has altered our relationship, worldwide, with sexual pleasure, as well as leaving a lasting, isolating stigma on the queer community. As for COVID-19, it will inevitably alter a generation's experience of social interaction and physical intimacy' (2022: 152). Historically, these collective traumas and calls towards social distancing and isolation left many people living with HIV hesitant to reach out to mutual aid groups or their neighbours, leaving them even more socially isolated (Savage and Lopez, 2020).

Researchers have seen in an increase in the life expectancy of people living with HIV, with 50 per cent of patients being over the age of 50 (Wing, 2016). This increase in life expectancy combined with stigma towards HIV and other chronic disease could account for the higher rates of isolation and depression among people living with HIV (Webel et al, 2014). Mazonson et al (2021) found that 50.8 per cent of people living with HIV aged 50 or older self-identified as lonely at higher rates than the general population. The authors suggest that some possible reasons for this could be having lost friends or loved ones to the original AIDS epidemic, higher rates of estrangement from their birth families (Serovich et al, 2011), being less likely to marry, and higher rates of substance abuse (Galvan et al, 2002; Williams et al, 2016). An additional factor in higher rates of isolation and loneliness can also be attributed to stigma related to a positive HIV diagnosis which can further impact health outcomes (both physical and psychosocial), including mental health and support-seeking behaviours (Earnshaw and Chaudoir, 2009). When people living with HIV internalise aspects of HIV stigma, they are less likely to disclose their status and withdraw from their social support networks (Derlega et al, 2004; Earnshaw and Chaudoir, 2009).

So far, we have focused on a descriptive aspect of isolation that is data-driven. We recognise that this is only half of the picture, as Osmundson (2022) reminds us when elaborating on how detrimental this isolation has been: '[At] the peak of the COVID-19 crisis, we couldn't even hold hands. The very gestures of care we need to resist American atomization, aloneness, isolation were the gestures that put us most at risk' (2022: 288). This isolation is the result of a series of structures influenced by late-stage capitalism that requires us to be hyper-individual, predicated on the same politics of disposability that enable environmental degradation that further enacts intersectional harm.

Queer peoples' experiences of isolation further exemplify how socioenvironmental harm can result in victimisation that arises from disasters. The ways those who have contracted HIV and/or COVID-19 are isolated, both physically and metaphorically, are clearly damaging to the victims as individuals but also to their communities. It is in this ripple effect of communal harm where the environmental degradation becomes more evident. Eco-feminist literature cites the importance of coalition-building in order to dismantle the systemic harms, and the isolation that queer peoples experience. These disasters fissure essential communal bonds that act as coalitions for working to mitigate the harms of racism, human-caused climate change and heteronormative patriarchal domination (Gaard, 1993). This coalition building can help counteract the victimisation of queer peoples and nature when exposed to such disasters because we are stronger when working against disasters as a coalition. Furthermore, in the COVID-19 pandemic, the isolation that has permeated many parts of society has resulted in dramatic increases in practices that further excessive consumption through online shopping, which has a much larger environmental impact through drastically more materials used in shipping and packaging, and also fossil fuels to make rapid deliveries (Boudreau, 2021). This demonstrates that the lockdowns mandated by the COVID-19 pandemic were harmful to queer peoples and also furthered environmental degradation. An eco-feminist approach teaches us that none of these systemic harms are taking place in in separated silos, and therefore the socioenvironmental solutions cannot be born in isolation either.

Queering green criminology in practice

Capitalism, especially the current trend of a neoliberal or 'disaster' capitalism (Solis and Klein, 2020), is built on the intensifying concept of intense hyper-individualism where humans, particularly those in the Global North, see themselves as individuals that are in competition with other individuals instead of part of a community. This conception of the hyper-individual is one that is inherently isolating given the combined effects of being an individual while calling upon the collective for legitimacy. This is most evident in the rise of individuals protesting mask and vaccine mandates and lockdowns designed to limit the spread of COVID-19. Here, individuals were engaging in calls for individual liberty, while relying on a homogeneous *whiteness* (and, arguably, cisheteronormativity) which is articulated through a disregard of others (Bratich, 2021). As Bratich stated in the title of his article, 'Give me liberty or give me COVID'. We claim that this call for liberty at the expense of others can be seen as an extension of necrocapitalism by relying upon individuals marking themselves as individuals worth saving because of their engagement with larger power structures, while *simultaneously* marking the other as disposable. The history of the HIV pandemic shows similar politics of disposability when marking gay men as a whole (and gay men of colour specifically) as not being worth saving and, in fact, deserving of the 'gay cancer' that was decimating their communities.

We are steadfast that the state's response to zoonotic disease pandemics is a green crime because of how it exacerbates the harms experienced by queer peoples, particularly trans women of colour, and needs to be studied through a lens of queer-feminist green criminology. The victimisation of these marginalised populations and nature can no longer be neglected by the field. The state continues to prioritise profit over both nature and marginalised peoples, and this is a green crime. These disasters, such as climate change, extreme weather or pandemics, manifest in the victimisation of queer peoples and nature. Despite these harms, the state continues to disregard both nature and queer peoples, or worse, perpetuate the extractive practices that result in widespread devastation and disempowerment.

Gordon and Green (2021) discuss how the state's early responses to the COVID-19 pandemic have undermined civil rights and threatened democracy. This is exemplified through the deregulation of various polluting industries and the outsourcing of tasks to private corporations that are traditionally covered by the public sector. The government's response to the pandemic demonstrates both overreach and underreach that have manifested in structural violence. The government overreach is characterised by the use of heightened security and surveillance measures through the guise of preventing the spread of the disease, while the underreach is exemplified by the inaction and racist, classist and homophobic exclusion of life-saving resources.

When considering the HIV/AIDS pandemic in this frame we claim that an example of government overreach is found in HIV criminalisation. While the Reagan administration did fail to evoke the name of the pandemic for four years (Bennington-Castro, 2020), that did not prevent local municipalities from enacting varying forms of HIV criminalisation designed to punish sexual deviants through faulty assumptions about transmission (Novak, 2021). HIV criminalisation usually takes the form of laws designed to criminalise the transmission of HIV. Thirty-four states have active laws criminalising HIV exposure (CDC, 2019), many of which (25 states) have little to no risk of exposure (Lehman et al, 2014), where HIV-positive sex is reduced to individuals being irresponsible and antisocial (Tomso, 2017). Because criminalisation does not affect society at random, it further burdens stratified groups, including racial minorities, sexual minorities, sex workers and rural populations (Herdt, 2001).

We argue that the failure of the Reagan administration to evoke the name of the pandemic for four years is one of many examples of government underreach. Government underreach does not simply mean the absence of the government, but rather 'denotes forms of intervention designed to implement certain kinds of deregulation and austerity measures that have led to the evisceration of welfare policies and the erosion of the social safety net' (Gordon and Green, 2021: 4). In the COVID-19 pandemic, we have seen examples of government underreach in the management of prisons. Widra and Hayre (2020) describe how when prison populations are swelling, the state prisons demonstrate underreach in their mismanagement of the situation; only five states completed adequate testing of inmates, only three complied with testing of correctional staff, and even when states seemed to be checking the appropriate boxes there were still very high death rates from COVID-19. We echo the calls of Widra and Hayre (2020) that these are the results of systemic failures that were preventable.

An example of the systemic failure of the state's response to the COVID-19 pandemic is not just the lack of attention specifically given to how the virus is impacting this marginalised community, but also that the sexual orientation of victims was never recorded. While there are stringent records of the death tolls of victims, we will never know exactly how many queer people have died from COVID-19 because patients were not asked their sexual orientation nor if their gender identity was beyond a cisgender binary (Nowaskie and Roesler, 2022). In this instance, we apply the necrocapitalist theory of 'body counts' to understand why this is so harmful. Banerjee uses 'body counts' when discussing US war intervention in Iraq, and he says that while there was no precise number of Iraqi citizen casualties, the body count of foreign soldiers was exact. Banerjee states: 'When General Tommy Franks infamously said "We don't do body counts", he of course meant that they only do the bodies that count' (2008: 1553). There are parallels to the understanding of how queer bodies are valued here, because in 1991 when HIV/AIDS death toll reached 100,000 Americans, most of them queer, that news was relegated to page 18 of the *New York Times*. But when 100,000 Americans died of COVID-19, that made the front page. The identity of the victims in the former made them less valued, and their stories were buried in the publication, meanwhile when a disease is spreading that is equally transmissible to *all* Americans it is highlighted as the most prominent story of importance. This stance is further supported since death tolls are not taking sexual orientation into account, therefore we will not know how many queer people are lost in the COVID-19 pandemic, and the lives of queer people are not deemed worthy of counting. The necrocapitalist theory of 'body counts' is one way to demonstrate that the state's response to zoonotic disease pandemics, in both overreach and underreach, has failed many of its most marginalised constituents.

This chapter is putting environmental degradation in conversation with zoonoitic diseases, while centring the experience of queer peoples. According to Lawler et al, minimising the risks of future zoonotic disease pandemics demands that we confront the anthropogenic factors that result in the proliferation of zoonotic diseases, which include 'land-use change, wildlife trade, intensive livestock production, and climate chang' (2021: 847). The lack of attention to queer peoples in the state's response to both HIV and COVID-19 characterise these disasters as green crimes because of how these responses have further marginalised this community and perpetuated environmental degradation. In the earlier sections of this chapter, we highlighted the ways in which queer peoples and nature were uniquely impacted by these zoonotic disease pandemics (houselessness, employment discrimination, incarceration and isolation) through the lenses of eco-feminism and necropolitics We agree with Lynch's (2018) call for green criminology to adopt a feminist approach and more acutely consider gender, but also call for this approach to include queer peoples and for that ontological understanding of gender to operate beyond a binary. This is essential because the current literature in the field of green criminology not only has an underdeveloped approach to gender but is neglecting queer peoples.

We argue that green criminology must explore the role that heteronormativity plays in perpetuating green crimes. For this field to be more intersectional, and to therefore be a stronger force of analytical critique, it is essential that the nuanced perspectives of queer peoples be considered. Intersectional and eco-feminist literatures teach us that oppression of peoples and nature is both overlapping and interconnected, and therefore an approach to green criminology that fails to consider how queer peoples are unique victims of disasters is incomplete and ill-equipped to achieve its goals.

We posit that this queering of green criminology can draw on the work of Gaard (2004), which has made immense progress towards queering eco-feminism. A queer eco-feminist perspective demonstrates how queer peoples are pejoritised through their feminisation, animalisation and eroticisation. The Western lens has historically understood queerness based on mainstream religious morality and thus has characterised queer sexuality and gender identity as being unnatural, while at the same time devaluing both queer lives and nature. Eco-feminists identify this as a contradiction in the dominant narrative, but this actually 'indicates that the "nature" queers are urged to comply with is none other than the dominate paradigm of heterosexuality – an identity and practice that is itself a cultural construction, as both feminists and queer theorists have shown' (Gaard, 2004: 28). Further historical analysis of sexuality alongside North American colonialism reveals that imperial rule was partly justified by both the domination of nature and the effort to stamp out gender-role deviance and nonheterosexual erotic practices among Indigenous peoples. Green criminology heavily identifies colonialism as a site of extensive green crimes (Crook et al, 2018; Rodríguez Goyes, 2019), but little to no attention has been given to the role that heteronormativity played and continues to play in furthering the victimisation of nature and queer peoples.

Berila (2004) postulates that environmental justice movements that are critical of heteronormativity are better equipped to navigate the ways inequality along lines of 'race', class, gender and sexuality impact communities. We maintain that green criminology can benefit from critiquing heteronormativity, but as Gaard (2004) demonstrates in the queering of eco-feminism, it is not as simple as adding heterosexism to the list of 'isms' the field is already considering. We note this to draw attention to the care and intentionality that must be taken when incorporating a more diverse and nuanced approach.

Green criminology centres on ontological questions of crime and the environment and *doing theory*. Brisman suggests that:

> [W]e consider *theory* as encompassing ideals and tools for describing and analyzing *what and how things are as they are; who engage in various behaviors, patterns and practices*; and *how do we—or how might we—interpret and ascribe meaning to those behaviors, patterns and practices, and to the consequences thereof.* (Brisman, 2014: 23; emphasis in original)

In this, we seek to contribute to the theoretical toolbox that green criminologists use. Historically, criminologists have left the study of environmental harms and laws and regulations to other fields (Lynch and Stretsky, 2003). While this is no longer the position with the expansion of the canon in green criminology, it is still the case that the connections between environmental crimes and queerness have been under-represented in the literature.

One of the main areas of attention in green criminology is the understanding of 'lawful but awful' (Passas, 2005; Brisman, 2015; Wyatt and Brisman, 2017). This essentially means that harms that, while not breaking any laws, have long-lasting societal impact. The state's (mis)management of zoonotic diseases such as COVID-19 and HIV are examples of this. In the early days of the HIV pandemic, the Reagan administration did not engage in any practice to curb the spread of the disease or help the queer communities affected by HIV. This was a lawful practice; there were no laws requiring Reagan to order his administration to tend to the disease or its victims. Ignoring HIV was certainly a 'lawful but awful' practice when measuring the cost to generations of queer people that are still being felt today.

The question becomes then, how should we queer green criminology? It is not enough to ask green criminologists to simply incorporate queer people into their analyses. While this inclusion is important for creating a better understanding of how queer people are affected by environmental disasters, it is only the first step. Instead of relying on an additive approach where queer people are included on the list of identities (for example, 'race', class, gender, sexuality, ability, and so on), a truly queer green criminology would prioritise the complex intersections of these facets to understand the nuanced lived experiences of queer people from the beginning.

A few ways this queering of green criminology can be accomplished is through adjusting research questions, drawing on more diverse perspectives and incorporating critiques of heteronormativity that are based in Western colonialism. This heteronormativity is the same as that which results in harms to the queer community and nature through laws that limit our civil rights but also through homophobic, lesbophobic and transphobic attacks. These harms are not limited to this, they also help to justify the destruction of the environment based in racialised and gender norms that assume that the environment is a feminine creature that needs to be dominated and controlled by a heterosexual white man.

Our analysis of the impacts of zoonotic diseases on queer peoples' experience with housing, employment, incarceration and isolation serves as an example of how green crimes victimise queer peoples. This should act as a model or stepping stone on which further research can build and further examine these connections.

This work mostly relies on eco-feminism and necropolitics to consider how the state's unjust response to zoonotic diseases, in this case, HIV/AIDS and COVID-19, result in green crimes that uniquely impact queer peoples. However, we argue that queer feminist green criminology can examine many other aspects of how queer peoples uniquely experience disasters and green crimes. There is opportunity to draw on diverse perspectives for interpreting the health disparities that result from disasters; work coming out of environmental humanities, visual cultural studies, affect studies, and more are also making strides to understand how queer peoples' victimisation need to be centred in research. For example, Straube (2019) discusses how trans peoples are framed as parasites of society, much like ticks that spread Lyme disease, through transphobic and racist ideologies. Straube's analysis of visual culture reveals that in a Western lens, queer peoples, notably trans people of colour, are depicted as parasitic because they are a threat to cisheteronormativity, despite being the common victim of hate crimes and many forms of structural inequality. Queerness is constructed as 'degenerate', 'deviant' and thus characterised as unnatural (Straube, 2019). This intimate intertwining of queerness and nature is also explored in eco-feminist literature, notably by Gaard (2004), who pushes the field to move towards a queer eco-feminism. Thus, we argue that a queer feminist green criminology should draw on a plethora of fields and methods to strive to understand the experiences and victimisation of queer peoples and nature.

A queer feminist green criminology should draw on a plethora of fields and methods to strive to understand the experiences of queer peoples in relation to the environment and social harm. The impacts of environmental disasters exacerbate inequality and existing forms of social marginalisation. We drew on the literatures of green criminology, eco-feminism and necropolitics to consider what a queer green criminology could look like in order to address the gap in green criminological literature regarding the LGBT community. Throughout this chapter, we aimed to draw connections between the treatment of queer peoples and green criminology by highlighting how the changes brought on by the COVID-19 pandemic mirrored many of the same failures and consequences of the HIV pandemic.

To fully understand these green crimes, we argue that green criminology needs a queer lens. However, we view this as a stepping stone for more research and praxis based in a green criminology that incorporates queerness into its theoretical lens. This will need to be an ongoing project, but one that will be worthwhile because both queerness and green criminology ask ontological questions that challenge the dominant narratives that excuse and enable the politics of disposability that harm both the environment and queer people. Only by actively rejecting these politics of disposability can we work to begin to mitigate the effects of the climate disaster.

Notes

[1]   Serostatus refers to the presence or absence of certain markers within the blood. For our work, a positive serostatus refers to the presence of antibodies to HIV, marking them as HIV+.

[2]   A Superfund site refers to the name for locations marked under the Comprehensive Environmental Response, Compensation and Liability Act of 1980. These locations are polluted with hazardous materials, many of which act as a significant harm for the health of people living near these sites.

[3]   Kyriarchy is an intersectional feminist extension of traditional understandings of patriarchal subordination forwarded by Schüssler Fiorenza (1992) which pushes understandings of patriarchal oppression beyond gender to include (but not limited to), racism, ableism, antisemitism, classism, homophobia, transphobia, speciesism and colonialism.

References

Alakunle, E.F. and Okeke, M.I. (2022) Monkeypox virus: A neglected zoonotic pathogen spreads globally. *Nature Reviews Microbiology*, 20(9), 507–8.

Alexander, M. and West, C. (2012) *The New Jim Crow: Mass Incarceration in the Age of Colorblindness*, revised edn. New York: New Press.

Annequin, M., Lert, F., Spire, B. and Dray-Spira, R. (2015) Has the employment status of people living with HIV changed since the early 2000s? *AIDS*, 29(12), 1537–47.

Barreca, A.I. (2012) Climate change, humidity, and mortality in the United States. *Journal of Environmental Economics and Management*, 63(1), 19–34.

Barry, D. and Hoyne, S. (2021) Sustainable measurement indicators to assess impacts of climate change: Implications for the New Green Deal era. *Current Opinion in Environmental Science & Health*, 22, 100259.

Bartlow, A.W., Manore, C., Xu, C., Kaufeld, K.A., Del Valle, S., Ziemann, A., Fairchild, G. and Fair, J.M. (2019) Forecasting zoonotic infectious disease response to climate change: Mosquito vectors and a changing environment. *Veterinary Sciences*, 6(2).

Beirne, P. (2021) Wildlife trade and COVID-19: Towards a criminology of anthropogenic pathogen spillover. *The British Journal of Criminology*, 61(3), 607–26.

Bennington-Castro, J. (2020) How AIDS remained an unspoken – but deadly – epidemic for years. Available at: www.history.com/news/aids-epidemic-ronald-reagan (accessed 11 May 2021).

Bergin, A. and Allen, R. (2008) *The Thin Green Line: Climate Change and Australian Policing*. Barton: Australian Strategic Policy Institute.

Berila, B. (2004) Toxic bodies? ACT UP's disruption of the heteronormative landscape of the nation. In R. Stein (ed) *New Perspectives on Environmental Justice: Gender, Sexuality, and Activism*. New Brunswick: Rutgers University Press, pp 21–44.

Bernd, C., Loftus-Farren, Z. and Mitra, M.N. (2017) *America's Toxic Prisons: The Environmental Injustices of Mass Incarceration*. Sacramento, CA: Truthout and Earth Island Journal.

Bezgrebelna, M., McKenzie, K., Wells, S., Ravindran, A., Kral, M., Christensen, J. et al (2021) Climate change, weather, housing precarity, and homelessness: A systematic review of reviews. *International Journal of Environmental Research and Public Health*, 18(11).

Boudreau, C. (2021) Shopping online surged during COVID. Now the environmental costs … Available at: www.politico.com/news/2021/11/18/covid-retail-e-commerce-environment-522786 (accessed 11 May 2022).

Bratich, J. (2021) 'Give me liberty or give me COVID!': Anti-lockdown protests as necropopulist downsurgency. *Cultural Studies*, 35(2–3), 257–65.

Brent, R.J. (2016) The value of reducing HIV stigma. *Social Science & Medicine*, 151, 233–40.

Brisman, A. (2014) Of theory and meaning in green criminology. *International Journal for Crime, Justice and Social Democracy*, 3(2), 21–34.

Brisman, A. (2015) Environmental harm as deviance and crime. In E. Goode (ed) *The Handbook of Deviance*. Chichester: John Wiley & Sons, pp 471–87.

Bromwich, R.J. (2022) Mother Earth in environmental activism: Indigeneity, maternal thinking, and animism in the Keystone pipeline debate. In J. Gacek and R. Jochelson (eds) *Green Criminology and the Law*. Cham: Springer International, pp 285–305.

Chambers, L.A., Greene, S., Watson, J., Rourke, S.B., Tucker, R., Koornstra, J., Sobota, M., Hwang, S., Hambly, K., O'Brien-Teengs, D., Walker, G. and The Positive Spaces Healthy Places Team (2014) Not just 'a roof over your head': The meaning of healthy housing for people living with HIV. *Housing, Theory and Society*, 31(3), 310–33.

Cheng, J.-F., Juhasz, A. and Shahani, N. (eds) (2020) *AIDS and the Distribution of Crisis*. London: Duke University Press.

Clements-Nolle, K., Marx, R., Guzman, R. and Katz, M. (2001) HIV prevalence, risk behaviors, health care use, and mental health status of transgender persons: Implications for public health intervention. *American Journal of Public Health*, 91(6), 915–21.

Critical Resistance (2022) What is the PIC? What is abolition? *Critical Resistance*. Available at: http://criticalresistance.org/about/not-so-common-language/ (accessed 11 May 2022).

Crook, M., Short, D. and South, N. (2018) Ecocide, genocide, capitalism and colonialism: Consequences for indigenous peoples and glocal ecosystems environments. *Theoretical Criminology*, 22(3), 298–317.

Davis, A.Y. (2000) Masked racism: Reflections on the prison industrial complex in the USA. *Lola Press*, 12, 1-6.

Davis, A.Y. (2003) *Are Prisons Obsolete?* New York: Seven Stories Press.

Davis, M. and Wertz, K. (2009) When laws are not enough: A study of the economy health of transgender people and the need for a multidisciplinary approach to economic justice. *Seattle Journal for Social Justice*, 8, 467–95.

Decker, M., Littleton, H.L. and Edwards, K.M. (2018) An updated review of the literature on LGBTQ+ intimate partner violence. *Current Sexual Health Reports*, 10(4), 265–72.

Derlega, V.J., Winstead, B.A., Greene, K., Serovich, J. and Elwood, W.N. (2004) Reasons for HIV disclosure/nondisclosure in close relationships: Testing a model of HIV–disclosure decision making. *Journal of Social and Clinical Psychology*, 23(6), 747–67.

Earnshaw, V.A. and Chaudoir, S.R. (2009) From conceptualizing to measuring HIV stigma: A review of HIV stigma mechanism measures. *AIDS and Behavior*, 13(6), 1160–1177.

Edelman, E. (2014) 'Walking while transgender': Neccropolitical regulations of trans feminine bodies of colour in the nation's capital. In J. Haritaworn, A. Kuntsman, and S. Posocco (eds) *Queer Necropolitics*. New York: Routledge, pp 172–90.

Farmer, P. (2004) An anthropology of structural violence. *Current Anthropology*, 45(3), 305–25.

Ferreira, H., Manuel Baptista, M. and Alves de Almeida, A.R. (2020) Prison as a solution? The principles of eco-feminism applied to the social reintegration of women detainees'. In M. Manuel Baptista and A.R. Alves de Almeida (eds) *Gender Performativities in Democracy under Threat*. Coimbra: Gracio Editor, pp 245–55.

Flores, A.R., Langton, L., Meyer, I.H. and Romero, A.P. (nd) Victimization rates and traits of sexual and gender minorities in the United States: Results from the National Crime Victimization Survey, 2017. *Science Advances*, 6(40), eaba6910.

Foucault, M. (1977) *Discipline and Punish: The Birth of the Prison*. New York: Pantheon Books.

Friedman, B. (2021) Unveiling the necrocapitalist dimensions of the shadow carceral state: On pay-to-stay to recoup the cost of incarceration. *Journal of Contemporary Criminal Justice*, 37(1), 66–87.

Gaard, G. (ed) (1993) *Ecofeminism: Women, Animals, Nature*. Philadelphia: Temple University Press.

Gaard, G. (2004) Toward a queer ecofeminism. In R. Stein (ed) *New Perspectives on Environmental Justice: Gender, Sexuality, and Activism*. New Brunswick: Rutgers University Press, pp 21–44.

Gaard, G. (2015) Ecofeminism and climate change. *Women's Studies International Forum*, 49, 20–33.

Galvan, F.H., Bing, E.G., Fleishman, J.A., London, A.S., Caetano, R., Burnam, M.A., Longshore, D., Morton, S.C., Orlando, M. and Shapiro, M. (2002) The prevalence of alcohol consumption and heavy drinking among people with HIV in the United States: Results from the HIV Cost and Services Utilization Study. *Journal of Studies on Alcohol*, 63(2), 179–86.

Gibson, A. (2019) Climate change for individuals experiencing homelessness: Recommendations for improving policy, research, and services. *Environmental Justice*, 12(4), 159–63.

Gil, R.M., Freeman, T.L., Mathew, T., Kullar, R., Fekete, T., Ovalle, A., Nguyen, D., Kottkamp, A., Poon, J., Marcelin, J.R. and Swartz, T.H. (2021) Lesbian, gay, bisexual, transgender, and queer (LGBTQ+) communities and the coronavirus disease 2019 pandemic: A call to break the cycle of structural barriers. *The Journal of Infectious Diseases*, 224(11), 1810–20.

GNP and ILO (2012) *Stigma and Discrimination at Work: Findings from the People Living with HIV Stigma Index*. Amsterdam: GNP and ILO.

Goffman, E. (1990) *The Presentation of Self in Everyday Life*. New York: Doubleday.

Goldsmith, L., Raditz, V. and Méndez, M. (2022) Queer and present danger: Understanding the disparate impacts of disasters on LGBTQ+ communities. *Disasters*, 46(4), 946–73.

Gordon, N. and Green, P. (2021) State crime, structural violence and COVID-19. *State Crime Journal*, 10(1).

Griggs, G. (2021) Rising seas in California: An update on sea-level rise science. In *World Scientific Encyclopedia of Climate Change: Case Studies of Climate Risk, Action, and Opportunity*, Volume 3. Singapore: World Scientific, pp 105–11.

Habibullah, M.S., Haji Din, B., Tan, S.-H. and Zahid, H. (2022) Impact of climate change on biodiversity loss: Global evidence. *Environmental Science and Pollution Research*, 29(1), 1073–86.

HALSA (nd) *HIV/AIDS Discrimination: Are You Breaking the Law?* Los Angeles: City of Los Angeles Department on Disability, AIDS Coordinator's Office.

Herdt, G. (2001) Stigma and the ethnographic study of HIV: Problems and prospects. *AIDS and Behavior*, 5(2), 141–9.

High, K.P., Brennan-Ing, M., Clifford, D.B., Cohen, M.H., Currier, J., Deeks, S.G. et al (2012) HIV and aging: State of knowledge and areas of critical need for research. A Report to the NIH Office of AIDS Research by the HIV and Aging Working Group. *Journal of Acquired Immune Deficiency Syndromes*, 60(Supplement 1), S1–18.

HRI (nd) *Population At-Risk: Homelessness and the COVID-19 Crisis.* Washington, DC: Homeless Research Institute: National Alliance to End Homelessness.

Irvine, A. (2010) We've had three of them: Addressing the invisibility of lesbian, gay, bisexual, and gender nonconforming youths in the juvenile justice system. *Columbia Journal of Gender and Law*, 19(3). doi: 10.7916/CJGL.V19I3.2603

Jarrell, M.L., Lynch, M.J. and Stretesky, P.B. (2013) Green criminology and green victimization. In B.A. Arrigo and H.Y. Bersot (eds) *The Routledge Handbook of International Crime and Justice Studies.* New York: Routledge.

Konapala, G., Mishra, A.K., Wada, Y., and Mann, M.E. (2020) Climate change will affect global water availability through compounding changes in seasonal precipitation and evaporation. *Nature Communications*, 11(1), 1-10.

Lawler, O.K., Allan, H.L. Baxter, P.W.J., Castagnino, R., Corella Tor, M., Dann, L.E. et al (2021) The COVID-19 pandemic is intricately linked to biodiversity loss and ecosystem health. *The Lancet Planetary Health*, 5(11), e840–50.

Lehman, J.S., Carr, M.H., Nichol, A.J., Ruisanchez, A., Knight, D.W., Langford, A.E., Gray, S.C. and Mermin, J.H. (2014) Prevalence and public health implications of state laws that criminalize potential HIV exposure in the United States. *AIDS and Behavior*, 18(6), 997–1006.

Lynch, M.J. (2018) Acknowledging female victims of green crimes: Environmental exposure of women to industrial pollutants. *Feminist Criminology*, 13(4), 404–27.

Lynch, M.J. and Stretsky, P.B. (2003) The meaning of green: Contrasting criminological perspectives. *Theoretical Criminology*, 7(2), 217–38.

Maulsby, C.H., Ratnayake, A., Hesson, D., Mugavero, M.J. and Latkin, C.A. (2020) A scoping review of employment and HIV. *AIDS and Behavior*, 24(10), 2942–55.

Mazonson, P., Berko, J., Loo, T., Kane, M., Zolopa, A., Spinelli, F., Karris, M. and Shalit, P. (2021) Loneliness among older adults living with HIV: The 'older old' may be less lonely than the 'younger old'. *AIDS Care*, 33(3), 375–82.

Meyer, I.H., Flores, A.R., Stemple, L., Romero, A.P., Wilson, B.D.M. and Herman, J.L. (2017) Incarceration rates and traits of sexual minorities in the United States: National inmate survey, 2011–2012. *American Journal of Public Health*, 107(2), 267–73.

Mies, M. and Shiva, V. (2014) *Ecofeminism*. London: Zed Books.

Moore, S.E., Wierenga, K.L., Prince, D.M., Gillani, B. and Mintz, L.J. (2021) Disproportionate impact of the COVID-19 pandemic on perceived social support, mental health and somatic symptoms in sexual and gender minority populations. *Journal of Homosexuality*, 68(4), 577–91.

Mock, J. (2014) *Redefining Realness: My Path to Womanhood, Identity, Love and So Much More*. New York: Atria Books.

Morse, N. and Orias, D. (2020) No one is disposable: Ecofeminism and climate change. In J. Shayne and N. Carter (eds) *Persistence is Resistance: Celebrating 50 Years of Gender, Women and Sexuality Studies*. Bothell, Washington: University of Washington Libraries.

Novak, A. (2021) Toward a critical criminology of HIV criminalization. *Critical Criminology*, 29(1), 57–73.

Osmundson, J. (2022) *Virology: Essays for the Living, the Dead, and the Small Things in Between*. New York City, NY: W.W. Norton.

Passas, N. (2005) Lawful but awful: Legal corporate crimes. *The Journal of Socio-Economics*, 34(6), 771–86.

Pellow, D.N. (2018) *What Is Critical Environmental Justice?* Cambridge: Polity.

Perry, R.W. (2018) Defining disaster: An evolving concept. In H. Rodríguez, W. Donner, and J.E. Trainor (eds) *Handbook of Disaster Research*. Cham: Springer International Publishing, pp 3–22.

Quarantelli, E.L. (2000). Disaster research. In E. Borgatta and R. Montgomery (eds) *Encyclopedia of Sociology*. New York: Macmillan, pp 682–8.

Robinson, B.A. (2020) The lavender scare in homonormative times: Policing, hyper-incarceration, and LGBTQ youth homelessness. *Gender & Society*, 34(2), 210–32.

Rodríguez Goyes, D. (2019) *Southern Green Criminology: A Science to End Ecological Discrimination*. Bingley: Emerald Publishing.

Rubin, L.H. and Maki, P.M. (2019) HIV, depression, and cognitive impairment in the era of effective antiretroviral therapy. *Current HIV/AIDS Reports*, 16(1), 82–95.

Savage, R. and Lopez, O. (2020) Isolation and HIV memories hit LGBT+ elderly hard in lockdowns … Available at: www.reuters.com/article/us-health-coronavirus-lgbt-idUSKBN21H0BO (accessed 11 May 2022).

Schüssler Fiorenza, E. (2005) *But She Said: Feminist Practices of Biblical Interpretation*. Boston: Beacon Press.

Serovich, J.M., Grafsky, E.L. and Craft, S.M. (2011) Does family matter to HIV-positive men who have sex with men? *Journal of Marital and Family Therapy*, 37(3), 290–8.

Singla, R.K., Singla, S. and Shen, B. (2022) Biased studies and sampling from LGBTQ communities created a next-level social stigma in monkeypox: A public health emergency of international concern (PHEIC). *Indo Global Journal of Pharmaceutical Sciences*, 12, 205–8.

Solis, M. (2020) Naomi Klein on Coronavirus and disaster capitalism. Available at: www.vice.com/en/article/5dmqyk/naomi-klein-interview-on-coronavirus-and-disaster-capitalism-shock-doctrine (accessed 11 May 2022).

Spinelli, M.A., Hessol, N.A., Schwarcz, S., Hsu, L., Parisi, M.-K., Pipkin, S., Scheer, S., Havlir, D. and Buchbinder, S.P. (2019) Homelessness at diagnosis is associated with death among people with HIV in a population-based study of a US city. *AIDS (London, England)*, 33(11), 1789–94.

Stanley, E.A., Spade, D. and Queer (In)Justice (2012) Queering prison abolition, now? *American Quarterly*, 64(1), 115–27.

Straube, W. (2019) Toxic bodies. *Environmental Humanities*, 11(1), 216–38.

Tomar, A., Spadine, M.N., Graves-Boswell, T., Wigfall, L.T. (2021) COVID-19 among LGBTQ+ individuals living with HIV/AIDS: Psycho-social challenges and care options. *AIMS Public Health*, 8(2), 303–8.

Tomso, G. (2017) HIV monsters: Gay men, criminal law, and the new political economy of HIV. In D. M. Halperin and T. Hoppe (eds) *The War on Sex*. Durham, NC: Duke University Press.

Treisman, R. (2022) As monkeypox spreads, so do concerns about stigma: NPR. Available at: www.npr.org/2022/07/26/1113713684/monkeypox-stigma-gay-community (accessed 11 May 2023).

Van Ness, J. (2022) Jonathan Van Ness: We are still not taking monkeypox seriously … Available at: https://time.com/6206159/jonathan-van-ness-monkeypox-response/ (accessed 11 May 2023).

Wacquant, L. (2000) The new 'peculiar institution': On the prison as surrogate ghetto. *Theoretical Criminology*, 4(3), 377–89.

Webel, A.R., Longenecker, C.T., Gripshover, B., Hanson, J.E., Schmotzer, B.J. and Salata, R.A. (2014) Age, stress, and isolation in older adults living with HIV. *AIDS Care*, 26(5), 523–31.

White, R. (2011) *Transnational Environmental Crime: Toward an Eco-Global Criminology*. London: Routledge.

White, R. (2022) Green criminology, environmental harms and eco-justice. In D. Nelken and C. Hamilton (eds) *Research Handbook of Comparative Criminal Justice*. Cheltenham: Edward Elgar Publishing, pp 315–31.

White, R. and Heckenberg, D. (2014) *Green Criminology: An Introduction to the Study of Environmental Harm*. Abingdon: Taylor & Francis.

Widra, E. and Hayre, D. (2020) *Failing Grades: States' Responses to COVID-19 in Jails and Prisons*. Northampton, MA: Prison Policy Initiative.

Williams, E.C., Hahn, J.A., Saitz, R., Bryant, K., Lira, M.C. and Samet, J.H. (2016) Alcohol use and Human Immunodeficiency Virus (HIV) infection: Current knowledge, implications, and future directions. *Alcoholism: Clinical and Experimental Research*, 40(10), 2056–72.

Wilson, B., O'Neill, K. and Vasquez, L. (2021) *LGBT Renters and Eviction Risk*. Los Angeles: Williams Institute.

Wing, E.J. (2016) HIV and aging. *International Journal of Infectious Diseases*, 53, 61–8.

Wyatt, T. and Brisman, A. (2017) The role of denial in the 'theft of nature': Comparing biopiracy and climate change. *Critical Criminology*, 25(3), 325–41.

8

Women and the Structural Violence of 'Fast-Fashion' Global Production: Victimisation, Poorcide and Environmental Harms

Sandya Hewamanne and Nigel South

Introduction

As Sollund (2020: 520) points out, '[w]omen, children and nature are typically positioned in contrast to man and masculinity, which creates the dualistic hierarchy so central to the organization of patriarchal thought and culture' – and, importantly, also central to the organisation of work and labour. In this chapter we examine how the profit motivation of private corporations and the insatiable demand of Western consumerism for cheap goods result in the oppression of marginalised women in the Global South and a variety of effects leading to environmental damage. Our starting point is that textile production is a globally 'outsourced' industry based on manufacturing in the Global South to supply markets of the Global North, creating an industry that is structured in such a way that the local labour laws of hosting countries are either routinely ignored or do not apply to the subcontracted factories operating in special economic/manufacturing zone territories. This situation represents the inextricable linking of structural and slow violence (Galtung, 1969; Nixon, 2011; Davies, 2019) characteristic of the processes of production in the 'fast fashion' industry and the incubation of varied forms of disaster that lead to suffering, death and injury. In particular and applying the concepts of poorcide (Udayakumar, 1995) and victimisation, we discuss their impacts on two groups of women factory workers in Bangladesh and Sri Lanka.

The chapter first presents a review of relevant literature, then describes the situation of garment, textile and assembly line workers in Bangladesh and Sri Lanka who work within contexts of unregulated labour conditions and where the effects of resultant disasters and the adverse consequences of environmental damage are disproportionately borne by poor women. In particular, we highlight the negligence and denial that can lead to what Turner (1976) called the 'incubation of disasters', in this case the examples of the collapse of the eight-storey Rana Plaza garment factory in April 2013 in Bangladesh and its aftermath, and the three waves of the COVID-19 pandemic affecting global factory workers in Sri Lanka in 2020/21. Linking these cases to the everyday environmental degradation caused by the fashion industry at the lowest rung of global value chains, we demonstrate how women, and the environment, are victims of the structural violence of global production.

Gender: work-labour-poverty-violence

According to Heckenberg and Johnston (2012: 150–1), while '[g]ender relations place men and women differently at risk in the face of climate change, extreme weather events and natural disaster', such relations, norms and values 'can evolve over time' and analysis should acknowledge that women have been both 'passive and active partners in the world's factories, fields and farms'. Whether or not they are *equal* partners is, however, a very different matter and although the aftermath of a disaster may bring some changes to gendered norms and roles, these are not necessarily long-lasting. Most predictably, the pressures of labour market expectations (for example, gendered pay) and the weight of cultural traditions can reimpose the previous status quo very rapidly. In this respect, feminist approaches to political economy (Bradshaw, 2013) and political ecology (Masse et al, 2021) have paid 'particular attention to gendered divisions of labour, access, and participation', to understand, for example, 'resource use and the differential effects of changes in land and natural resource governance, tenure and access' (Masse et al, 2021: 208). Building upon such analysis, Masse et al argue that '[g]endered discourses and expectations "set in motion differentiated and unjust life opportunities and exclusions" (Elmhirst, 2011: 130) and create different "rules of the game" for men and women (Kandiyoti, 1988: 274)' (Masse et al, 2021: 208). In addition – if this needs to be said – the 'agency' of local women workers in the face of the demands of global capitalism is also likely to be fairly limited.

The ability of women to resist decision-making by others and 'reset the rules of the game' is weakened by the institutional and cultural 'weight' of global market operations and demands but also as a result of the structurally embedded nature of the forms of violence that women face. The force of violence is usually seen as immediate and direct. The point made by Galtung (1969) and Nixon (2011) is that structural or slow violence is not sharply defined by effect or timing but may be hard to isolate. Such slow violence is cumulative and possibly leads to the 'incubation' of disasters (Turner, 1976). Originating in 'organisation theory' and 'administrative science', the idea of 'incubation' has a wider application in relation to the consequences of the wilful or negligent failures of regulators, corporations or states. In this sense, it refers to the period of accumulation of generally unnoticed (certainly unremedied) factors that contribute to or precipitate a damaging event that disproportionately affects workers and communities who 'don't count'. Such 'incubation' is a form of 'slow violence' and is built into the very premises of transnational production and its assumptions about the disposability of cheap labour: women are factored into production costs as secondary or supplementary earners who will be docile and voiceless in the face of gross exploitation. Moreover, the gendered devaluing of their labour creates a status of them as 'disposable' and a label for their work as 'unskilled' (Elson and Pearson, 1981; Wright, 2007; Hewamanne, 2020). This is the situation of poor, marginalised women workers, seen only as secondary citizens under the patriarchal systems prevalent in both Sri Lanka and Bangladesh.

Entwined here are the processes whereby the assignment of a lesser value to a particular group of people – racial and ethnic minorities, women, the poor – makes it easier to justify oppression and exploitation. This will also mean these individuals and groups are more likely to be victims of environmental injustice (Bullard, 1990; Brulle and Pellow, 2006). In relation to environmental damage related to climate change, for example, Wonders and Danner (2015) argue that women are particularly vulnerable because they are more likely to be poor. Udayakumar's concept of 'poorcide' (1995: 342) is therefore suggestive here, for although 'race, ethnicity, gender, generation and political powerlessness all contribute considerably to poverty' it is the 'economic worth' factor that underpins 'poorcide' – the genocide of the poor.

Nixon illustrates this well in his discussion of 'Slow violence, gender, and the environmentalism of the poor' by noting how rural women in Kenya had

> suffered the perfect storm of dispossession: colonial land theft; the individualizing and masculinizing of property; and the experience of continuing to be the primary tillers of the land under increasingly inclement circumstances, including soil erosion and the stripping of the forests ... women inhabited the betrayals of successive narratives of development that had brutally excluded them. The links between attritional environmental violence, poverty, and malnutrition was a logic they lived. (Nixon, 2011: 140)

Similarly, Csevár (2021: 1) reports how in West Papua, 'Indigenous women are disproportionately affected by environmental degradation, caused by resource extraction and increasingly compounded by climatic changes. This in turn exacerbates other vulnerabilities, including sexual and gender-based violence and other forms of marginalization' (see also Shiva, 2010). In contrast to this context of rurality and the legacy of colonialism, Deb (2021) traces the legacy of urban industrialisation and operations of transnational corporations under India's market liberalisation, using the case of the Bhopal disaster as reflected in the experiences of women and quoting activists who point to long-lasting suffering that no one wants to hear about.

The occurrence of structural violence and of the incubation of disaster are global phenomena and women around the world suffer the impacts. Furthermore, when the relentless pursuit of neoliberal forms of development endangers the lived environment, the consequences are experienced alongside already existing inequalities. Environmental damage only deepens these inequalities and as Simončič (2021: 360) notes with regard to the production processes involved in the clothing, garment and textile industries, '[e]missions of chemicals, wastewater, solid waste generation, excessive fresh water, minerals, fossil fuel and energy consumption are only some of the devastating environmental consequences of clothing production' (see also Allwood et al, 2006).

Global fast fashion

Western investors and multinational corporations have taken full advantage of the structural adjustment policies that the World Bank and political neoliberalisation ushered in from the 1990s to address high debt burdens affecting various economies of the Global South. The main investments have been in establishing factories in tariff-free export-oriented production zones, utilising local labour paying very low wages, and adopting strategies to reduce production costs to the minimum. Countries such as Vietnam, Thailand, Indonesia, Mexico, Honduras and China established demarcated zones for transnational manufacturing of garments, electronics, toys, rubber products and medical equipment. Despite the evident exploitation, manufacturing jobs in global assembly lines provided much-needed sources of income for unemployed young people. The operation of these global assembly lines is structurally gendered, with women doing assembly line work, and men filling managerial positions. The role of women has been based on patriarchal stereotypes of women being 'nimble fingered', docile and supplementary earners – as opposed to men who are considered the main 'breadwinners' for a family. Notwithstanding reports of labour exploitation, poor working conditions, human rights abuses and industrial disasters, these factories continue to operate with varied levels of compliance to international labour codes (Ong, 1991; Hewamanne, 2008, 2016; Prentice and De Neve, 2017; Siddiqi, 2017).

Assembly line work in Bangladesh and Sri Lanka

The ready-made garment industry in Bangladesh

In recent decades, Bangladesh has 'experienced rapid industrialisation, enormous urbanisation and massive digitalisation, with serious environmental consequences resulting from the associated increases in consumption and waste and the use of energy and resources' (Faroque and South, 2022: 3). It has also become known as one of the world's major subcontracting nations in the ready-made garment industry. In 1996 Bangladesh was home to 2,353 factories with 1.29 million workers and a US$2.55 billion turnover; by late 2019, just before the COVID-19 pandemic, the industry's export turnover had reached US$33.07 billion, with more than 4,500 factories and 4.2 million workers (Strumpell and Ashraf, 2021). This rapidly growing sector in Bangladesh is also well known for frequent industrial disasters, bad working conditions and exploitation of women and children.

Assembly line work in Sri Lanka

The first Sri Lankan Free Trade Zone (FTZ) was established in 1978 in Katunayake, after a newly elected government began pursuing structural adjustment programmes. Now Sri Lanka boasts of 13 such locations, officially known as Export Processing Zones. The newly created assembly line jobs were quickly filled by unemployed rural women. As in other transnational factories around the world, the Export Processing Zone assembly lines demand maximum output for minimal wages in exploitative working conditions and practice a distinctively late-capitalist form of gendered working relations. In 2020, over 40 years after the first FTZ was established, the basic worker's salary was Rs.14,000 (about US$80) per month, although women could earn about Rs.25,000 by working overtime. About 45,000 rural women from economically and socially marginalised groups work as machine operators in the Katunayake FTZ's 92 factories and a similar number work for subcontracting factories located around the zone. Altogether close to 300,000 workers engage in this kind of assembly line work with most being female, young, unmarried and well-educated, often with 10–12 years of schooling (Hewamanne, 2016). There are few state- or factory-run hostels so the women who migrate from rural areas rent hastily built rooms from area residents. The difficult work and living conditions are compounded by the sexual harassment workers face on city streets and the shop floor (Hewamanne, 2012, 2016).

Social harm and state-corporate crime

The garment and similar industries are gendered whether in Bangladesh, Sri Lanka, or elsewhere. Simončič notes that:

> There is a saying among young girls in Bangladesh: 'If you're lucky you'll be a prostitute, if not, a worker in the garment industry'. … One of these girls is Anju, a garment factory worker in Bangladesh, who earns $900 a year, works twelve hours a day and skips meals because she cannot afford to buy food … Anju is also more exposed to the negative effects of climate change than those living in the Global North, to which the fashion industry … which produces 8% of all global greenhouse gas emissions … and one-fifth of global wastewater … contributes extensively. In another part of the world, in Spain, the CEO of the fashion conglomerate Inditex … works hard too and rarely takes a vacation. However, he owns a private plane in his backyard and has more than $60 billion in his bank account. (Simončič, 2021: 343)

In Sri Lanka, especially during the protracted civil war (1983–2009), there was a different popular saying: 'garment for (rural) girls and military for (rural) boys'. While the dangers of death and injury to foot soldiers at the time was well recognised and compensated for via various incentives, such as risk area allowances, the violence of fast fashion committed on women's bodies, minds and spirit was never recognised or acknowledged. While the soldiers were celebrated as 'war heroes', women who suffered to bring much-needed dollars to the country were stigmatised as women working 'only' to buy fancy goods until they could earn their dowry (De Mel, 2007; Hewamanne, 2008).

Social harms and environmental impacts

The mass-production of cheap, disposable clothing, and frequent introduction of new trends has the (intended) effect of making consumers feel constantly out-of-date and encouraging more buying (Brisman and South, 2014: 55–60). As Large (2019: 95) notes, citing Bauman (2007), 'this "buy it, enjoy it, chuck it out cycle" (p 98) coupled with the "constant pressure to be someone else" (p 100) reinforces the constant demand in neoliberal capitalist societies for the consumption of fashion'. In the United States alone, 35kg of textile waste per person per year is generated because, on average, a garment is worn only seven times. As the Sustain Your Style[1] fashion impact report (2022) notes, instead of the classic 'two seasons' of the past, fast-fashion brands now release 52 mini collections, and 80 billion garments are made each year. This represents an increase in production of more than 400 per cent compared to just 20 years ago.

These trends in the fashion industry lead to harmful impacts on the environment on many fronts. For example: water pollution from toxic waste water released from textile factories; excessive water consumption for producing cotton and fabric dyeing; micro-fibre pollution; non-biodegradable material waste; and the extreme use of chemicals in the fashion industry (Bick et al, 2018). The Sustain Your Style report elaborates that the fashion industry accounts for 10 per cent of global carbon emissions and due to the massive use of chemicals the industry contributes to soil degradation. Rainforest destruction is a consequence of the tree-felling required to produce wood-based fabric, threatening ecosystems and the livelihoods of Indigenous communities, especially women, around the world. Every year, 70 million trees are cut to make clothes with 30 per cent of wood-based fabric being produced from endangered ancient forests. Of the chemicals produced worldwide, 23 per cent are used in the textile industry (1kg of chemicals is needed to make 1kg of textiles). Unsurprisingly, 63 per cent of the fabric tested from major brands contained harmful chemicals and 27 per cent of the weight of a 100 per cent 'natural' fabric is made up of chemicals (Sustain Your Style, 2022).

The burden of these environmental harms will be disproportionately borne by women. The devaluation of poor, gendered communities in the Global South as expendable resources in the process of lowering manufacturing costs, corresponds with the denial by corporate stakeholders in the global value chain that the fast-fashion industry causes any harm to the environment of the countries and regions that host subcontracting factories. In Bangladesh, the bulk of such environmental harm occurs through water pollution. For example the garment factories near Dhaka regularly dump waste in the river Turag, causing the river water to change colour and emit a foul odour (Faroque and South, 2022). Bansal river and the irrigation canals surrounding the industrial arena of Narayanganj are also the victims of Bangladeshi ready-made garment production. The tests on upstream water sources and even deep tube wells have shown the presence of heavy metals and trace radioactive elements. Toxic industrial waste products have negatively impacted the fish population of these rivers, affecting the quality of fruits and vegetables grown with water from the rivers (Sakamoto et al, 2019).

Sri Lanka has only seven textile and raw material factories as opposed to 300 apparel manufacturing facilities, so the environmental impact of the global garment industry in Sri Lanka has been less than in Bangladesh. According to the knittingindustry.com website (2021), the championing of commitment to sustainable manufacturing in Sri Lanka has seen changes in manufacturing processes, sourcing strategies and product innovation, attracting brands that highlight their concern about the environmental impact of their business. Nonetheless, the establishment of the textile industry in Sri Lanka was rapid and has meant that the generation of waste at all levels of the production process – from sourcing raw material through the processes of design and logistics to end-disposal – has been high (Udara et al, 2019). Environmental harms still follow but in Sri Lanka it is particularly the exploitation, harsh work and living conditions of the labour force that expose how the corporations, global capitalist hierarchies and local politicians have colluded through certain policies to incubate large disasters and crises that bubble up and explode periodically, occasionally forcing the Global North to take stock of the environmental and human cost of fast fashion.

Workers in Sri Lanka are exploited in many ways – the conditions of very low wages, very long hours, forced labour, child labour, and bad health and safety conditions at these factories are well known. As unionisation within global garment production is prohibited, there is no way for workers to collectively organise. The two case studies included in the following sections are prime examples of all these labour and human rights violations and consequences.

Incubating disaster and COVID-19

On 24 April 2013, the eight-storey Rana Plaza building in Dhaka, Bangladesh, which housed five garment factories, collapsed, killing at least 1,135 people and injuring more than 2,500 workers (Chowdhuri, 2017). The previous day, serious cracks had appeared in key structural positions and the employees of a bank and several shops which operated in the lower floors were evacuated due to the dangers of working in a building that was obviously buckling. The garment workers were also evacuated but the next day were ordered to return to work. The cracks were rapidly widening and they were reluctant to enter the factory but were pressured into resuming work or face losing their jobs. Some were threatened and beaten to force them to return (Parveen, 2014).

In this specific case, the responsibility for immediate decisions about safety and return to work lies with the local supervisors and managers. However, it is the very structuring of global fashion value chains – which forces local factories to deliver within limited time-frames or risk losing orders – that is at the root of the call that sent garment workers back to work in a buckling building. This incident epitomises the human cost of fast fashion where competitively obtained orders have to be completed within a very short period of time. Some of the fashion brands manufacturing in the Rana Plaza garment factories during the collapse were Walmart, the Children's Place, Zara and Primark.

The biggest tragedy was how easily this could have been prevented were it not for the highly unequal economic relationships between the buyers of the Global North and the manufacturers in the Global South (Lim and Prakash, 2017). The eight-storey building was constructed on swamp land, and the top three floors were built without official approval and had no supporting walls. Crowded with workers and heavy machinery, the already structurally unsound building was at high risk of causing a disaster. Yet, even after significant cracks appeared, the local factory owners, who feared loss of favour with buyers and of subsequent orders, forced workers back to the building, demonstrating the devaluation of the lives of the poor – the process of poorcide referred to earlier.

Rana Plaza was one of the most devastating and well-known industrial disasters to occur in recent history. However, it was not a one-off or simple incidental event. It was the outcome of a system based on negligence, exploitation and the oppression of poor, marginalised women and children within the global fashion industry in Bangladesh. For example, in 2005, a substandard building housing the Spectrum sweater company collapsed, resulting in 64 deaths and 80 people being injured and, in 2012, a devastating factory fire in the Tazreen fashions factory resulted in the deaths of more than 112 workers and injuries to 200 more (Sumon et al, 2017). These cases were in addition to incidents of fire and collapsing buildings that resulted in smaller numbers of deaths and injuries among the predominantly female workforce of these global assembly line factories. Most workers died during these incidents due to being trapped inside factories with locked doors and windows that made it difficult for them to escape when fires broke out, and injuries of survivors occurred when jumping out of windows that offered the only means of escape.

As Western companies try to escape the costs of the strict environmental and other safety measures that are in place in affluent countries, by outsourcing manufacturing to countries like Bangladesh, they squeeze these factories to the lowest cost possible in order to increase profits. In turn the local factories squeeze the workers to widen their profit margin. The government of Bangladesh itself has been willing to overlook general building regulation violations in the fear that factories will lose orders to competing poor countries. Lim and Prakash (2017) demonstrate how corporate social responsibility policies are ineffective in the context of such unequal global power relations, noting that an auditing of the Rana Plaza building just a few months before the collapse failed to report its structural weaknesses.

Dying for fashion: Bangladesh

The shocking scale of the Rana Plaza disaster created a public relations crisis for the garment industry and led to a legally binding agreement that is commonly called the Accord (IndustryAll, 2022). The Accord holds brands responsible for ensuring safety in factories and liable in court if safety violations are not resolved in a timely manner. The legally binding nature of the Accord resulted in measurable safety improvements and fewer disasters. Nazma Akter, founder and executive director of the Awaj Foundation, a labour rights organisation in Bangladesh, and co-chair of IndustriALL Global Union's Asia Pacific Women's Committee, notes: 'Before the accord, every year there were so many accidents, fire incidents, every year people were dying, got severely injured. … After Rana Plaza, there are also incidents, but our workers are not dying. Our factories are safe' (Vogue Business, 2022). Victims of the Rana Plaza incident are often called 'fashion victims' who died for fast and cheap fashion and any improvements in safety are a memorial to how 1,134 workers' lives and a further 2,500 workers' futures were sacrificed to win the most basic labour and human rights within the ready-made garment industry in Bangladesh.

Unfortunately, these improvements come with costs that the brands/buyers are not willing to continue to bear without compensation. Thus the Accord forced most brands to factor the cost of safety into their overall production costs and subsequently into their pricing. For this reason, according to some critiques, there is now a move away from engagement with renegotiating the Accord and such brands are favouring alternative bipartite or tripartite initiatives that do not have the legally binding character of the Accord that led to its success. As Ashraf and Prentice (2018) show, although the Accord addresses important issues regarding workers' health and safety, it does not challenge the fundamental structures of exploitation on which the industry rests. Since the Rana Plaza disaster, 109 accidents have been recorded with at least 35 being in textile factories. The evident human cost of the Rana Plaza collapse was simply a highly visible demonstration of the consequences of the slow violence, injury and death that the fast-fashion industry inflicts upon the poor, marginalised and gendered populations and environment of the Global South. The Accord may have improved safety and ensured there are better health facilities for workers but the structural conditions that make workers faint at their machines or work in a crumbling building to meet production targets remain unchanged.

There is, however, one interesting gender-related structural change that has followed post-Accord improvements in safety and wages. This is the change in worker demographics. As Strumpell and Ashraf (2021) argue, due to the increase in wages, technological up-skilling, and some other organisational and locational developments, the share of men in the workforce in late 2019 had almost doubled compared to 2015 (from roughly 20 per cent to at least 40 per cent). These authors note there is some irony in male workers now benefiting from improved conditions that women workers won through industrial action and, of course, after suffering through events like the Rana Plaza tragedy.

In the Sri Lanka case in the next section, we note many parallels to the Bangladeshi workers' experiences and again how fast fashion is not only detrimental to the environment but underpins an invisible process of gendered poorcide. In addition, we consider here how the COVID-19 pandemic unravelled global production networks and adversely affected local factories and their workers in a way that was similar for most of the subcontracting nations of the Global South.

Health disaster and gendered poorcide: Sri Lanka

The Rana Plaza disaster prompted some comparisons with conditions (especially regarding fire and building safety) in the global garment industry in Sri Lanka, which, in its 35-year history, has not had an industrial disaster that killed even one worker. But does this obvious difference in the numbers of incidents and deaths translate to better work practices and living conditions for workers or mean there is less environmental harm? In what follows we explore how the processes of devaluation of poor women's labour and lives within global assembly line production have familiar outcomes and result in unrecognised and untold suffering for women. This is done via an analysis of how the pandemic was experienced within two FTZs in Sri Lanka that hosted global garment factories.

Having no reported industrial disasters does not mean that the dynamics surrounding such production do not lead to gendered and classed suffering and environmental harm. The intertwined factors of unplanned development, structuring of work to produce fast fashion and the devaluation of rural, poor women's labour all work together to inflict violence on working women. Rather than incubating visually and physically transparent disasters such as Rana Plaza, this violence becomes embodied, and its cost is then borne by workers, families, communities and the national health system.

The advent in 1978 of the Katunayake FTZ, and the subcontracting factories surrounding it, attracted young rural women to take up employment. The arrival of more than 100,000 temporary migrants and the growth of the FTZ saw the area developing from a semi-rural area to a city in a short time; itself a considerable environmental impact with most of this development being unplanned. For example, although factories were well built, the accommodation for the rural-to-urban migrants was not planned alongside this, resulting in people in neighbouring areas building unauthorised rows of rooms in their gardens for workers. These rooms were small, badly ventilated, dusty and hot, and the electricity was illegally obtained through cables stretched from the landlord's house. Each small room would be occupied by several workers, sometimes up to six, making them very crowded and most such boarding compounds only had one well and very few toilets (Hewamanne, 2008, 2016, 2019).

The consequences of such unplanned development have been varied but include environmental and public health impacts such as the women workers experiencing flooding of their boarding houses, and resulting loss of all their belongings, and exposure to unsanitary conditions in their living arrangements, leading to frequent bouts of communicable diseases on top of the long-term health hazards of the relentless work pace in their factories. And then – in 2020 – the COVID-19 pandemic arrived into these already compromised work and living arrangements.

Precarity and the pandemic

Global factory work has always been precarious and also had an element of a 'labour clock' – meaning women workers were expected to work only for a certain period in their lives before moving on to meet traditional expectations about their roles in marriage and family. From around 2010 in Sri Lanka, this precarity doubled as factories started to cut their regular work force by half and hire casual daily-paid workers through employment agencies. Factories now encourage casual labour to the extent of asking regular workers to quit and re-enter as daily-hired labourers (Dabindu Collective, 2017; Hewamanne, 2021). This allows the factories to escape some of the new responsibilities that came with 'corporate social responsibility' policies regarding their regular workforce, such as the Employees Provident Fund and Employees Trust Fund payments.

On the one hand, daily-hire work allows women to refuse being enslaved by factory production regimes. On the other hand, they lose out on many forms of benefits. It is important to consider the context in which women come to prefer casual work. Global factory work is characterised by stressful, back-breaking, target-oriented production leading to forced overtime and weekend work. All of this leads some women to prefer casual work that allows them a little more agency in 'when and where' to work. Ultimately, however, the exploitative labour regimes ensure the continuation of a 'race to the bottom'. The pandemic put this situation in perspective.

Sri Lanka reported its first local COVID-19 case in late February 2020. The country went into full lockdown on 19 March 2020 and an island-wide curfew was imposed, lasting until early June. While lockdown and curfew prevented the virus spreading, the abrupt declaration of the curfew affected thousands of FTZ workers adversely. The factories shut down without paying workers for the month of March while the cancellation of public transport services made it difficult for many to return to their village homes. Workers soon found out that these immediate concerns were the least of their worries. Between March and 30 May Sri Lanka's apparel exports fell 50 per cent against the same period of the previous year. This was a loss of close to US$680 million. By April, the apparel industry was facing losses of over US$1.5 billion (*Financial Times*, 2020). By October, overall exports to the United States alone had fallen by 22.15 per cent, while exports to the

European Union fell by 21.36 per cent (De Silva, 2020). The same bleak picture faced all other countries that depended on global assembly line work for most of their export income. When COVID-19 struck, Western brands cancelled orders worth billions of dollars, leaving shipments of clothing with no takers. Hundreds of factories closed across Asia, including Vietnam, China and Bangladesh. Millions of garment workers in Asia, the vast majority of them women, have been suspended or laid off, depriving them of US$6 billion in wages (Nilsson, 2020).

As global production chains unravelled, the closing of shops in Europe and North America translated into more and more closed factories in the Global South (Nilsson and Terazono, 2020). In the West, companies such as JCPenney and Neiman Marcus filed for bankruptcy protection. The chain reaction caused 20–30 per cent of companies along the value chain – from brands to wholesalers and department stores – to cease business, thereby affecting numerous other jobs across the industry. Many factory owners in the Global South are now finding it difficult to recover payments from major brands and retailers who refuse to pay or demand deep discounts for goods that were already in production or ready for shipment (Anner, 2020). Many factories found it difficult to stay afloat without steady orders. As a result, factories are closing indefinitely, leading to a massive loss of employment. For example, at least 110 garment factories closed between January and September in Cambodia alone, leaving over 55,000 workers unemployed (Clean Clothes Campaign, 2020).

By July 2021, Asian workers had lost US$12 billion in wages (Nilsson, 2021). Workers who migrated for work in global subcontracting factories were especially affected. The Clean Clothes Campaign reported that in August 2021 around 2,000 migrant workers from Myanmar were stranded in Laos without work or money for rent or food (Clean Clothes Campaign, 2021). Perhaps more alarmingly, the reduced orders and subsequent cutting of jobs created a context where there were many more workers waiting for jobs than there was work available, thus providing an environment that advantaged employers. In this climate of financial struggle, many governments of the Global South are ignoring violations of rules and of best practice, such as cutting wages (sometimes up to half of pre-pandemic pay levels) and cancelling bonuses and incentives (Dabindu Collective, 2020; Hewamanne, 2022). The Asia Floor Wage Alliance, an Asian labour-led global social alliance across garment producing countries, carried out an in-depth six-country regional study surveying over 2,000 (mostly women) workers in Bangladesh, Cambodia, India, Indonesia, Pakistan and Sri Lanka, and found massive wage theft in the global garment industry during the pandemic. Reports on all countries highlighted similar processes of wage reductions and slashing of benefits (AFWA, 2021).

The pandemic led to forms of modern slavery within global assembly lines, while many countries slowly moved towards authoritarian politics under the pretext of addressing pandemic-related difficulties (Lorch and Sombatpoonsiri, 2020; DeVotta, 2021; Hewamanne, 2021). Non-vaccinated workers toiled within crowded factory conditions and became infected and quarantined at alarming rates in Sri Lanka and other subcontracting countries in the Global South. For example, in mid-August 2021 Bangladesh alone had 2 million global garment workers still waiting for vaccinations (Clean Clothes Campaign, 2021). The companies that profit from the labour of these workers offloaded their setbacks to subcontracting companies, thereby exacerbating difficulties faced by those at the bottom-most rungs of the global production networks – the assembly line workers. The disproportionate suffering that female global factory workers endured during the second and third waves of the pandemic in Sri Lanka was in fact very much a part of the unjust structuring of fast fashion.

During the second wave, local labour regimes, shaped by global production pressures, led to infected and exposed workers staying on the shop floor. Years of being mocked and sent back to the lines when they complained of illness have conditioned FTZ workers to put up with bodily pains and exhaustion. The routine discouragement of reporting health problems, and the normalisation of working while sick, meant many workers experiencing symptoms of COVID-19 did not seek on-site healthcare. Many worked for days while sniffling, coughing and exhausted until one worker tested positive at the hospital. Before she was diagnosed, she along with others who were showing symptoms, were treated at the factory clinic but told they just had flu. When several workers started fainting on the production floor, managers and nurses sprinkled water on their faces and sent them back to work after a short break (Bandara, 2020).

As the factories are overcrowded, infections spread fast. Workers returned to their multiple-occupancy, poorly ventilated boarding houses and shared toilets with many others, leading to the virus being spread throughout the FTZ area and then to the Gampaha district. Since workers were not given paid leave until they themselves were tested or their boarding house was declared as a quarantining facility (which took close to a week after someone was diagnosed) the workers continued to go to work, exposing others at the factories. Public health inspectors and the military removed women workers who were suspected contacts of infected persons to quarantine centres considerable distances away without warning. Others who remained in the boarding houses experienced being treated with contempt and fear by neighbours (Perera, 2020): 'These women are being treated like criminals and "lepers"', lamented Chamila Thushari, leader of a non-governmental organisation (Thushari, 2020).

All of this represented an intertwined process of class, gender and, to a lesser extent, region-based marginalisation of this group of workers and the devaluation of rural women's work built into the global garment industry.

Fears of contamination have always centred on the intersections of race and class (Preston and Firth, 2020) and the fears and anxieties of the pandemic in Sri Lanka seem to have been expressed in ways producing further ostracisation of already stigmatised and marginalised communities: the poor Muslims in congested living circumstances in the first wave, and the garment workers (predominantly poor women) in the second wave.

The Clean Clothes Campaign (2021: 7) has accused factory owners of avoiding their responsibility towards worker health by failing to set up bipartite (employer and employee) health committees to deal with COVID-19 as mandated by the Labour Ministry task force. Only two factories had complied with this requirement by February 2021. Unions argue that establishing such committees could have minimised the intensity of the second wave which broke out in global garment factories. In addition, the lack of responsibility within the supply chain was clearly evident. Instead of monitoring how workers were treated during the pandemic, brands and retailers turned away and focused squarely on protecting their profits (AFWA, 2021; Clean Clothes Campaign, 2021). All this can be characterised as a case of necrocapitalism, allowing people to die in the name of making a profit, and a form of poorcide.

Discussion

As outlined in this chapter, the slow violence committed by the fast-fashion industry and affecting global factory workers, incubates varied disasters which cause suffering, death and injury to female workers. Rather than learning from these disasters and restructuring the industry to minimise human suffering and harms to the environment, the multinational corporations use these disasters (as demonstrated during the COVID-19 pandemic) to increase their profit margins further.

The process of textile production, especially from natural fibres, is complex, and despite the idea that using 'natural' materials sounds 'green', it can actually create many environmental problems. All stages of textile production – fibre production, processing, spinning, yarn preparation, fabric production, bleaching, dyeing, printing and finishing – create polluted waste water, chemical discharge and solid waste matter. Multinational corporations that are frustrated with environmental safety measures in Western, affluent countries happily outsource this process to poorer countries in the Global South to escape their responsibilities towards safeguarding the environment. Subcontracting factories operate with little to no oversight from the Western companies outsourcing their production and while governments may protest about the significant effects of unchecked textile production on water sources and other increased levels of pollution, they are in a powerless position in the global capitalist hierarchy.

Thus women at the bottom-most rung of the global fashion industry not only suffer within bad working and living environments but also suffer the consequences of industry-related environmental damage more acutely. Specifically, as the providers of water for their families, women are now experiencing much more anxiety about gathering clean water for drinking, cooking and washing. Poor men affected by economic disruption in the ready-made garment factory zones, specifically, farmers and fishermen, also face precarity and hardships related to their livelihoods, with many having to join the surplus of casual labourers waiting for day-labour jobs in the city and the suburbs.

The Rana Plaza disaster was a culmination of intertwined factors produced by a global capitalist hierarchy whereby the most environmentally damaging aspects of textile manufacturing are outsourced to poorer countries. According to the Clean Clothes Campaign (2021), when the Rana Plaza disaster happened, the factory buildings in the industry were notoriously unsafe with no regard to building or fire safety, making them potential 'death traps' for the 2 million women workers in about 1,600 factories subcontracting for around 200 global fashion brands. As seen in the case of Sri Lanka, the waves of the COVID-19 pandemic inflicted further unnecessary and excessive suffering on global garment workers, as the global labour regimes in place had already normalised working with illness and pain, and also suppressed awareness and concern about environmentally damaging processes and contaminants that are also damaging to workers and wider communities.

Conclusion

There is some growing awareness of the environmental costs of 'fast fashion' and cheap clothing, with journalists, international agencies and even parts of the fashion industry itself drawing attention to the price paid by the environment as a result of over-use of water, chemical run-off, encouragement of waste and landfill, intensive energy use and carbon emissions. This chapter has discussed this form of social harm in connection with another – the exploitation of women workers in the textile and garment industries. The focus here has been to provide examples of labour exploitation and the forms in which this occurs, mainly (but not exclusively) in the Global South which also suffers environmental problems more acutely than the Global North. The chapter has aimed therefore, to show how the adverse effects of environmental damage, unregulated labour conditions and resultant disasters are *disproportionately* borne by poor women in the Global South and how this is intimately connected to the lower socioeconomic status of women in these particular contexts.

Around the world – in the Global North and the Global South – cheap fashion is still a significant driver of real and digital footfall in stores and online, regardless of growing ethical and environmental concerns in some quarters. The idea that this is truly 'cheap' and without much cost will, however, inevitably be challenged by reality as environmental harms move across borders. As Thanhauser (2022) remarks, just because the environmental problems of fast fashion seem to be 'more acute elsewhere', doesn't mean they are 'going to stay elsewhere' (Morris, 2022).

Note

1 Sustain Your Style is an independent platform created by those who are concerned about the current practices of the fashion industry. Their website is a resource that provides tools to make informed, sustainable fashion choices.

References

AFWA (Asia Floor Wages) (2021) Money heist: COVID-19 wage theft in global garment supply chains. *AFWA*. Available at: https://asia.floorwage. org/wp-content/upl oads/2021/07/Money Heist_Book_Finalcompressed. pdffbclid=IwAR2S6dE SEv37HGydzd24ngjKa6cbWiamnU81If41oVIIr (accessed 7 July 2021).

Allwood, J.L., Laursen, S.E. and de Rodriguez, C.M. (2006) *Well Dressed? The Present and Future Sustainability of Clothing and Textiles in the United Kingdom.* Cambridge: University of Cambridge.

Anner, M. (2020) Leveraging desperation: Apparel brands' purchasing practices during COVID-19. Available at: www.workersrights.org/wp-cont ent/ uploads/2020/10Leveraging_Desperation.pdf (accessed 5 May 2022).

Ashraf, H. and Prentice, R. (2018) Beyond factory safety: Labor unions, militant protest, and the accelerated ambitions of Bangladesh's export garment industry. *Dialectical Anthropology*, 43(1), 93–107.

Bandara, N.K. (2020) Workers who fainted had water sprinkled on their faces and had to work again. *Daily Mirror*, 14 October. Available at: www.dail ymirror.lk/newsfeatures/Workers-who-fainted-had-water-sprinkled-on-their-faces-and-had-to-workagain/131-197861 (accessed 21 January 2021).

Bauman, Z. (2007) *Consuming Life*. Cambridge: Polity.

Bick, R., Halsey, E. and Ekenga, C. (2018) The global environmental injustice of fast fashion. *Environmental Health*, 17, 92.

Bradshaw, S. (ed) (2013) *Gender, Development and Disasters*. Cheltenham: Edward Elgar.

Brisman, A. and South, N. (2014) *Green Cultural Criminology: Constructions of Environmental Harm, Consumerism, and Resistance to Ecocide.* London: Routledge.

Brulle, R. and Pellow, D. (2006) Environmental justice: Human health and environmental inequalities. *Annual Review of Public Health*, 27(1), 103–24.

Bullard, R.D. (1990) *Dumping in Dixie: Race, Class, and Environmental Quality.* London: Routledge.

Chowdhuri, R. (2017) The Rana plaza disaster and the complicit behavior of elite NGOs. *Organization*, 24(6), 938–49.

Clean Clothes Campaign (2020) Un(der) paid in the pandemic: An estimate of what the garment industry owes its workers. *Clean Clothes Campaign*. Available at: https://cleanclothes.org/filerepository/underpaid-in-the-pandemic.pdf/view (accessed 15 July 2021).

Clean Clothes Campaign (2021) COVID-19 pandemic: A pretext to roll back Sri Lankan garment workers' rights. *Clean Clothes Campaign*. Available at: https://file:///C:/Users /Sandya/Downloads/Sri_Lanka_Brief_March_2021%20(4).pdf (accessed 15 July 2021).

Csevár, S. (2021) Voices in the background: Environmental degradation and climate change as driving forces of violence against indigenous women. *Global Studies Quarterly*, 1, 1–11.

Dabindu Collective (2017) *Living for the Day: Contract Workers in Sri Lanka's Free Trade Zones*. Nugegoda: NEO Graphics.

Dabindu Collective (2020) NEXT strike. Available at: www.facebook.com/chamila. thushari.18/videos/387335592729044 (accessed 22 December 2020).

Davies, T. (2019) Slow violence and toxic geographies: 'Out of sight' to whom? *Environment and Planning C*, 40(2), 409–27.

De Mel, N. (2007) *Militarizing Sri Lanka: Popular Culture, Memory and Narrative in the Armed Conflict*. London: SAGE.

De Silva, C. (2020) Apparel exports drop by 22% to 3.1 b in 9 months. *Financial Times* (Lanka). Available at: ft.lk/front-page/Apparel-exports-drop-by-22-to-3-1-bin-9-months/44-707531 (accessed 15 October 2020).

Deb, N. (2021) Slow violence and the *Gas Peedit* in neoliberal India. *Social Problems*. https://doi.org/10.1093/socpro/spab058

DeVotta, N. (2021) Ethnoreligious nationalism and autocratization in Sri Lanka. In S. Widmalm (ed) *Routledge Handbook of Autocratization in South Asia*. London: Routledge, pp 285–97.

Elmhirst, R. (2011) Introducing new feminist political ecologies. *Geoforum*, 42(2), 129–32.

Elson, D., and Pearson, R. (1981) 'Nimble fingers make cheap workers': An analysis of women's employment in third world export manufacturing. *Feminist Review*, 7(1), 87–107.

Faroque, S. and South, N. (2022) Water pollution and environmental injustices in Bangladesh. *International Journal for Crime, Justice and Social Democracy*, 11(1), 1–13.

Financial Times (2020) Apparel appeals to government for swift support. *Financial Times*. Available at: ft.lk/front-page/Apparel-appeals-to-Govt-for-swiftsupport/44-698740 (accessed 10 April 2020).

Galtung, J. (1969) Violence, peace, and peace research. *Journal of Peace Research*, 6(3), 167–91.

Heckenberg, D. and Johnston, I. (2012) Climate change, gender and natural disasters: Social differences and environment-related victimisation. In R. White (ed) *Climate Change from a Criminological Perspective*. London: Springer, pp 149–72.

Hewamanne, S. (2008) *Stitching Identities in a Free Trade Zone: Gender and Politics in Sri Lanka*. Philadelphia: University of Pennsylvania Press.

Hewamanne, S. (2009) Duty bound? Militarization, romances and new spaces of violence among Sri Lanka's Free Trade Zone garment factory workers. *Cultural Dynamics*, 21(2), 153–84.

Hewamanne, S. (2012) Negotiating sexual meanings: Global discourses, local practices and Free Trade Zone workers on city streets. *Ethnography*, 13(3), 352–74.

Hewamanne, S. (2016) *Sri Lanka's Global Factory Workers: (Un)disciplined Desires and Sexual Struggles in a Post-colonial Society*. London: Routledge.

Hewamanne, S. (2019) From global workers to local entrepreneurs: Former global factory workers in rural Sri Lanka. *Third World Quarterly*, 41(3), 547–64.

Hewamanne, S. (2021) Invisible bondage: Mobility and compulsion within Sri Lanka's global assembly line production. *Ethnography*. doi:10.1177/1466138121995843

Hewamanne S. (2022) Wither labor and human rights? Precarious work and informal economies in the post COVID-19 Global South. In S. Hewamanne and S. Yadav (eds) *Political Economy of Post-COVID Life and Work in the Global South: Pandemic and Precarity*. London: Palgrave.

IndustryAll (2022) Rana plaza from tragedy to an international accord to make garment factories safe. Available at: www.industriall-union.org/rana-plaza-from-tragedy-to-an-international-accord-to-make-garment-factories-safe (accessed 30 May 2022).

Kandiyoti, D. (1988) Bargaining with patriarchy. *Gender & Society*, 2(3), 274–90.

Knittingindustry.com (2021) Transforming sustainable textile manufacturing in Sri Lanka. Available at: www.knittingindustry.com/intimate-apparel/transforming-sustainable-textiles-manufacture-in-sri-lanka/ (accessed 20 May 2022).

Large, J. (2019) *The Consumption of Counterfeit Fashion*. London: Palgrave.

Lim, S. and Prakash, A. (2017) Four years after one of the worst industrial accidents ever what have we learned. *Washington Post*. Available at: www.washingtonpost.com/news/monkey-cage/wp/2017/04/24/four-years-after-one-of-the-worst-industrial-accidents-ever-what-have-we-learned/ (accessed 10 June 2022).

Lorch, J. and Sombatpoonsiri, J. (2020) Southeast Asia between autocratization and democratic resurgence. Available at: https://carnegi eeurope.eu/2020/12/07/sou theast-asia-between-autocratization-and-democratic-resurgence-pub-83139 (accessed 22 December 2020).

Masse, F., Giva, N. and Lunstrum, E. (2021) A feminist political ecology of wildlife crime: The gendered dimensions of a poaching economy and its impacts in Southern Africa. *Geoforum*, 126, 205–14.

Morris, S. (2022) Fashion's dirty laundry. *The I*, 27 January, pp 36–7.

Nilsson, P. (2020) Asia's garment workers lose out on $6bn after pandemic cuts. *Financial Times*, p 10. Available at: www.ft.com/content/cc430a78-59ff-4c36-9994-cde265301a3f (accessed 10 August 2020).

Nilsson, P. (2021) Pandemic deprives Asia's garment workers of almost $12bn in wages. *Financial Times*, 19 July, p 4. Available at: www.ft.com/content/22007eb9-440d-48c7-b3dc-fce62c735e1e (accessed 30 July 2022).

Nilsson, P. and Terazono, E. (2020) Can fast fashion's $2.5tn supply chain be stitched back together? *Financial Times*. Available at: www.ft.com/cont ent/62d c687e-d15f-46e7-96df-ed7d00f8ca55 (accessed 17 May 2020).

Nixon, R. (2011) *Slow Violence and the Environmentalism of the Poor*. Cambridge, MA: Harvard University Press.

Ong, A. (1991) The gender and labor politics of postmodernity. *Annual Review of Anthropology*, 20, 279–309.

Parveen, S. (2014) Rana Plaza factory collapse: Survivors struggle one year on. *BBC News*. Available at: www.bbc.co.uk/news/world-asia-27107860 (accessed 11 May 2023).

Perera, A. (2020) Government's extreme measures to contain the garment factory outbreak put workers at a great disadvantage. *ADNASIA*. Available at: https://adnasia.org/ 2020/11/09/in-tatters-sri-lankas-excessive-pandemicresponse/?fbclid=IwA R1L5VRWA62WDVZe9rvIjI5312yp4lihybIas-bwKbebM0TdEHuk-82KEjk (accessed 11 November 2020).

Prentice, R. and De Neve, G. (eds) (2017) *Unmaking the Global Sweatshop: Health and Safety of the World's Garment Workers*. Philadelphia: University of Pennsylvania Press.

Preston, J. and Firth, R. (2020) *Coronavirus, Class and Mutual Aid in the United Kingdom*. London: Palgrave.

Sakamoto, M., Ahmed, T., Begum, S. and Huq, H. (2019) Water pollution and the textile industry in Bangladesh: Flawed corporate practices or restrictive opportunities? *Sustainability*, 11(1951). Available at: sustainability-11-01951.pdf (accessed 20 August 2022).

Shiva, V. (2010) *Staying Alive: Women, Ecology and Survival in India*. Boston: South End Press.

Siddiqi, D. (2017) Before Rana Plaza: Towards a history of labour organizing in Bangladesh's garment industry. In V. Crinis and A. Vickers (eds) *Labour in the Clothing Industry in the Asia Pacific*. London: Routledge, pp 60–79.

Simončič, K. (2021) Fast fashion: A case of social harm and state-corporate crime. *Howard Journal of Crime and Justice*, 60(3), 343–69.

Sollund, R. (2020) The victimisation of women, children and nonhuman species through trafficking and trade: Crimes understood through an ecofeminist perspective. In A. Brisman and N. South (eds) *The Routledge Handbook of Green Criminology*, 2nd edn. Abingdon: Routledge, pp 512–28.

Strumpell, C. and Ashraf, H. (2021) Of 'nimble fingers' and 'jacquard's soldiers': Up-scaling, up-skilling, and the re-masculinization of labor in Bangladesh's garment. Available at: https://trafo.hypotheses.org/30448 (accessed 23 June 2022).

Sumon, M., Shifa, N., and Gulruk, S. (2017) Discourses of compensation and the normalization of negligence: The experience of the Tazreen factory fire. In G. De Neve and R. Prentice (eds) *Unmaking the Global Sweatshop: Health and Safety of the World's Garment Workers*. Philadelphia: University of Pennsylvania Press.

Sustain Your Style (2022) What's wrong with the fashion industry. Available at: www.sustainyourstyle.org/en/whats-wrong-with-the-fashion-industry (accessed 11 May 2023).

Thanhauser, S. (2022) *Worn: A People's History of Clothing*. London: Allen Lane.

Thushari, C. (2020) Condemn the uncivilized media. *YouTube*. Available at: www.youtube.com/watch?v=Tb7d4fDAC6E&fbclid=IwAR2pltI abAdHuF_uPtj2aNWUKM_MYJtdnCPKUfGyk3L-o8F02CyDWt8URPE (accessed 15 December 2020).

Turner, B.A. (1976) The organizational and interorganizational development of disasters. *Administrative Science Quarterly*, 21, 378–97.

Udara, S.P., Arachchige, R., Vithanage, K.D, Wadanambi, R. and Wan, T. (2019) Environmental impacts of textile industry in Sri Lanka. *International Journal of Scientific and Technology Research*, 8(9), 251–3.

Udayakumar, S.P. (1995) The futures of the poor. *Futures*, 267(3), 339–51.

Vogue Business (2022) The Bangladesh accord is set to expire: here's what's at risk. *Vogue Business*. Available at: www.voguebusiness.com/sustainability/the-bangladesh-accord-is-set-to-expire-heres-whats-atrisk#:~:text=The%20accord%2C%20put%20in%20place,why%20brands%20are%20pursuing%20alternatives.&text=An%20agreement%20that%20led%20to,after%20less%20than%20a%20decade (accessed 25 May 2022).

War on Want (2020) Garment workers on the frontline of the pandemic: Outbreak in Sri Lanka. *War on Want*. Available at: https://waronw ant.org/news-analysis/garment-workers-frontline-pandemicoutbreak- sri-lanka (accessed 24 May 2022).

Wonders, N. and Danner, M. (2015) Gendering climate change: A feminist criminological perspective. *Critical Criminology*, 23(4), 401–16.

Wright, M. (2007) *Disposable Women and Other Myths of Global Capitalism*. London: Routledge.

9

Green Victims of the International Waste Industry: An Analysis from a Gender Perspective

María-Ángeles Fuentes-Loureiro

Introduction

The waste management industry has undergone a strong internationalisation process in recent decades. As a result, an export-import flow from countries of the Global North to specific regions and countries of the Global South has been established, which externalises waste management processes and, by extension, the risks and consequences. In many cases, receiving areas are not equipped with appropriate means for handling the waste and protecting both the population and the environment from the severe effects that may be caused by waste pollutants. Thus, despite the efforts of international institutions to prevent those hazards, the transfer of waste from the Global North to certain areas of the Global South still causes great ecological damage and has a major impact on the quality of life and health of the population of the receiving areas.

The international transfer of waste has been extensively studied, especially within the fields of green criminology and green victimology. These disciplines have addressed in depth a wide range of related issues, such as the dynamics of the criminal phenomenon, its relationships with criminal organisations and companies, its consequences from a perspective of ecological justice, social justice and species justice. However, they have overlooked gender.

In general terms, women experience green victimisation differently than men (Lynch, 2018: 408) and victimisation arising from waste crime is not an exception. The way in which waste crime affects women differs from the way in which it affects men for several reasons. First, the social dynamics and gender roles are also reflected in some irregular waste industries, causing different forms of victimisation for men and women. As long as their roles and participation in the waste industries are different, their degree of victimisation also varies. Second, the exposure to pollutants causes different illnesses among men and women. In this sense, the exposure to waste pollutants might have special implications for women's health, especially in terms of reproductive diseases and maternity (Dolk and Vrijheid, 2003; García-Pérez et al, 2015; Kresovich et al, 2019; Melody et al, 2020; Kanner et al, 2021).

The aim of this chapter is to critically analyse waste crime victimisation from a gender perspective in the context of the international waste industry. This will involve making visible the different forms of victimisation of men and women resulting from the transfer and irregular management of waste in the Global South. However, it is not possible to analyse the consequences of waste crime for women on an overall level, as each type of industry and each region has its own characteristics. For that reason, two of the most striking cases of irregular waste management activities in the Global South were selected for analysis: the ship breaking industry and the e-waste irregular industry.

For this purpose, in the first part of the chapter, I explain how the international waste industry has emerged as the ideal context for waste crime. It is only by understanding the dynamics and working of the international waste industry that is it possible to understand waste crime and its characteristics. These characteristics are also explained in the first part of this chapter, where I give a general overview of how this criminal phenomenon works. On this basis, the second part of this chapter analyses the waste crime consequences focusing on two case studies: the ship breaking industry and the e-waste recycling irregular dumps in the Global South. To do so, I explore the literature and other secondary sources related to three consequences: environmental damage, human victimisation and social harm. As a result, this chapter elucidates the different forms of green victimisation experienced by women and men in the context of waste crime and reflects on the social harm and inequalities generated and maintained by the international waste industry.

International waste industry as a perfect scenario to waste crime

This section illustrates the international waste industry as a setting for irregular waste management activities. Thus, the first subsection provides an overview of how the waste phenomenon became an international environmental problem promoted by the countries of the Global North but whose consequences mainly affect countries in the Global South. The second subsection explains how waste crime is carried out in the context of the global waste industry.

International waste industry dynamics

The enormous amount of waste annually generated may be subjected to treatment and disposal procedures that are more or less complex depending on the legislation of each region or country. In many cases, these activities must be carried out in an environmentally sound manner, complying with the regulation created to protect the environment and ensure the safety of citizens. It is in this context that the waste management industry has emerged as an important sector, both economically and socially.

There are no data on the monetary and social impact of the international waste industry on a global level. However, the data available regarding the European Union (EU) shines some light on the size of this industry. Researchers estimate that the waste management sector in the EU generates around 100 billion euros per year (Sahramäki et al, 2016: 2). Likewise, the European Environment Agency has pointed out that the waste management sector in the EU creates more than 1 million jobs (2021: 24).

A few decades ago, the waste management industry was limited to operations within the countries where the waste was produced, mainly countries of the Global North. However, between the 1970s and 1980s, there was an increase in the control and regulation of waste treatment activities, so that undertaking this activity within the country became increasingly costly. Waste disposal, is, in the end, a business, and waste-management corporations are profit-oriented (Lynch, 2017: 473). Therefore, in light of the tightening conditions for waste disposal in the countries that generated it, other ways of getting rid of it were sought.

Since the 1980s, we have been witnessing the increasing internationalisation of waste management activities. Treatment and disposal activities are no longer limited to the territorial boundaries of the state in which the waste is produced, but exporting waste to other countries has become the general rule. In this context, toxic colonialism (Puckett, 1992) emerges, reproducing the colonial dynamics of dependency and exploitation, now in the field of waste (Panjabi, 2010: 420). Indeed, a large part of the waste generated in countries of the Global North is transferred to economically deprived countries in the Global South, where both legal and illegal waste management plants are set up. In this way, the risks of waste management are externalised, transferring the consequences of these activities, and their victims, to third countries (White, 2011: 77).

The concepts of Global North and Global South that I use here are circumscribed within the more general concept of 'world–system' or 'world–economy', developed by Wallerstein since the 1970s (Wallerstein, 1974, 1983,

1991, 2004; Arrighi et al, 1989). This scheme reflects the dynamics of the economic system and the hierarchy in the distribution of tasks that exists in our world. Those tasks that require higher professional qualifications and more capital are carried out in the Global North. Conversely, the Global South is home to natural resource exploitation activities and activities that are 'undesirable' in the Global North. In this sense, the 'dirty industries' (Jänicke et al, 1997; Grether and De Melo, 2003) are mainly located in the countries of the Global South. Dirty industries are those that have a high degree of environmental impact, even if they are carried out with the necessary systems to minimise that impact. These include waste treatment and disposal activities.

Waste crime and its characteristics

With the aim of minimising the transnational shipment of toxic waste and ensuring that waste management activities are carried out in an environmentally sound manner, the Basel Convention on the Control of Transboundary Movements of Hazardous Waste was enacted in 1989. Later amended in 1995, it remains the global standard for waste shipments, although a few countries, notably the United States, have not signed it. With this regulation, transnational movements of waste are subject to a complex system of notifications and authorisations designed to ensure that waste is treated and disposed of while taking safety measures to protect the ecosystems and the population of the receiving areas (Cubel Sánchez, 2001). However, the Basel Convention has not succeeded in making international waste management completely safe for the environment and population. Loopholes in the established system, the lack of uniformity in environmental protection concepts and standards between signatory countries and economic interests have made waste crime one of the major forms of environmental crime (BAN and SVTC, 2002: 28–33; Pellow, 2008: 230).

There are different ways of committing waste crime. On the one hand, from a strictly legal point of view, waste crime can be considered to be a breach of the legal system established mainly in the Basel Convention, although each country or region will have its own regulations in this respect (Clapp, 2002). On the other hand, there are also completely illegal shipments, carried out clandestinely, bypassing the entire system and measures of the Basel Convention (Suvantola et al, 2017: 33–5). Likewise, we must not forget shipments that are technically legal, in practice cause pollution and damage to ecosystems, animals and plants, and to the health of human citizens. In this sense, waste crime is understood to be waste shipments or waste management activities carried out illegally, as well as activities carried out within the legal waste industry, but which cause environmental or social damage (Fuentes-Loureiro, 2017: 335–7).

The percentage of waste management business conducted illegally at the global level is unknown. However, again, the data available for the EU can help with initial estimations. Within the EU waste management industry there is a relatively high percentage of business that is conducted illegally. In this regard, a study carried out by Suvantola et al (2017: 21–2) revealed that an average of 13 per cent of non-hazardous waste generated between 2010 and 2014 disappeared from the legal market in EU member states. The figures for hazardous waste are even more alarming. The study indicates that on average 33 per cent of hazardous waste generated in the EU between 2010 and 2014 ended up in illegal industry.

Illegal waste treatment and disposal activities have three main characteristics, explained in the following paragraphs: the use of organised criminal networks; the absence of proper authorisations and licences; and the use of counterfeit documentation (Ruggiero and South, 2013: 366; Ruggiero, 2015: 80; Suvantola et al, 2017: 34).

Illegal waste management is often carried out within the framework of criminal networks or organisations engaged in illicit activities. Organised crime in general, and in the field of waste management in particular, is characterised by the use of violence and corruption as a means of securing its activities (Peluso, 2015: 15). Thus, violence is used to impose the organisation's will and to pressure the officials and authorities involved in the process – from the initial waste producer to intermediate actors such as transporters, employees of treatment facilities, and so on. Moreover, corruption is used to infiltrate waste economics and politics, so that the organisations secure control of the industry (Peluso, 2015: 16). In this way, the phenomenon of illegal waste management in the sphere of organised crime not only causes damage to the environment and people's health, but it also creates an environment of violence, monopoly and extortion.

The role of large companies and organisations as actors in the field of waste crime should also be highlighted. Indeed, due to the nature and scope of the activities of certain entities, their activities are more likely to cause significant environmental damage than those of private individuals. Ruggiero and South (2013) use the term 'dirty collar crime' to denote those offences related to waste disposal and treatment that are committed at the corporate level (see also Ruggiero, 2015: 80).

Regarding the means used to ship waste irregularly, forgery stands out as a common tactic used to violate the Basel Convention's system of authorisations and licences. In other cases, authorisations or licences to carry out treatment and disposal procedures or waste shipments simply do not exist (Suvantola et al, 2017: 34). Indeed, the system for transnational shipments of waste becomes more complicated as the number of states involved increases. Since authorisations have to be issued not only by the authorities of the states of origin and destination, but also by all transit states, the system is highly susceptible to forgery and corrupt practices.

Consequences of waste crime

As described already, much of the waste generated in the Global North is exported to the Global South for treatment or disposal. This flow has led to waste-receiving areas becoming the dumping ground of the planet, with all the subsequent environmental degradation and exposure of their ecosystems and population to the inherent risks in waste treatment and disposal procedures. The following subsections focus on the consequences of the two case studies – ship breaking and e-waste recycling industries – from a threefold perspective. To start, the first subsection addresses the damage to ecosystems in the waste-receiving areas. The second subsection offers an analysis of human victims from a gender perspective. Finally, the third subsection concludes with a reflection on damages of those activities in terms of social harm.

Environmental damage and destruction of ecosystems

Despite the efforts of the signatory states to the Basel Convention to ensure that waste treatment and disposal procedures are carried out in appropriate facilities and in an environmentally sound manner, criminal networks involved in waste activities manage to circumvent legal barriers and carry out waste shipments without following authorisation and prior consent procedures (White, 2011: 77). As a consequence, in many cases, the recipients of waste are not equipped with adequate means to handle the waste and to protect the environment from the severe effects that the substances in it can cause (Clapp, 1994: 505). Thus, many irregular shipments of waste result in the destruction of natural ecosystems in the receiving areas.

A clear example of outsourced waste management activities that lead to the destruction of ecosystems in the receiving areas is ship dismantling. Several locations in the Global South, namely Bangladesh, China, India, Turkey and Pakistan, receive hundreds of obsolete ships every year to be scrapped and disposed of on their beaches (Sonak et al, 2008: 144–5). These operations are carried out in a rudimentary manner, using the *beaching method*, which consists of grounding the ships on the sandy beaches and dismantling them there, with very few measures in terms of pollution prevention (Watkinson et al, 2016). It is common for scrapped ships to contain hazardous materials that pollute the sea and the beach surroundings, such as paint chips containing heavy metals, bilge and ballast water, glass wool, asbestos, oily sludge/rags/ sands, radioactive wastes as well as gases such as acetylene, halon, and so on (Anand et al, 2014: 161). Likewise, the resulting waste from the dismantling process is often recycled or sold, but it is not uncommon for it to be disposed of in an irregular manner in the immediate vicinity, buried without any measures or treatment to prevent soil contamination (Neşer et al, 2008: 354). These activities have left a disproportionately large environmental footprint in the ecosystems in the areas of Alang, in India, Aliaga, in Turkey and Sitakunda, in Bangladesh. In this respect, several studies have analysed the degree of degradation of the ecosystems surrounding the ship breaking yards in these localities and have shown high levels of maritime and air pollution that have persisted since the early 2000s (Tewari, 2001; Sponza and Karaoğlu, 2002; Vardar and Harjono, 2005; Watkinson et al, 2016; Kutub et al, 2017; Hasan et al, 2020).

Analysing the victims of waste crime from a gender perspective

Irregular waste management activities also have significant consequences for human life and health. Many of the irregular waste management industries in the receiving areas of the Global South are not equipped with adequate means to handle the waste, exposing people to waste pollutants and their severe effects. In this regard, workers are the most exposed to harm, both because of their direct contact with pollutants and because of poor working conditions in many of the irregular waste dumps and plants. However, the harms of irregular waste management also extend to the population in the surrounding areas.

Ship breaking

Returning to the example of ship breaking through the beaching method, dismantling and recycling activities traditionally took place without protection measures for workers' health and safety (Neşer et al, 2008: 354; Gunbeyaz et al, 2019). These activities put workers' health and integrity at serious risk, as they are directly exposed to the remaining materials of the ship's cargo and to the materials of the ship itself. In addition, due to the rudimentary way in which workers dismantle ships, manually, using physical force and generally without protective measures, workers are also at high risk of accidents such as falls, fires, explosions, blows, and so on (Sonak et al, 2008: 144; Chang et al, 2010: 1390–1; Zakaria et al, 2012: 96).

To prevent illness and accidents, an awareness-raising and training campaign for workers has been initiated in the ship breaking yard of Aliaga, Turkey. Furthermore, some security measures have been implemented, such as the implementation of personal protective equipment for workers, including gloves, gumboots, goggles and masks (Anand et al, 2014: 161). However, such occupational safety measures are still neglected in other locations, such as in Bangladesh (Gunbeyaz et al, 2019).

The study of the gender profile of the shipyard workers shows that there are no women among the workers. Only men work on the waste disposal beaches manually dismantling ships (Kutub et al, 2017: 41). As a consequence, men suffer from a greater victimisation in this field, since they are more exposed than women to the harm inherent in this activity, as they are directly exposed to the pollutants and the accidents resulting from the manual dismantling of the ships. The young age of the workers is also noteworthy. In the ship breaking yard of Sitakunda, workers over 30 years of age only account for 10 per cent of the total and most of the workers are between 19 and 22 years old (41.75 per cent of the total) (Kutub et al, 2017: 40–1). In addition, around 11 per cent of the labour force is male children and minors under 18 years old (Kutub et al, 2017: 40–1). Therefore, we can conclude that men and children have a higher victimisation than women in the ship breaking yards.

However, human consequences of ship dismantling are not limited to the risks and damage of the yard workers. The ship waste recycling network extends into the immediate vicinity, where dumping grounds of ship breaking remainings are established (Kutub et al, 2017: 39–40). In these facilities workers are generally women, who work choosing and separating recyclable elements without protective measures and exposed to the pollutants of the discarded materials. Therefore, women are not entirely exempt from the harm arising from ship breaking facilities.

Apart from the damage to people working in the yards and surrounding facilities, the surrounding population remains exposed to a polluted environment. Air pollution exposure and incidental ingestion of contaminated particles through the consumption of drinking water and food from the impacted area are some of the risks (Islam et al, 2022). Several studies show a high level of organic pollutants and metal levels in the water, soil and even food in nearby areas, with a subsequent risk to the health of the local inhabitants. Indeed, populations around ship recycling yards have reported high levels of various hazardous substances in their bodies (Zakaria et al, 2012; Islam et al, 2022) and are at average risk of developing different types of cancer (Nøst et al, 2015). In this case, however, no gender differences were found.

E-waste recycling

Another example of the great impact of irregular waste management on human health is the uncontrolled dumping sites of e-waste established in Asian and African countries, especially in some regions of Nigeria, Benin, Ghana, India, Vietnam and Pakistan (Basel Convention, 2012). Every year, these places receive tons of electronic waste, such as household appliances, mobile phones or obsolete computers. As a result, an irregular e-waste recycling industry has developed in several locations, in which the local population has been carrying out activities highly hazardous to their health for decades without any safety measures (BAN and SVTC, 2002; Greenpeace, 2009; World Health Organization, 2019). Such activities are often carried out without basic protective equipment such as masks, gloves and even appropriate footwear (Hull, 2010: 3; Lundgren, 2012: 18; Bisschop, 2015: 66–7; Park, 2019).

The situation in the informal e-waste recycling sector is different than the situation in ship breaking yards, since many of the workers in e-waste irregular dumps are traditionally women and children (BAN and SVTC, 2002: 16). Furthermore, in e-waste dumps there is a gender-based distribution of labour that victimises women more than men. Work in e-waste dumps includes hazardous activities ranging from manual searching for reusable materials among tons of electronic equipment to even more harmful activities, such as open air burning of components to recover valuable metals inside them. In this context, women take on the most undesirable and dangerous tasks, in particular, the use of acid baths to search for precious metals (McAllister et al, 2014: 172).

The discarded devices in e-waste dumps contain high levels of dangerous substances, such as lead in electronic circuits, beryllium in motherboards, brominated compounds in plastic cases, chlorofluorocarbons in cooling appliances, and lithium in batteries. As a result of the exposure to those materials, workers develop serious diseases such as silicosis and other respiratory diseases, skin lesions and cancer (BAN and SVTC, 2002; Hull, 2010: 8–15; Grant et al, 2013; Purchase, 2017). Among the health consequences of working in irregular waste dumps stand those related to the reproductive system, which places women in a particularly vulnerable situation. In this sense, several studies have shown that cancers of the reproductive system – such as ovarian, uterine and breast cancer – menstrual cycle irregularities or endometriosis are some common consequences of exposure to e-waste components (Hull, 2010: 13; McAllister et al, 2014: 172).

Regarding damage to the health of the population exposed to e-waste pollutants, several studies conducted in localities and regions where irregular dumpsites are located show reproductive complications for women. Indeed, the polluting materials that remain in women's bodies after exposure to e-waste pollutants are often passed on to their children. Elevated levels of various substances such as lead, dechlorane or organohalogen compounds – including polychlorinated biphenyls and brominated flame retardant Polybrominated Diphenyl Ethers – have been detected in the blood, breast milk, maternal serum, umbilical cord blood and placenta of women residing in localities where irregular e-waste dumps are located (BAN and SVTC, 2002: 9; Hull, 2010: 13–14; Leung et al, 2010; Devanathan et al, 2012; Xu et al, 2012; Ben et al, 2013). In addition, more birth complications have also been found in women exposed to e-waste pollution than in those who are not. Births in locations near e-waste disposal plants showed significantly higher rates of adverse birth outcomes, including low birth weight, low birth weight at term, lower Apgar scores – an examination made immediately after birth that assesses five aspects: colour, heart rate, reflexes, muscle tone and respiration – and a four times higher risk of stillbirth (Xu et al, 2012; Grant et al, 2013). Furthermore, it has also been found that pregnant women living near e-waste recycling centres are more likely to suffer from spontaneous abortion (Chan et al, 2007; Leung et al, 2008; Grant et al, 2013).

Consequences of waste crime in terms of social harm and inequalities

In the field of critical criminology, Hillyard and Tombs (2004, 2017) identify social harm as a wide range of events and conditions that affect people during their lives. The authors argue that social harm can manifest as physical harm, such as death or serious physical harm; as financial or economic harm, such as poverty or loss of purchasing power; as emotional or psychological harm; and even as cultural harm, such as restrictions on human autonomy or access to culture or information. Hillyard and Tombs explicitly include exposure to pollutants as a form of social harm in the form of physical harm to individuals (Hillyard and Tombs, 2004: 19). Such harms were discussed and analysed in depth in the previous section. However, irregular waste management can manifest into many types of social harm, affecting society as a system. In this regard, the following paragraphs discuss how the transfer of waste to countries in the Global South has negative consequences for society in terms of both economics and equality.

As noted, the waste management industry is an important sector that generates high economic benefits. Waste management companies obtain these revenues from different sources. On the one hand, these profits come from the price that waste generators pay waste managers to dispose of their waste. On the other hand, through the sale of by-products or recycled materials resulting from recycling processes, such as metals or plastics or energy generated through organic waste recovery processes. Therefore, illegal waste management behaviour generates social harm in the sense developed by Hillyard and Tombs (2004: 19–20), insofar as it causes financial losses to the people professionally engaged in these activities.

Likewise, illicit waste activities, especially the shipment of waste to third countries in the Global South for treatment or disposal, cause social harm in terms of inequality. This flow of exports-imports from the Global North to the Global South is a clear manifestation of the dynamics of the global economic system and contributes to maintaining the existing inequalities between the two areas. These differences, as discussed in what follows, are not only economic, but also materialise in 'access to education, health care, adequate food and shelter, effective political institutions and safe and secure living environments' (Carrington et al, 2016: 7).

First, in terms of the means of securing waste crime activities, the use of corrupt means, such as bribes to designated licensing authorities, influences the effectiveness and trust of political institutions. Similarly, the use of coercive means, threats and other types of violence used against operators involved in the waste shipment process compromise the safety of the living environments of people living in the areas of origin, transit and destination of the waste (Peluso, 2015: 15). Therefore, in this sense, illegal waste activities weaken public institutions and safety. Second, irregular waste management industries located in the Global South worsen the living conditions of workers and the general population (Kutub et al, 2017). Waste recycling activities carried out without environmental protection measures, as noted previously, pollute the ecosystems surrounding the facilities. This leads to impoverished living conditions of workers and the general population as well as problems related to the ingestion of contaminated particles through the consumption of drinking water and food from the impacted area (Islam et al, 2022). Third, insofar the existence of child labour has been confirmed in the two waste management industries analysed (BAN and SVTC, 2002: 16; Kutub et al, 2017: 40–1), we can maintain that the transfer of such waste to countries in the Global South leads to inequalities related to the right to the development of the personality and the right to education in childhood and adolescence.

To conclude, waste shipments to countries in the Global South generate environmental degradation and social harm, not only in terms of damage to people's health, but also in economic terms and the maintenance of inequalities between different areas of the planet, in terms of economic, health, work, education, trust in institutions and security.

Concluding reflections

The waste management industry has experienced a process of internationalisation in the last decades. As a result, nowadays, an increasing amount of waste of all kinds is shipped to low-income countries for treatment and disposal procedures. This export-import flow of waste is produced mainly from the countries of the Global North to countries of the Global South. This shift externalises the consequences of waste management to the most economically deprived countries, which have to deal with environmental degradation and health damage to their population.

Despite international legal attempts to ensure the safety of waste management activities in all countries, many waste-receiving areas lack the measures to manage waste in an environmentally sound manner and to protect human health. As a result, entire communities suffer the consequences of irregular waste management activities, which are not just limited to a major environmental degradation, but also translate into social damage in terms of poor working conditions, social exclusion and serious health problems of the local population. In this context, there are significant differences in the degree of victimisation depending on gender. However, it is not possible to draw general conclusions in this regard since the distinction on the grade of victimisation between men and women depends on the kind of waste under consideration.

In the case of ship breaking industry, men and teenagers suffer from a greater victimisation than women, as only they work directly in the ship breaking yards, manually dismantling the vessels. Although the entire population is exposed to pollution from this activity, yard workers are more directly exposed to hazardous waste and accidents resulting from the rudimentary way in which the ships are scrapped.

On the contrary, the irregular e-waste management industry victimises women to a greater extent. In e-waste irregular dumps there is a division of tasks in which women traditionally take on the most arduous and dangerous activities, exposing them to greater health risks. Diseases and complications related to the reproductive system are one of the most common consequences of the exposure to e-waste pollutants, which puts women in a particularly vulnerable situation. In addition, pregnant women in contact with e-waste have been shown to transmit hazardous materials in their bodies to their foetus and are more likely to suffer birth complications and miscarriages. In the end, both mothers and children face an increased health risk with damaging consequences for future generations.

References

Anand, M.H., Sachin, K.P., Dinesh, K. and Shyam, R.A. (2014) Ecological engineering, industrial ecology and eco-industrial networking aspects of ship recycling sector in India. *APCBEE Procedia*, 10, 159–63.

Arrighi, G., Hopkings, T.K. and Wallerstein, I. (1989) *Antisystemic Movements*. London: Verso.

BAN (Basel Action Network) and SVTC (Silicon Valley Toxic Coalition) (2002) *Exporting Harm: The High-Tech Trashing of Asia*.

Basel Convention (2012) *Where are WEee in Africa? Findings from the Basel Convention E-waste Africa Programme*. Available at: www.researchgate. net/publication/261242977_Where_are_WEEE_in_Africa (accessed 16 May 2023).

Ben, Y.J., Li, X.H., Yang, Y.L., Li, L., Di, L., Wang, W.Y., Zhou, R.Z., Xiao, K., Zheng, M.Y., Tian, Y. and Xu, X.B. (2013) Dechlorane Plus and its dechlorinated analogs from e-waste recycling center in maternal serum and breast milk of women in Wenling, China. *Environmental Pollution*, 173, 176–81.

Bisschop, L. (2015) *Governance of the Illegal Trade in E-Waste and Tropical Timber: Case Studies on Transnational Environmental Crime.* Farnham: Ashgate.

Carrington, K., Hogg, R. and Sozzo, M. (2016) Southern criminology. *British Journal of Criminology*, 1(56), 1–20.

Chan, J.K.Y., Xing, G.H., Xu, Y., Liang, Y., Chen, L.X., Wu, S.C., Wong, C.K.C., Leung, C.K.M. and Wong, M.H. (2007) Body loadings and health risk assessment of polychlorinated dibenzo-p-dioxins and dibenzofurans at an intensive electronic waste recycling site in China. *Environmental Sciences & Technology*, 41, 7668–74.

Chang, Y.C., Wang, N. and Durak, O.S. (2010) Ship recycling and marine pollution. *Marine Pollution Bulletin*, 60, 1390–6.

Clapp, J. (1994) The toxic waste trade with less-industrialized countries: Economic linkages and political alliances. *Third World Quarterly*, 3(15), 505–18.

Clapp, J. (2002) Seeping through the regulatory cracks. *SAIS Review*, 22(1), 141–55.

Cubel Sánchez, Z.P. (2001) *Comercio Internacional de residuos peligrosos (La regulación internacional de los movimientos transfronterizos de desechos peligrosos).* Valencia: Tirant lo Blanch.

Devanathan, B., Subramanian, A., Sudaryanto, A., Takahashi, S., Isobe, T. and Tanabe, S. (2012) Brominated flame retardants and polychlorinated biphenyls in human breast milk from several locations in India: Potential contaminants sources in a municipal dump site. *Environment International*, 39, 87–95.

Dolk, H. and Vrijheid, M. (2003) The impact of environmental pollution on congenital anomalies. *British Medical Bulletin*, 68, 25–45.

European Environment Agency (2012) *Movements of Waste across the EU's Internal and External Borders*, report no 7. Available at: www.eea.europa.eu/publications/movements-of-waste-EU-2012 (accessed 16 May 2023)

Fuentes-Loureiro, M.A. (2017) El traslado trasnacional de residuos peligrosos: origen y evolución normativa. *Rivista Giuridica dell'Ambiente*, 2, 319–46.

García-Pérez, J., Lope, V., López-Abente, G., González-Sánchez, M. and Fernábdez-Navarro, P. (2015) Ovarian cancer mortality and industrial pollution. *Environmental Pollution*, 205, 103–10.

Grant, K., Goldizen, F.C., Sly, P.D., Brune, M.N., Neira, M., Van der Berg, M. and Norman, R.E. (2013) Health consequences of exposure to e-waste: A systematic review. *The Lancet Global Health*, 1(6), 350–61.

Greenpeace (2009) Greenpeace undercover operation exposes illegal dumping of e-waste in Nigeria. *Greenpeace*. Available at: http://africa.gm/africa/nigeria/article/greenpeace-undercover-operation-exposes-illegal-dumping-of-e-waste-in-nigeria (accessed 16 May 2023).

Grether, J. and De Melo, J. (2003) Globalization and dirty industries: Do pollution havens matter? *National Bureau of Economic Research*, 9776.

Gunbeyaz, S.A., Kurt, R.E. and Baumler, R. (2019) A study on evaluating the status of current occupational training in the ship recycling industry in Bangladesh. *WMU Journal of Maritime Affairs*, 18(1), 41–59.

Hasan, A.B., Reza, A.H.M.S., Kabir, S., Siddique, M.A.B., Ahsan, M.D. and Akbor, M.A. (2020) Accumulation and distribution of heavy metals in soil and food crops around the ship breaking area in southern Bangladesh and associated health risk assessment. *SN Applied Sciences*, 2.

Hillyard, P. and Tombs, S. (2004) Beyond criminology? In P. Hillyard, C. Pantazis, S. Tombs and D. Gordon (eds) *Beyond Criminology: Taking Harm Seriously*. London: Pluto Press, pp 10–29.

Hillyard, P. and Tombs, S. (2017) Social harm and zemiology. In A. Liebling, S. Maruna and L. McAra (eds) *The Oxford Handbook of Criminology*, 6th edn. Oxford: Oxford University Press, pp 284–305.

Hull, E.V. (2010) Poisoning the poor for profit: The injustice of exporting electronic waste to developing countries. *Duke Environmental Law & Policy Forum*, 21, 1–48.

Islam, M.N., Ganguli, S., Tanvir, E.M., Rifat, M.A.H., Saha, N., Peng, C. and Jack, C.N. (2022) Human exposure assessment of mixed metal/loids at and near mega-scale open beaching shipwrecking activities in Bangladesh. *Exposure and Health*, 15, 69–84.

Jänicke, M., Binder, M. and Mönch, H. (1997) Dirty industries: Patters of change in industrial countries. *Environmental and Resource Economics*, 9, 467–91.

Kanner, J., Pollack, A.Z., Ranasinghe, S., Stevens, D.R., Nobles, C., Rohn, M.C.H., Sherman, S. and Mendola, P. (2021) Chronic exposure to air pollution and risk of mental health disorders complicating pregnancy. *Environmental Research*, 196.

Kresovich, J.K.K., Erdal, S., Chen, H.Y., Gann, P.H., Atgos, M. and Rauscher, G.H. (2019) Metallic air pollutants and breast cancer heterogeneity. *Environmental Research*, 177.

Kutub, M.J.R., Falgunee, N., Nawfee, S.M. and Rabby, Y.W. (2017) Ship breaking industries and their impacts on the local people and environment of coastal areas of Bangladesh. *HSS*, 6(2), 35–58.

Leung, A.O.W., Duzgoren-Aydin, N.S., Cheung, K.C. and Wong, M.H. (2008) Heavy metals concentrations of surface dust from e-waste recycling and its human health implications in southeast China. *Environmental Science and Technology*, 42(7), 2674–80.

Leung, A.O.W., Chan, J.K.Y., Xing, G.H., Xu, Y., Wu, S.C., Wong, C.K.C., Leung, C.K.M. and Wong, M.H. (2010) Body burdens of polybrominated diphenyl ethers in childbearing-aged women at an intensive electronic e-waste recycling site in China. *Environmental Science and Pollution Research*, 17(7), 1300–13.

Lundgren, K. (2012) *The Global Impact of e-waste: Addressing the Challenge*. International Labour Office Publications.

Lynch, M.J. (2017) Conceptualizing green victimisation, green criminology and political economy: A reply. *Critical Sociology*, 43(3), 473–8.

Lynch, M.J. (2018) Acknowledging female victims of green crimes: Environmental exposure of women to industrial pollutants. *Feminist Criminology*, 13(4), 404–27.

McAllister, L., Magee, A. and Hale, B. (2014) Women, e-waste and technological solutions to climate change. *Health and Human Rights*, 16(1), 166–78.

Melody, S.M., Wills, K., Knibbs, L.D., Ford, J., Venn, A. and Johnston, F. (2020) Maternal exposure to ambient air pollution and pregnancy complications in Victoria, Australia. *International Journal of Environmental Research and Public Health*, 17(7).

Neşer, G., Ünsalan, D., Tekoğul, N. and Stuer-Lauridsen, F. (2008) The shipbreaking industry in Turkey: Environmental, safety and health issues. *Journal of Cleaner Production*, 16, 350–8.

Nøst, T.H., Halse, A.K., Randall, S., Borgen, A.R., Schlabach, M., Paul, A., Rahman, A. and Breivik, K. (2015) High concentrations of organic contaminants in air from ship breaking activities in Chittagong, Bangladesh. *Environmental Science & Technology*, 49, 11372–80.

Panjabi, R.K.L. (2010) The pirates of Somalia: Opportunistic predators or environmental prey? *William and Mary Environmental Law and Policy Review*, 2(34), 377–491.

Park, M. (2019) Electronic waste in recycled in appalling conditions in India. *The Conversation*. Available at: http://theconversation.com/electro nic-waste-is-recycled-in-appalling-conditions-in-india-110363 (accessed 16 May 2023).

Pellow, D.N. (2008) The global waste trade and environmental justice struggles. In K.P. Gallagher (ed) *Handbook on Trade and the Environment*. Cheltenham: Edward Elgar, pp 225–33.

Peluso, P. (2015) Dalla terra dei fuochi alle terre avvelenate: lo smaltimento illecito dei rifiuti in Italia. *Rivista di Criminologia, Vittimologia e Sicurezza*, 9(2), 13–30.

Puckett, J. (1992) *Toxic Trade Update*. Greenpeace, No. 5.2, Second Quarter 1992.

Purchase, D. (2017) Electronic waste: An emerging global environmental and health challenge in the 21st century. Presentation at *SmART Cities and Waste Research Workshop. Urban Waste Streams & Flows Workshop*, held in London, 6 April.

Ruggiero, V. (2015) Creative destruction and the economy of waste. In R.A. Sollund (ed) *Green Harms and Crimes: Critical Criminology in a Changing World*. Basingstoke: Palgrave Macmillan, pp 79–96.

Ruggiero, V. and South, N. (2013) Green criminology and crimes of the economy: Theory, research and praxis. *Critical Criminology*, 21, 359–73.

Sahramäki, L., Favarin, S., Mehlbaum, S., Savona, E., Spapens, T. and Kankaaranta, T. (2016) *Crime Script Analysis of Illicit Cross-border Waste Trafficking: Key Findings.*

Sollund, R. (2013) The victimisation of woman, children and non-human species through trafficking and trade-crimes understood through an ecofeminist perspective. In N. South and A. Brisman (eds) *Routledge International Handbook of Green Criminology*. London: Routledge, pp 541–53.

Sonak, S., Sonak, M. and Giriyan, A. (2008) Shipping hazardous waste: Implications for economically developing countries. *International Environmental Agreements: Politics, Law and Economics*, 2(8), 143–59.

Sponza, D. and Karaoğlu, N. (2002) Environmental geochemistry and pollution studies of Aliağa metal industry district. *Environmental International*, 27(7), 541–53.

Suvantola, L., Favarin, S., Mehlbaum, S., Sahramäki, I., Spapens, T., Savona, E. and Kankaaranta, T. (2017) *Blocking the Loopholes for Illicit Waste Trafficking. Final Consolidated Report.*

Tewari, A.A. (2001) The effect of ship scrapping industry and its associated wastes on the biomass production and biodiversity of biota in in situ condition at Alang. *Marine Pollution Bulletin*, 6(42), 461–8.

Vardar, E. and Harjono, M. (2005) *Ships for Scrap V. Steel and Toxic Wastes for Asia, Greenpeace Report on Environmental, Health and Safety Conditions in Aliağa Shipbreaking Yards, Izmir, Turkey.*

Wallerstein, I. (1974) *The Modern Word-system: Capitalist Agriculture and the Origins of European World-economy in the Sixteenth Century*. New York: Academic Press.

Wallerstein, I. (1983) *Historical Capitalism*. London: Verso.

Wallerstein, I. (1991) *Geopolitics and Geoculture: Essays on the Changing World-system*. Cambridge: Cambridge University Press.

Wallerstein, I. (2004) *World-systems Analysis: An Introduction*. Durham, NC: Duke University Press.

Watkinson, R.V., Hossain, M.S., Sharifuzzaman, S.M. and Chowdhury, S.H. (2016) *Evaluation of Environmental Impacts of Ship Recycling in Bangladesh Final Report for International Maritime Organization*. Available at: wwwcdn. imo.org/localresources/en/OurWork/PartnershipsProjects/Documents/ Ship%20recycling/WP1b%20Environmental%20Impact%20Study.pdf (accessed 16 May 2023).

White, R. (2011) *Transnational Environmental Crime: Towards an Eco-global Criminology*. London: Routledge.

Wonders, N.A. and Danner, M.J.E. (2015) Gendering climate change: A feminist criminological perspective. *Critical Criminology*, 23, 401–16.

World Health Organization (2019) *Electronic Waste. Children's Environmental Health*.

Xu, X., Yang, H., Chen, A., Zhou, Y., Wu, K., Liu, J., Zhang, Y. and Huo, X. (2012) Birth outcomes related to informal e-waste recycling in Guiyu, China. *Reproductive Toxicology*, 33(1), 94–8.

Zakaria, N.M.G., Ali, M.T. and Hossain, K.A. (2012) Underlying problems of ship recycling industries in Bangladesh and way forward. *Journal of Naval Architecture and Marine Engineering*, 9, 91–102.

10

The Green Road Project and Women's Green Victimisation in Turkey

Halil Ibrahim Bahar

Introduction

There have been criticisms that green criminology neglects women's green victimisation (WGV); and feminist criminology has paid little attention to the victimisation of women by green crimes (Lynch, 2018). It is argued that 'the specific "green victimization experiences" varies across types/groups of victims, as well as across individuals in the same victim group' (Lynch, 2018: 406). For example, environmental toxins affect children and adults differently. 'Illustrating the importance of gender, medical studies of toxic exposure also indicate that women are affected differently than men' (Lynch, 2018: 406). Taking gender inequalities into consideration, feminist green criminology seeks to address this ignorance and offers an understanding of WGV and conceptualises WGV as part of the broader social structure of gender relationships (Lynch, 2018). To understand the nature and extent of victimisation, feminist green criminology provides the needed empathy towards women victims of green crime.

One instance where such an approach may provide valuable insight is in the case of Turkey. Paving the way for transportation of minerals and other extracted sources between Turkey, Russia and Europe, in the Eastern Black Sea (EBS) region of Turkey, the controversial 2,600km 'Green Road' is planned to be constructed. The project also aims to connect tourism centres throughout the highlands of the provinces of Artvin, Rize, Trabzon, Giresun, Ordu, Gumushane, Bayburt and Samsun in the EBS region; all previously protected as conservation areas and public land. Environmentalists and people in this region are greatly concerned that the project may have a potentially devastating impact on the environment; local people are under threats of green crime from extractive industries. There are fears that the 'Green Road' will cause erosion, forest loss, habitat fragmentation, stream pollution and other ecological destruction. The project also threatens the traditional, seasonal migrations of people who bring their livestock up into the highland pastures to graze each summer (Bayraktar, 2022). These devastating developments cause WGV and put women's livelihoods at risk, ultimately forcibly removing them from their traditional living spaces (Akay, 2020).

Relevant studies in the Turkish context have so far focused only on environmental politics, local opposition movements (Eryılmaz, 2016; Erensu, 2018), climate justice, energy politics (Kaya, 2016; Turhan et al, 2016; Mazlum, 2017; Turhan, 2017), eco-feminism (Seckin, 2016) and other similar issues. No academic study has so far been carried out to address WGV and green crime in the country.

Drawing attention to how the organisation of capitalism has accelerated ecological disorganisation and destruction, green political-economic criminological analysis has given importance to the Treadmill of Production (ToP) theory. The ToP emphasises that 'the accelerated pace of capitalism requires expanded raw material extraction, expanded production and hence increased pollution, and consumption to maintain economic expansion' (Lynch, 2016: 6). The ToP theory is a general model of capitalism's connection to ecological destruction. The relevance to the discussion of WGV and green feminist criminology is that women who are socially and economically marginalised by capitalism are victimised and exploited in the same way that capitalism victimises and exploits nature. Of course, this is true for some men as well. Because capitalism's primary concern is profit, it ignores how the pursuit of profit produces the ecological disorganisation of nature, and how the victimisation of nature feeds back and affects humans, but especially women as marginalised and exploited populations within capitalism (Salleh, 1995, cited in Lynch, 2016). The ToP theory can be used to frame the study of green crime, deviance and victims. One of the negative consequences of industrial production is the emergence of pollution and harmful substances as by-products. The ToP theory is an analogy used to describe how the economy drives society's continuous march towards ecological destruction (Schnaiberg, 1980). For Schnaiberg (1980) in the ToP approach, green crimes are identified by the fact that they cause 'ecological disorganization' (Schnaiberg, 1980). Ecological additions and ecological withdrawals are the two methods of ecological disorganisation. The former refers to adding pollution to the ecosystem; the latter refers to withdrawing resources from nature for production (Lynch, 2020).

Eco-feminists criticise patriarchal profit-oriented capitalist modes of production and they propose subsistence lifestyles as an alternative (see Lynch and Stretesky, 2003). An eco-feminist solution proposes that: 'patriarchy must be replaced with egalitarian forms of social organization in which men and women have equal power, and by a vision of social ecology wherein the natural environment is treated with respect in so far as it should be sustained rather than manipulated and destroyed' (Epstein, 1993: 145; cited in Lynch and Stretesky, 2003: 223). This is possible by reorienting cultural values, returning to local economies and grassroots democracy. A criminological perspective is necessary, however, the attempt to explain green crimes, victimisations and reactions through a narrow criminological lens is insufficient. Orthodox criminology fails to provide an adequate analysis of harm and victimisation in the ecosystem.

Extractive industries are not concerned about the commodification of nature and the damage to humans, other living things and the ecosystem. The use of chemicals in production processes and abandoning waste materials in natural sites increases the destruction of nature and WGV. For Jarrell et al (2013), the ToP theory provides a framework for understanding green crime, WGV and green feminist criminology. Utilising the ToP to analyse WGV, this study argues that although devastating effects are very clearly visible, WGV tends to persist unnoticed in Turkey. If there is a victim of crime, of course, there is also a perpetrator who committed this crime.

Interest-based relationships of companies involved in the Green Road Project with the Justice and Development Party (AKP) are already known. Company owners and the AKP appear not to hide these ecologically destructive and politically, economically and socially unjust ties. The devastating activities of companies against the habitat are not covered by the AKP-sponsored media. Local coverage of habitat destruction, inhabitants' resistance and collective actions are also very limited due to political pressure, fear of losing economic, political or social advantages, or other reasons. For those who bear the risk of challenging the authorities, social media provides different options and platforms to post about loss of habitat, WGV and resistance movements. Women are at the front of all kinds of resistance movements and collective actions.

In general, women's mobilisation is often constrained within cultural contexts limiting them to gendered spaces and roles. Drawing evidence from this research, the question emerges: Why has the burden of participating in collective actions and other resistance movements against WGV been placed mostly on the shoulders of women? This chapter specifically examines the relations between green crime victims, resistance movements, collective actions, the ruling AKP and extractive corporations. This chapter first briefly discusses green feminist criminology, WGV in Turkey, the Green Road

Project, and WGV in the EBS region of Turkey. Before the conclusion, women's resistance and collective actions are examined.

Green feminist criminology

Feminism brings significant perspectives to green crime and fills an important gap in criminology. For example, feminists have brought care theory to the philosophical debate over how humans should treat nonhuman animals (Donovan, 2006). Also, the concept of 'compassionate protection' reveals new ethical values about how people share nature with other living things (Ramp and Bekoff, 2015). For more than three decades a growing body of literature has focused on green criminology. It is the study of harm caused to humanity, the environment and nonhuman animals committed by extractive companies, industrial institutions and also by ordinary people (Lynch, 2020; South and Beirne, 2006; Ruggiero and South, 2013; Sollund, 2013, 2015a, 2015b, 2015c, 2017, 2019; South, 2014; Sollund et al, 2016; Sollund and Brisman, 2017; Beirne et al, 2018; Brisman et al, 2018; Goyes, 2018; White, 2018; Brisman and South, 2019; and many others).

Because of excessive consumption, pollution and other environmental destruction, the earth has entered into a global ecosystem collapse (Sato and Lindenmayer, 2018) and the victims are humans, nonhumans and ecosystems. Under these circumstances, it is argued that corporate crimes and state crimes do more harm than street crime victimisation (Lynch, 2020). Affecting women badly and causing WGV, the ToP damages the environment and raises concerns over deforestation, river damming and loss of habitat (Schnaiberg et al, 2002).

In terms of WGV, and a feminist approach to green crime, there is a gap in the literature. However, some encouraging signs have also been emerging, raising the issue of WGV. As was witnessed during the initial delay in the acceptance of feminist criminology, the green feminist approach to green crime has not yet become widespread. For Sollund (2017: 246), 'conventional criminology is insufficient for understanding the complexities of human–animal relationships and nature, and insights produced by the feminist movement are vital'. Women are most vulnerable to patriarchal and profit-oriented capitalist modes of production, which can give rise to a long list in terms of WGV. It includes the environmentally destructive enterprises that threaten women in terms of physical wellbeing, degradation of agricultural land and forests, and other instances of water being poisoned with chemicals, harmful gases from factories and power plants, the prevalence of counterfeit medicines, and questionable beauty and cosmetic products. Beauty salons, aesthetic surgery clinics and hairdressers engage in activities that threaten women's health (Lynch and Stretesky, 2001).

Although green criminology has drawn attention to the widespread forms of green victimisation, it has neglected WGV (Lynch, 2016). Green feminist criminology has the potential to contribute to understanding, which is currently missing in the criminological literature, just as it is ignored in real life. As happened during the process of the emergence of feminist criminology in the context of domestic violence in the 1970s (Lynch, 2016), now there is a great need for the development of green feminist criminology through WGV. The status of women has not changed in terms of victimisation of crime, since Walklate (1990: 27) proposed that feminist criminology emphasised that women's victimisation resulted from 'the inescapable material reality of women's … relative structural position'.

Green crimes that threaten ecological sustainability and harm people, nonhuman animals and the ecosystem are often not recognised as harmful and remain outside the law in Turkey. Taking patriarchy and widespread gender inequality into consideration in Turkey (Bahar, 2018), green feminist criminology can add significantly to the understanding of WGV in the country. Green feminist criminology has great potential in developing social awareness and resistance to tackle WGV in Turkey. To truly address the root causes and the scope and scale of WGV in Turkey, a green feminist approach can create an opportunity to fight historical oppressions. It also proposes a transformative feminist agenda that puts the leadership of women at the centre.

Methodology

In the absence of sufficient data, analytic methods and, perhaps most importantly, gender-responsive policies and practices, feminist criminology argues that gender-neutral mainstream explanations fail to provide an adequate account for offending and victimisation (Wattanaporn and Holtfreter, 2014). Drawing upon 'cyber ethnography' and the analyses of secondary sources, this study suggests that the incorporation of green and feminist criminology is needed to understand the gendered dynamics involved in WGV, which is rooted in patriarchal power structures. Such a power structure causes green crime, affecting women's social, economic and political survival. An online community has been developed in Turkey among local, national and international environmentalists, residents whose habitat is at risk, journalists, politicians and other interested parties. This community enables people to interact with one another (Hallett and Barber, 2014) and gives researchers opportunities to carry out cyber-ethnographic research. News, photos, videos, opinions and comments about the collective actions of the activists, victims of green crime, women victims of green crime and perpetrators are shared through different news and social media accounts. 'Cyber ethnography' enables the observation of women's green victimisation and political activism of women through the Internet.

Within this non-participant cyber-ethnographic study, the data were collected mainly from tweets of national and international journalists, activists, victims of green crime, online news sites, environmental groups, politicians, as well as videos posted on YouTube and some personal blogs. The intent was that the combination of data would provide a fuller and more comprehensive account. The data were documented through field notes. The study took more than one year, starting from the beginning of July 2020 to August 2021. In this study, the virtual space is used as online fieldwork and has made it possible to 'follow' the victimisation, actions and opinions of environmental organisations, together with the publishing of ideas and interactions between different actors such as the activists, victims and ruling authorities. Other sources include a selection of literature and online newspaper articles, including updated information on recent political developments. Environmental movements, women, nature, the EBS region, the Green Road Project, Carattepe Resistance, women resistance, green crime, eco-feminism and women victimisation were the main search terms that were used in this study. When it comes to ethical considerations, all data were collected from the public domain.

In terms of positionality, I was born in a village in the EBS region of Turkey and spent almost half of my childhood there. I have regularly visited the region for family visits and holidays. I remember how I spent my childhood and remember clearly the traditional way of life in this mountainous, green land with its rainy climate. My position as an 'outsider within' who hails from this area, my memories, human relations, observations and visits played a significant role in conducting this research and collecting the data. Since my family, my relatives and our neighbours in my village have experienced similar life and green victimisation patterns, I now understand what they feel in terms of green crime and WGV. I also understand the nature and extent of WGV and their struggle to cope with WGV. It is argued that there is a two-way relationship between feeling and thinking. We cannot think without feeling; likewise, we can't feel without thinking. As Sollund (2017: 248) argues: 'Feeling is fundamental for thinking and questioning "taken for granted truths". Intuition and sensitivity for positions and suffering of others, whether human or nonhuman, are requirements both for initiating research on oppressed groups and for reporting their experiences.'

Being from the EBS region has facilitated my detailed observations about the traditional family structure of the region, the relationships between men and women, and the roles and status of women in the public and private sphere. My experience with the culture and the geography of the EBS region has enabled me to obtain data from direct and indirect empirical insights. This helped to yield interesting results in revealing and analysing WGV and women's responses. For example, when looking at the reasons, discourses, forms, culture and patterns of solidarity in women's collective actions that are at the forefront of collective actions, or only carried out by women, it can be said that women are actually in an unnamed feminist movement.

The Green Road Project and women's green victimisation in the Eastern Black Sea region of Turkey

The 2,600km 'Green Road' Project will connect the EBS region's plateau to Russia and Europe, paving the way for the transportation of minerals and other extracted resources between Turkey, Russia and European countries. As mentioned, the project also aims to connect tourism centres throughout the once-protected and conserved highlands of the provinces of Artvin, Rize, Trabzon, Giresun, Ordu, Gumushane, Bayburt and Samsun in the EBS region (Canefe, 2016; DOKAP, 2016). Environmentalists and people in this region are greatly concerned that the project may have a potentially devastating impact on the environment. The Green Road and similar infrastructure and environmental projects paved the way for developing resistance movements against environmental harm and other habitat destruction. For example, Carattepe, a district of Artvin located in the northeastern part of Turkey, turned out to be one of the largest protest sites and has become a significant location of the contemporary environmental movement in Turkey (Doğu, 2017). In July 2015, Mrs Rabiye Bekar, known as *Havva Ana* (Mother Eve) by local Carettepe residents, sat down in front of the bulldozers, and she was forcibly dragged along by gendarmerie forces during a protest against the 'Green Road' Project in Carettepe (Hürriyet TV, 2015; Canefe, 2016). She later became a heroine and a symbol of resistance against the project and authorities.

The data obtained from this study suggests that the economies of rural regions, forests or seaside are destroyed by ToP by means of hydroelectric and coal power plants, mining, mass tourism, construction of mass housing estates and other extractive and infrastructure work resulting in habitat destruction, water and air pollution, deforestation and other forms of destruction. Losses of biodiversity are shocking and many are irreversible (@yeşilartvinder, 2021; @temavakfi, 2021). Women and men are running out of habitat (140journos, 2015). They are also under oppression by arrogant, menacing, marauding, patriarchal capitalism. The takeover of their lands drives the rural people from being local producers to merely being consumers, or in some places, people are forced to work in low-income jobs without any economic and social rights. In the cities, house prices are increasing and the cost of living is expensive. Social and economic harms heavily burden women who are doubly victimised, in that they fear financial and social insecurity (Evrensel, 2015) whereas men only experience financial insecurity. In the EBS region generally, the status of women is secondary to that of men, as is the norm elsewhere in Turkey because of the strong patriarchal social structure (Gazioğlu, 2013). Interestingly, from my long-lived experience in the EBS region, it could be argued that some women play the 'backseat driver' role in family decision-making processes. They do not directly challenge authority. When the time comes, and if it is desirable, they want to realise the plans that they have designed in their minds; step by step, with great courage, cleverness and also solidarity with other women. This pattern of solidarity among women is developed in socialisation processes and with the effects of the unique character of the geographic climate of the region, which is subjected to change very rapidly, sometimes being unpredictable even within a short time-frame. Therefore, solidarity among rural women is a necessity in the EBS region, where some elements of solidarity and a traditional division of labour continue to exist. Because women cannot fight alone against obstacles and harsh geographical conditions, they have developed the ability to organise things very easily and quickly for any kind of collective action.

The traditional culture of the region shows how intertwined the relations are between people and nature. In traditional folk songs, relationships between lovers are defined by reference to natural phenomena. For example, exuberant love is symbolised by a waterfall in the river or the high waves of the Black Sea. Lovers who do not have the opportunity to meet are described as high mountains standing side by side. Some love stories are compared to the falling and rising of waves in the sea. Lovers who cannot physically meet wish to be as rain or clouds with the ability to meet elsewhere. Couples share their whole life. They may even share a hazelnut: 'I don't eat a hazelnut without you.' Generally, women carry agricultural products or other loads in baskets on their backs. Men do not carry any weight on their backs. In traditional folk songs, men in love have the opportunity to show their love: 'Let me be the porter of the basket on your back.' Trees, cows, sheep, horses, chickens and plants are very valuable in this region. Even an inch of land is very valuable. For this reason, nobody wants to sell their family lands. If someone sells her or his family land, she or he will feel shame, and sometimes even humiliation.

Family lands in the region are generally small, mountainous and not suitable for agriculture except for growing tea and hazelnuts. Inhabitants migrate from the lowland regions to the highlands during the summer season, doing both small-scale agricultural work and grazing their animals. That is why even the destruction of a tree cannot be tolerated for economic and cultural reasons. The physical environment – forests, plateaus, lakes, streams, rivers, trees, farmland together with agricultural products and animals – are indispensable for the people. The elders want their lives to end on their lands and to be buried in their village's cemetery. They fear dying elsewhere, away from their local home. Some want to continue their bond with their land even after death and wish to be buried in their gardens, instead of a cemetery.

Both tea-picking and nut-harvesting activities are considered to be an indispensable family tradition that unites family members across the generations every summer. It is also not only locals who participate in the highland festivals held in the mountains during summer but also those who have migrated from the local region to other cities and even to foreign countries are keen to participate in these summer festivals. Highland festivals seem very functional in terms of maintaining family and social ties, kinship, identity, culture and tradition. Different generations come together at these festivals. In all these efforts to maintain traditional bonds, the weight of the burden is once again placed on the shoulders of women. Women are regarded as the pillars and/or cement of the family structure. For this reason, it is not just economic and/or other concerns that make women fearlessly defend their habitat from authorities and other interest groups. If their habitats are destroyed, it means their homeland, maternal hearth, past, present and future, and their identities will disappear. That is why they believe that they have to save their lands from destruction and/or other risks and threats.

If the environment is destroyed, community, family and kinship bonds will come to an end. Saving the environment means saving the right to live and the future generations. Grandparents would plant trees, tea and nut gardens not for themselves, but for their grandchildren and great-grandchildren. They are more afraid of the disintegration of their ancestral hearth and their past than they are afraid of death. Women, in particular, take their own, sometimes traditional, sometimes modern, ways of fighting against destructive change. They resist authorities and participate in collective actions and other forms of resistance. For Canefe (2016: 10), 'these movements underline the right to life, to dignity, to the desire of leaving a liveable environment to future generations as well as for the long-term well-being of the national populace'.

For the elderly living in the villages, ageism also matters in terms of not being willing to go to cities, because they feel they are unable to make new friends or establish a proper social environment. Despite all kinds of health risks, loneliness and harsh winter conditions, some elderly people prefer living in their homes in the villages all year round to living in the cities. Some people describe living in apartments in cities as 'living in iron cages'. For those who have domestic animals, it is a very strong excuse to stay in the village even during harsh winters. Their adult children, who live in cities, mostly fail to persuade their old parents to live with them for some part of the year. For the elderly, the old worry of 'what will happen to my children' in the past is now replaced by worry over what will happen to their house, garden, trees and animals (Akay, 2020; Bayraktar, 2021). They have a very strong bond with the trees, vegetables, nature, mountains, rivers, seas, gardens and animals. For this reason, they never listen to their children's genuine offers or suggestions, and they decline invitations to live with them in the cities. Those who prefer to spend winter seasons in cities start to miss their villages, friends, homes and gardens from the moment of arrival in the city. Their eyes will always be on the way back to their villages. For these reasons, even capitalist companies operating in the region cannot easily separate them away from their natural habitat, because no matter how old they are, they are willing to fight to the end.

Collective actions of rural women differ from urban women's actions in the streets of the cities, both in the number of participants and in their mode of action. Just as was experienced in the Carettepe resistance protests, which began at the end of the 1980s, even the collective actions of a small number of women in rural areas can make a lot of noise both in the local community and across the country. Rural women are even likely to become symbols of resistance against green crime. Intimidation practices applied to women participants in collective actions in the streets of cities cannot be applied to women engaged in collective actions in rural areas. Rural women reconstruct their individual and social relations with their resisting status and roles through collective actions. They are sometimes on the tops of the mountains, sometimes on the banks of rivers, or in the plains. Just like women who demand gender equality through collective actions in the streets of cities, they defend their right to live in their habitat.

Women's resistance and collective actions

Turkey's political regime might be considered populist authoritarian, possibly now even purely authoritarian. Therefore, the reason why men show a low profile compared to women, and why women are at the forefront in collective environmental actions and resistance movements in rural areas (Kaçar, 2015; Show TV, 2015), could be explained by the fear of violence of the authoritarian regime. This disturbing fact has not been expressed in academic studies on collective environmental actions and resistance movements. For example, Seckin (2016: 4) claims that 'men do not participate in collective actions, because they work far from the region, or they just do not participate because they spend time in the village coffee house'. This is not true. Findings from this study suggest that the reason men maintain a low profile or never turn up is politically a strategic decision made within the family (Özbucak, 2019). Men generally show a low profile in collective actions in order not to face the economic, physical and social violence of the regime. It is not easy to challenge the brutal violence and/or other social and economic sanctions of this regime.

In Turkey, individual and collective actions are determined by power relations, such as the 'strong–weak' distinction. Being right does not have meaning in the eyes of the authorities. For this reason, people have to consider this bitter reality before engaging in individual and/or collective actions. The names of people who participate in collective actions in different parts of the country are recorded, and punitive and/or deterrent practices, especially economic and social sanctions, are likely to be put into effect against them. This political reality is known by everybody. Activists' families and children may be harmed. For example, their chances of getting jobs in public and/or private companies may be reduced or eliminated. Those who have a job may be faced with its loss. The social media accounts of the activists are each targeted. For all these reasons, people are afraid to fight for their rights; even writing a couple of sentences on social media or pressing the 'like' button for a comment with which they agree. They do not dare to challenge capitalist companies that have turned their lives upside down. Even though some people are victimised, they appear to be staying on the side of the AKP. Possibly because the AKP either intimidates people or offers them money.

The burden of participating in collective actions and other resistance movements has also been placed on the shoulders of women. Men try to reduce the tension between themselves and the AKP and defend themselves by saying, 'what can we do, we can't talk to our wife or mother. They do not listen to anybody. In fact, we do not want them to participate in collective actions'. There are also some women from the different local and national women's associations who participate in some rural collective actions. These urban women wear modern outfits, such as miniskirts or shorts and modern tops during collective actions and the AKP seems to try to humiliate these women in the eyes of society by saying that 'feminists are demonstrating in miniskirts!'

Legitimacy and rightfulness are the sources of power of women who are victims of green crime and whose response is to take part in collective actions. Interesting scenes of young and old women have emerged in the countryside, standing fearlessly in front of excavation vehicles, or surrounding trees that are planned to be cut down by the authorities (Karadeniz İsyandadır, 2021). In some resistance actions in other regions of Turkey, we have witnessed the occupation of the highways by women. Violence against women victims of green crime is not easily legitimised, as it is applied to those who, for example, protest violence against women by performing the traditional '*las tesis*' dance in the streets of Istanbul. Taking into account the social legitimacy of using force, the authorities calculate in terms of when, how and against whom. Of course, it does not mean that rural women are not subjected to physical violence by the authorities. In collective actions against WGV, the AKP first applies the tactic of reducing tension in the early stages (Human Right Association, 2020). Later, as mentioned, it tries to buy or discourage people who resist, by giving money or discouraging them through different intimidation tactics. When the 2,600km length of the Green Road Project is considered, including Carattepe, there are many locations in the EBS region where collective action movements against WGV have been in process. The project has been continuing in a piecemeal fashion. It does not proceed in a single line from beginning to end.

Women are free in their local habitat and they contribute to local agricultural production. It will be easier for a family to survive if they have a cow and a garden to grow vegetables and other agricultural products, such as tea and nuts. When we consider collective actions against WGV, it can be argued that the victimisation of women in this context is tripled. The first victimisation emerges as a result of structural patriarchy and gender inequality. The second victimisation is due to green crime. The third one is that, as mentioned previously, it is mainly women who are left alone to resist WGV. Under these heavy burdens, an old peasant woman, who saw excavator scoops and gendarmes ranged against her in the Carattepe resistance, utters the curse with her famous local accent: "F★★★ the mine!" (Hürriyet TV, 2015). This action of an old woman has become a symbol of resistance against WGV. Even in the countryside, women dare to risk exclusion, insult and other forms of demonisation.

Rural women do not identify themselves as feminists. They may not know what a feminist is, although they act as if they are feminists. This shows how rural women activists engage in 'implicitly feminist practices', which is defined as a 'strategy practiced by feminist activists within organisations that are operating in an anti- and postfeminist environment in which they conceal feminist identities and ideas while emphasising the more socially acceptable angles of their efforts' (Giffort, 2011: 569). Women's collective actions are mostly spontaneous. However, it should be said that some environmentalist and feminist associations engage in joint efforts in the realisation of collective actions. It is observed that some local and national representatives of the opposition parties have also participated in collective actions which were ignored by the AKP-funded media.

In Turkey, while populist authoritarianism has been getting stronger, the ToP activities have been increasing. These two negative developments have great effects on social inequality, oppression and collective actions against WGV. Those living in these regions, especially children and women, face more risks of being a victim of green crime. As a result, it is not difficult to predict that those who take part in collective action against green crime and WGV in different areas of Turkey will likely be neutralised by excessive force used by law enforcement officers and private security guards of private companies.

Conclusion

Ecological disorganisation increases social, economic and political inequalities in Turkey. A way of life has emerged which has resulted in green crime and WGV. Mostly women and children are victimised by the ToP. Women are politically, economically and socially marginalised. As a result of this turmoil, even women over 85 years old find themselves at the top of a mountain, defending trees in the forests, by the sea, or a river, resisting security forces to protect their right to life, their habitat and the ecosystem in general (Hürriyet TV, 2015). With the ToP, the destruction of local production opportunities ends policies of sustainable economic growth and regional development. Within this context, individuals are reduced to the status of 'consumers'. This negative development, in turn, destroys local economies and leaves local communities vulnerable to exploitation by national and international capitalist companies. The takeover of their lands and the destruction of women's traditional productive roles further emphasise the issue of gender inequality. Furthermore, WGV is not recognised as a crime in the criminal justice system.

In Turkey, it is essential to develop feminist criminology that will recognise women's green victimisation and raise awareness that WGV is a crime. To realise this objective, it is necessary to introduce methods to deal with patriarchy and to develop relations at local, national and international levels. It is a fact that the EBS region's culture has relational and collective characteristics rather than individualistic. This results in important implications in terms of the dynamics of resistance movements and collective actions to tackle WGV. It is also necessary to deal with privileged capitalist companies and their arrogant approaches that ignore WGV.

'Environmental justice can probably not be properly discussed without a parallel discussion of power' (Wyatt and Arroyo, 2018: 11). Environmental damage is inflicted by powerful actors, companies and governments. This reality has important consequences in terms of green feminist criminology and WGV in Turkey. Transnational activism and cooperation are a must. Taking the concept of transnational resistance into consideration, regional, national and transnational resistance networks against dams, mines and other environmental threats have the potential to raise awareness, especially among women, and tackle WGV.

To explain transnational activism, Tarrow (2005) looks at the structures through which globalisation is arbitrated. He argues that horizontal and vertical relations between state and non-state actors acting at international, national and subnational levels achieve networks of formal and informal institutions as internationalisation proceeds. These institutions, he claims, create 'an opportunity space into which domestic actors can move, encounter others like themselves, and form coalitions that transcend their borders' (Tarrow, 2005: 25). Tarrow goes on to ask where transnational activism happens, and, who are the transnational activists? For Tarrow (2005: 42), transnational activists are, for the most part, not much different from their non-global counterparts. While they may hold transnational connections, and sometimes engage in transnational conflict, they are entrenched in local situations and interests and the 'social networks … resources, experiences, and opportunities that place provides them'. Tarrow (2005: 43) is right to argue that they are distinctive in their potential to 'shift their activities among levels, taking advantage of the expanded nodes of opportunity of a complex international society'.

As Turkey's Ecology Union Women's Assembly argues, the patriarchal capitalist system destroys both nature and women. The Ecology Union Women's Assembly says that domestic and foreign companies that covet land, air and water with all kinds of unlawfulness receive their power from the government. However, the AKP's pressure and oppression will not be able to deter women. Women are determined to continue to save their habitat (Ekoloji Birliği, 2021).

In Turkey, feminist and green criminology should be integrated to adequately explain WGV. Academic studies need to consider how to integrate feminist criminology and green criminology literature. Feminist criminology has the potential to contribute significantly to the debate on WGV of which a whole range has tended to be neglected (Lynch, 2018). Green crime results in not only habitat destruction and WGV, but also results in what Goyes et al (2021) describe as 'the erosion of a way of life and memory'. As at Carattepe, local women's collective action against WGV is a minor contribution to resistance movements against a major problem in Turkey's ToP construction, mining extraction and environmental destruction issues. Because there is resistance, there is hope, and every effort should be valued and encouraged no matter how large or small.

Acknowledgements

I am grateful to Dr Pamela Davies and Dr Tanya Wyatt for their valuable comments and suggestions, which helped me in improving the quality of the chapter.

References

140journos (2015) Yeşil yol projesi ve iklim üzerine olası etkisi. *140journos*. Available at: https://140journos.com/ye%C5%9Fil-yol-proj esi-ve-i%CC%87klim-%C3%BCzerine-etkisi-3c9b01f66fed (accessed 3 June 2022).

@temavakfi (2021) 740 Dönüm orman alanı yok edilmek isteniyor (Twitter post), 29 July. Available at: https://twitter.com/temavakfi/status/14207915 11403405315 (accessed 16 August 2022).

@yeşilartvinder (2021) *Artvin ili maden ruhsat haritası* (Twitter post), 4 July. Available at: https://twitter.com/yesilartvinder (accessed 16 August 2022).

Akay, S. (2020) *Yeşil Yol Nedir?* (Video). YouTube, 20 December. Available at: www.youtube.com/watch?v=KMQwbLTUSSM (accessed 17 August 2022).

Bahar H.I. (2018) Patriarchy, gender inequality and criminal victimization of women in Turkey. In K. Carrington, R. Hogg, J. Scott and M. Sozzo (eds) *The Palgrave Handbook of Criminology and the Global South*. Cham: Palgrave Macmillan, pp 931–45.

Bayraktar, T. (2021) *Anzer'deki Yaylacı Kadın* (Video). YouTube, 18 July. Available at: www.youtube.com/watch?v=txMvcQR9spY (accessed 17 August 2022).

Bayraktar, T. (2022) *Yayladaki Hayat* (Video). YouTube, 3 July. Available at: www.youtube.com/watch?v=mC5MUuMI_kw (accessed 17 August 2022).

Beirne, P., Brisman, A., Sollund, R. and South, N. (2018) Editors' introduction to the special issue: 'For a green criminology'—20 years and onwards. *Theoretical Criminology*, 22(3), 295–7.

Brisman, A. and South, N. (2019) Green criminology and environmental crimes and harms. *Sociology Compass*, 13(1), e12650.

Brisman, A., South, N. and Walters, R. (2018) Climate apartheid and environmental refugees. In K. Carrington, R. Hogg, J. Scott and M. Sozzo (eds) *The Palgrave Handbook of Criminology and the Global South*. Cham: Palgrave Macmillan, pp 301–21.

Canefe, N. (2016) The politics of public protests against extractivism in Turkey. *Peace Review*, 28(1), 10–19.

Doğu, B. (2017) Environment as politics: Framing the Cerattepe protest in Twitter. *Environmental Communication*, 13(5), 617–32.

DOKAP (2016) Yeşil Yol Projesi. *DOKAP*, 15 December. Available at: www. dokap.gov.tr/projeler/yesil-yol-projesi/1/Detay (accessed 7 July 2021).

Donovan, J. (2006) Feminism and the treatment of animals: From care to dialogue. *Signs: Journal of Women in Culture and Society*, 31(2), 305–29.

Ekoloji Birliği (2021) Kadın Meclisi. *Ekoloji Birliği*. Available at: https://eko lojibirligi.org/kategori/kadin-meclisi/ (accessed 9 March 2021).

Erensu, S. (2018) The contradictions of Turkey's rush to energy. *Middle East Report*, 288, 32–5. Available at: file:///C:/Users/Sony/Downloads/ The_Contradictions_of_Turkeys_Rush_to_En.pdf (accessed 7 July 2021).

Eryilmaz, Ç. (2016) Koşu bandı üretimi kuramının yeşil yol karşıtı yerel harekete uygulanması. *Journal of Black Sea Studies*, Kış(52), 119–37.

Evrensel (2015) Yeşil yol dediğin ranttır. *Evrensel*, 13 November. Available at: www.evrensel.net/haber/265066/camlihemsinli-kadinlar-yesil-yol-dedi gin-ranttir (accessed 3 June 2022).

Gazioğlu, E. (2013) Doğu Karadeniz Bölgesinin Toplumsal Cinsiyet Rejimi ve Kadınların Toplumsal Konumları. *Karadeniz Araştırmaları*, 40(40), 95–108.

Giffort, D.M. (2011) Show or tell? Feminist dilemmas and implicit feminism at girls' rock camp. *Gender & Society*, 25(5), 569–88.

Goyes, D.R. (2018) Green criminology as decolonial tool: A stereoscope of environmental harm. In K. Carrington, R. Hogg, J. Scott and M. Sozzo (eds) *The Palgrave Handbook of Criminology and the Global South*. Cham: Palgrave Macmillan, pp 323–46.

Goyes, D.R., South, N., Abaibira, M.A., Baicué, P., Cuchimba, A. and Ñeñetofe, D.T.R. (2021) Genocide and ecocide in four Colombian indigenous communities: The erosion of a way of life and memory. *The British Journal of Criminology*, 61(4), 965–84.

Hallett, R.E. and Barber, K.B. (2014) Ethnographic research in a cyber era. *Journal of Contemporary Ethnography*, 43(3), 306–30.

Human Right Association (2020) *Turkey, a Perpetual Emergency: Attacks on Freedom of Assembly in Turkey and Repercussions for Civil Society*. Available at: www.fidh.org/IMG/pdf/obs_turkeyweb.pdf (accessed 3 June 2022).

Hürriyet TV (2015) *Yeşil Yol'a Hava Ana İsyanı: Halkım Ben* (Video). YouTube, 12 July. Available at: www.youtube.com/watch?v=bUpG2fL7 Cic (accessed 17 August 2022).

Jarrell, M.L., Lynch, M.J., Stretesky, P.B., Arrigo, B. and Bersot, H. (2013) Green criminology and green victimization. In B.A. Arrigo and H.Y. Bersot (eds) *The Routledge Handbook of International Crime and Justice Studies*. London: Routledge, pp 423–44.

Kaçar, M. (2015) Yeşil Yol Direneişinin sembolü Kadınlar. Mücadelemiz Sürecek. *Evrensel*, 12 July. Available at: www.evrensel.net/haber/255807/yesil-yol-direnisinin-sembolu-kadinlar-mucadelemiz-surecek (accessed 20 August 2022).

Karadeniz İsyandadır (2021) Devler oldu Cengiz, hukuku alicengiz. *Karadeniz İsyandadır*. Available at: http://karadenizisyandadir.net/devlet-oldu-cengiz-hukuku-alicengiz/ (accessed 3 June 2022).

Kaya, A.Y. (2016) Sermaye-emek kutuplaşmasının yeniden üretimi: Acele kamulaştırma kararlarında HES'ler. In C. Aksu, S. Erensü and E. Erdem (eds) *Sudan Sebepler: Türkiye'de su-enerji politikaları ve direnişler*. Istanbul: İletişim, pp 65–92.

Lynch, M.J. (2016) Green criminology and social justice: A re-examination of the Lynemouth plant closing and the political economic causes of environmental and social injustice. *Critical Sociology*, 43, 449–64.

Lynch, M.J. (2018) Acknowledging female victims of green crimes: Environmental exposure of women to industrial pollutants. *Feminist Criminology*, 13(4), 404–27.

Lynch, M.J. (2020) Green criminology and environmental crime: Criminology that matters in the age of global ecological collapse. *Journal of White Collar and Corporate Crime*, 1(1), 50–61.

Lynch, M.J. and Stretesky, P.B. (2001) Toxic crimes: Examining corporate victimization of the general public employing medical and epidemiological evidence. *Critical Criminology*, 10, 153–72.

Lynch, M.J. and Stretesky, P.B. (2003) The meaning of green: Contrasting criminological perspective. *Theoretical Criminology*, 7(2), 217–38.

Mazlum, S.C. (2017) Turkey and post-Paris climate change politics: Still playing alone. *New Perspectives on Turkey*, 56, 145–52.

Özbucak, C. (2019) Siyanüre Direnen Kadınlar. *Ekoloji Birliği*, 18 August. Available at: https://ekolojibirligi.org/siyanure-direnen-kadinlar/ (accessed 20 August 2022).

Ramp, D. and Bekoff, M. (2015) Compassion is a practical and evolved ethic for conservation. *BioScience*, 65(3), 323–7.

Ruggiero, V. and South, N. (2013) Green criminology and crimes of the economy: Theory, research, and praxis. *Critical Criminology*, 21(3), 359–73.

Sato, C.F. and Lindenmayer, D.B. (2018) Meeting the global ecosystem collapse challenge. *Conservation Letters*, 11(1), e12348.

Schnaiberg, A. (1980) *The Environment: From Surplus to Scarcity*. New York: Oxford University Press.

Schnaiberg, A., Pellow, D.N. and Weinberg, A. (2002) The treadmill of production and the environmental state. In A.P.J. Mol and F.H. Buttel (eds) *The Environmental State Under Pressure*. Bingley: Emerald Group Publishing Limited, pp 15–32.

Seckin, E. (2016) *An Analysis of Selected Cases of Environmental Movements in Turkey Through an Ecofeminist Approach*, unpublished master of science thesis, Middle East Technical University, Graduate School of Natural and Applied Sciences, Ankara, Turkey.

Show TV (2015) *Karadenizli Kadınların Yeşil Yol Direnişi* (Video). YouTube, 12 July. Available at: www.haberturk.com/video/haber/izle/karadenizli-kadinlarin-yesil-yol-direnisi/144619 (accessed 10 August 2022).

Sollund, R. (2013) Crimes understood through an ecofeminist perspective. In A. Brisman and N. South (eds) *Routledge International Handbook of Green Criminology*, 2nd edn. Milton Park: Routledge, pp 317–31.

Sollund, R. (ed) (2015a) *Green Harms and Crimes, Critical Criminology in a Changing World*. London: Palgrave.

Sollund, R. (2015b) Introduction: Critical, green criminology-an agenda for change. In R. Sollund (ed) *Green Harms and Crimes, Critical Criminology in a Changing World*. London: Palgrave, pp 1–26.

Sollund, R. (2015c) Michael Lynch and Paul Stretesky: Exploring green criminology. Toward a green criminological revolution. *Punishment and Society*, 19(4): 514–18.

Sollund, R. (2017) Doing green, critical criminology with an auto-ethnographic, feminist approach. *Critical Criminology*, 25(2), 245–60.

Sollund, R. (2019) Wildlife management, species injustice and ecocide in the Anthropocene. *Critical Criminology*, 28(3), 351–69.

Sollund, R. and Brisman, A. (2017) Editors' introduction to the Special Issue, researching environmental harm, doing green criminology. *Critical Criminology*, 25(2), 159–63.

Sollund, R., Stefes, C. and Germani, A.R. (eds) (2016) *Fighting Environmental Crime in Europe and Beyond: The Role of the EU and Its Member States*. London: Palgrave.

South, N. (2014) Green criminology: Reflections, connections, horizons. *International Journal for Crime, Justice and Social Democracy*, 3(2), 5–20.

South, N. and Beirne, P. (eds) (2006) *Green Criminology*. Aldershot: Dartmouth.

Stretesky, P.B., McKie, R., Long, M.A., Lynch, M.J. and Barrett, K.L. (2018) Where have all the Falcons gone? Saker exports in a global economy. *Global Ecology and Conservation*, 13, e00372.

Tarrow, S. (2005) *The New Transnational Activism*. Cambridge: Cambridge University Press.

Turhan, E. (2017) Right here, right now: A call for engaged scholarship on climate justice in Turkey. *New Perspectives on Turkey*, 56, 152–8.

Turhan, E., Mazlum, S.C., Şahin, Ü., Şorman, A.H. and Gündoğan, A.C. (2016) Beyond special circumstances: Climate change policy in Turkey 1992–2015. *WIREs*, 7(3), 448–60.

Walklate, S. (1990) Researching victims of crime: Critical victimology. *Social Justice*, 17(3): 25–42.

Wattanaporn, K.A. and Holtfreter, K. (2014) The impact of feminist pathways research on gender-responsive policy and practice. *Feminist Criminology*, 9(3): 191–207.

White, R. (2018) The global context of transnational environmental crime in Asia. In K. Carrington, R. Hogg, J. Scott and M. Sozzo (eds) *The Palgrave Handbook of Criminology and the Global South*. Cham: Palgrave Macmillan, pp 281–300.

Wyatt, T. and Arroyo, I.Q. (2018) Green harms and crimes in Mexico. In I. Arroyo-Quiroz and T. Wyatt (eds) *Green Crime in Mexico*. Cham: Palgrave Macmillan, pp 1–22.

11

'Daughters of Dust': An Eco-Feminist Analysis of Debt-for-Nature Swaps and Underage Marriage in Indonesia

Delon Alain Omrow

Introduction

An eco-feminist analysis of debt-for-nature (DFN) swaps and underage marriage highlights what philosopher Karen J. Warren (1990) refers to as the 'logic of domination'. DFN swaps are agreements through which select developing nations have foreign debt forgiven in exchange for investments in environmental conservation measures (Cassimon et al, 2011), while underage marriage is the formal marriage or informal union between a child under the age of 18 and an adult or another child. Underage marriage is a human rights violation, preventing women and girls from living their lives free from all forms of violence globally (United Nations, 1993). The 'logic of domination' represents an intersectional vision of how the oppression of nature mirrors the oppression of women and girls. While some criminologists have attempted to subsume eco-feminist thought into the discipline of criminology (Lane, 1998; Lynch and Stretsky, 2003; Gaarder, 2013; Davies, 2014; Lynch, 2017, 2018; Sollund, 2020; Varona, 2020), these publications do not utilise the 'logic of domination' to draw concrete parallels between the eco-violent commodification of nature and women's bodies. In this chapter, I offer an invitation to recapture the theoretical promise of eco-feminism, explaining how all types of oppression are ultimately connected to this same logic. The purpose of this chapter is fourfold. I begin with a brief analysis of human and environmental security through the prism of gender-based violence (GBV). This is followed by an examination of eco-feminist theory as it applies to

Indonesia's DFN transaction and prevalence of underage marriage, expanding the conceptual terrain of debt. Finally, I offer some concluding thoughts on the interconnections of DFN swaps and various forms of violence against women in the context of human and environmental security.

Positionality

I offer the following author's note: at the time of this writing, I celebrate the five-year anniversary of the death of my grandmother, Aisha Omrow. Born in the hinterlands of Guyana, she often spoke of the institution of indentured labour – an insidious replacement for slavery in British Guiana that recruited servants from small towns in and around Kolkata, India – and the impact it had on my family. Being Amerindian – the Indigenous peoples of Guyana – she had witnessed the onslaught of ecologically destructive activities on ancestral lands, and the concomitant eco-violence against women and children. Mahdia, one of Guyana's major 'boom towns', had always haunted my grandmother. These makeshift towns catered to the influx of miners entering the region in search of gold and other resources to eke out a comfortable living. Replete with eateries and bars, hotels and brothels, 'boom towns' were defined by anomic conditions – a shining example of lawlessness where men exploited the remoteness of their mining operations, starved political infrastructure and lack of government support for the country's Indigenous communities. She recalled numerous accounts of young girls being deceived into leaving their communities to work in these towns and who, eventually, were forced into sex work and various forms of control, manipulation and violence. She knew of three families who had sold their daughters to one well-known miner in the area: she referred to these young girls as 'daughters of dust'.

It was these stories which served as an impetus for my doctoral studies, exploring both social and environmental injustice in Guyana (Omrow, 2017). My work with Red Thread, a non-governmental organisation (NGO) dedicated to advocating for the rights of women and young people, forced me to draw connections between human security and environmental security through the 'logic of domination'. It was through this framework that I recognised the larger pattern across the globe: the oppression of nature is directly related to the oppression of women. I cannot do justice in this space to the range and depth of injustice women across the globe have faced, and how this is connected to environmental injustice. I can, however, provide a cursory review of international efforts to record and mitigate these injustices. In 1993, the United Nations General Assembly adopted the Declaration on the Elimination of Violence against Women, defining violence against women as 'any act of gender-based violence that results in, or is likely to result in, physical, sexual or psychological harm or suffering to women, including threats of such acts, coercion or arbitrary deprivation of liberty, whether occurring in public or in private life' (United Nations, 1993: 48).

Almost 30 years later, not much has changed to ameliorate the status of women. In fact, a 2021 report by the World Health Organization (WHO) and a UN interagency working group on gender inequality provides the most accurate estimates of the prevalence of GBV. From 2000 to 2018 across 161 countries and areas, nearly one in three, or 30 per cent, of women have been subjected to physical and/or sexual violence by an intimate partner or non-partner or both (WHO, 2021). Over a quarter of women between the ages of 15 and 49 have been subjected to physical and/or sexual violence by their intimate partner at least once in their lifetime (WHO, 2021). Moreover, prevalence estimates of lifetime intimate partner violence range from 20 per cent in the Western Pacific, 22 per cent in high-income countries and Europe, 25 per cent in the WHO regions of the Americas to 33 per cent in the WHO African region, 31 per cent in the WHO Eastern Mediterranean region and 33 per cent in the WHO Southeast Asia region (WHO, 2021: 26).

The violence of gender inequality establishes (and reinforces) patterns of gender-based abuse across environmental contexts, hindering the security and wellbeing of nations, communities and individuals. The International Union for Conservation of Nature (IUCN) lays bare the links between environmental pressures and GBV, documenting how the degradation of nature, competition over increasingly scarce resources and environmental crime exacerbate GBV (Camey et al, 2020). GBV is a human and environmental security issue, signalling a paradigmatic shift from reductionist approaches of violence towards more comprehensive approaches which include many manifestations of violence – direct and indirect, interpersonal and structural, physical, emotional, economic and others (Jakobsen, 2014). It is through the interstices of GBV that the 'logic of domination' can be applied, invoking Johan Galtung's definition of violence. For Galtung, 'violence is present when human beings are being influenced so that their actual somatic and mental realizations are below their potential realizations' (Galtung, 1969: 168). In other words, to get at the core of Galtung's definition of violence, we must ask: why are some human beings unable to fully actualise their potential? Galtunian violence is the act of preventing persons from reaching their physical and psychological potential – what Galtung refers to as the 'somatic realization' of a human being's lifespan. In the context of GBV, gender inequality and oppression inhibit the lifespan of young girls and women, limiting both their physical and psychological potential.

The key to understanding Galtunian violence is the principle of avoidance. In *Cultural Violence* (1990: 292), Galtung explains: 'I see violence as avoidable insults to basic human needs, and more generally to *life*, lowering the real level of needs satisfaction below what is potentially possible. Threats of violence are also violence.' Suffice to say, if there is *something* preventing young girls and women from achieving basic human needs, that *something* is called violence – or, more accurately, GBV. GBV can be applied to Galtung's theory because it can be prevented through organised campaigns, and education and awareness on the important intersections between climate change, livelihoods and violence against women.

The 'logic of domination' pervades GBV, bringing questions about environmental and human security to the fore. The UN Commission on Human Security define human security in the following manner:

> [H]uman security means protecting the vital core of all human lives in ways that enhance human freedoms and human fulfillment. Human security means protecting people from critical (severe) and pervasive (widespread) threats and situations. It means using processes that build on people's strengths and aspirations. It means creating political, social, environmental, economic, military and cultural systems that together give people the building blocks of survival, livelihood and dignity. (Commission on Human Security, 2003)

Environmental security, on the other hand, has various meanings because it merges two capacious concepts: the environment and security. Belluck et al (2006) interpret environmental security as measures guarding against environmental degradation in order to preserve or protect human, material and natural resources. Zurlini and Müller (2008) offer a comprehensive overview of the environmental security field, noting that the environment is a transnational issue, and its security is imperative for securing peace, national security and human rights. In similar fashion, Barnett (2009) defines environmental security as the mitigation of threats to the integrity of the biosphere and to its interdependent human component. These definitions of environmental security explore resource scarcity, the contamination of inputs into human systems, biodiversity loss, land degradation, water pollution and scarcity, ecological degradation, the contamination of people, plants and animals by hazardous substances, and climate change. In a globalised world, we need to think about security in the context of globalisation and the manner in which one nation's security occurs at the expense of another (Barnett, 2009).

Environmental security is also linked to environmental consciousness and the framing of environmental problems as security issues. High-profile studies of environmental problems in the 1960s and the creation of international, environmental NGOs, such as the World Wildlife Fund, Friends of the Earth and Greenpeace signalled a growing awareness of potential threats to the security of the planet. This was followed by international summits on environmental issues and international agreements on how the global community could work together to protect the environment. In 1987, the World Commission on Environment and Development introduced the concept of sustainable development and the term environmental security. The United Nations Conference on Environment and Development soon followed and used environmental security as the crux of its multilateral environmental treaties on climate change and biodiversity (Bruggeman, 2008).

Today, the social sciences use the term environmental security to describe the multitudinous risks all people across the globe must manage. In particular, research conducted by human geographers such as Steve Lonergan (1999), Michael Redclift (2001, 2003) and Michael Watts (2013, 2014) have informed the study of environmental security. They offer five principal interpretations. First, environmental security refers to the impacts of anthropogenic activities on the planet and potential effects on human health. The second interpretation of environmental security is the impact of the military–industrial complex on the environment, while the third interpretation views environmental change as a threat to national security. Similarly, the fourth interpretation records environmental change as a possible cause of violent conflict. Finally, the fifth interpretation sees environmental change as a risk to human security (Barnett, 2009). For the purpose of this chapter, Barnett (2009) and Belluck et al's (2006) definition of environmental security will be employed, as both definitions draw upon one key principal interpretation: environmental change as a risk to human security.

This chapter argues that underage marriage and DFN swaps constitute threats to human security and environmental security, respectively. Eco-feminism can improve our understanding and application of these concepts because it recognises and affirms the crucial role that women play in environmental protection, ecological development and sustainable use of natural resources (Leahy, 2003). Both human security and environmental security can be enhanced by improving the status of women, empowering them to design and participate in the processes of the protection of natural systems on which all life depends (United Nations Development Programme, 1994). I will now discuss in further detail eco-feminism's 'logic of domination' and its promise in theorising human insecurity and environmental insecurity.

The 'logic of domination'

How might we begin to theorise human (in)security and environmental (in)security through an eco-feminist lens? For Warren (1988a), critical thinking fosters creative thinking, passion and empathy, all of which play important roles in 'reasonable reflection' about what to do or believe. Women are situated in a socially constructed 'frame of reference', or what Warren refers to as an oppressive conceptual framework, which acts as a set of basic beliefs, values, attitudes and assumptions which explain, shape and reflect views of ourselves and our world. Warren identifies three typical features of oppressive conceptual frameworks in Western societies for theorising women's oppression. First, oppressive conceptual frameworks feature value-hierarchical thinking, whereby men, culture and reason or intellect are 'up' and women, nature and emotion are 'down', respectively, creating systems of patriarchal domination.

Second, and relatedly, oppressive conceptual frameworks support 'either-or' thinking. This creates misleading or harmful value dualisms in which disjunctive terms are pitted against each other and viewed as exclusive (rather than inclusive) and oppositional (rather than complementary). Warren notes, also, that higher value is attributed to one disjunct, creating social and cultural binaries. Finally, the third and most important feature of an oppressive conceptual framework is its 'logic of domination'. Warren describes this as 'a structure of argumentation which explains, justifies, and maintains the subordination of an "inferior" group by a "superior" group on the grounds of the (alleged) superiority and inferiority of the respective groups' (Warren, 1988b: 32).

Oppressive conceptual frameworks can be patriarchal. They can assign greater value to that which has been identified as 'male' than to that which has been identified as 'female', or they can be anthropocentric whereby they assign greater value to that which has been identified as 'human' than to that which has been identified as 'nature'. The alleged superiority of one justifies the subordination of the 'Other'. This is further unpacked in Warren's 1990 publication, *The Power and the Promise of Ecological Feminism*. Consider the following logic of argumentation:

(A1) Women are identified with nature and the realm of the physical; men are identified with the 'human' and the realm of the mental.
(A2) Whatever is identified with nature and the realm of the physical is inferior to ('below') whatever is identified with the 'human' and the realm of the mental: or, conversely, the latter is superior to ('above') the former.
(A3) Thus, women are inferior to ('below') men; or, conversely, men are superior to ('above') women.
(A4) For any X and Y, if X is superior to Y, then X is justified in subordinating Y.
(A5) Thus, men are justified in subordinating women. (Warren, 1990: 133)

Clearly, both nature and women are the 'Other' in this schema, outlining what Warren (1994) refers to as the twin and interconnected dominations of nature and women. The question is: which form of domination precedes the other? Such tautology is beyond the scope of this chapter. However, we can look to earlier feminist inquiry for a clearer understanding of how power is exercised over nature. For example, eco-feminists Gourish Chandra Mondal and Palash Majumder (2019) and Susan Buckingham (2015) argue that the domination of women and the degradation of the environment are rooted in patriarchy and capitalism, and that strategies to address one must take into account its impact on the other. Women's equality should not be achieved at the expense of environmental degradation, and neither should environmental improvements be gained at the expense of women. Carolyn Merchant (1980) and Val Plumwood (1993) once advocated 'the feminine principle' as an antidote to environmental destruction, highlighting the oppressive structures of capitalism and commodification. The 'feminine principle' is an awareness of nature as a living, interdependent force, and the inclusion of women as sources of expertise and knowledge when solving society's current ecological crises (Garrity-Bond, 2018).

While there is a substantial scholarship dedicated to analysing the commodification of nature through the onslaught of neoliberal policies in the latter half of the 20th century – namely, privatisation, marketisation, monetary valuation and other associated processes (Akerman, 2003; Blomley, 2006; Erickson, 2010; Benjaminsen and Bryceson, 2012; Fairhead et al, 2012; Wynne-Jones, 2012; Green and Adams, 2015; Bigger et al, 2018; Missemer, 2018) – we know very little about the conceptual frameworks driving the human economy's relationship with natural environments. Ecological economics comes close as an exciting vista of inquiry in the critical social sciences, evolving into critical geography (Robertson, 2002, 2004, 2006; Bakker, 2003, 2005, 2007; Castree, 2003; Mansfield, 2004; McCarthy and Prudham, 2004; Heynen et al, 2007). Feminist critical geographers, too, have made a case for grounding geography in feminist, anti-racist and postcolonial theory, foregrounding questions of race, colonialism and history in critical geographies of development (Kwan, 2002; Lahiri-Dutt, 2017; Jones, 2019; Thorpe and Chawansky, 2021). What is missing in these works, I suggest, is the application of Warren's 'logic of domination' as an ideological force behind the neoliberal ethos of the commodification and subordination of both nature and women's bodies. An inquiry into such a relationship raises more issues and questions than can be answered by any one author in one submission, and this chapter does not try. I can, however, draw upon select works to demonstrate the thoughtful synthesis of eco-feminism and green criminology.

Lane (1998) uses eco-feminism to reconceptualise feminist political protest and women's relationship with nature. This has informed green criminology's critique of environmental law as anthropocentric and law breaking as a symbolic act that seeks to challenge dominant ideas and values in our society. Lynch and Stretsky (2003) unpack the commonalties among environmental justice movements, highlighting diverse strategies for redressing environmental harms. The authors explain that eco-feminist thought critiques capitalist, profit-growth orientations and its patriarchal nature. Concomitantly, patriarchy must be dismantled and replaced with more egalitarian forms of social organisation wherein women and the natural environment are treated with respect and dignity. This gender-based analysis of domination and exploitation of nature and women lends itself to wider discussions of green criminology and various meanings of the term 'green'.

More recently, Taylor and Fitzgerald (2018) use eco-feminism and feminist intersectional theories in green criminological debates to theorise animal abuse. The authors argue that eco-feminism and green criminology can bridge the divide between environmentalism, animal advocacy and their associated areas of academic study, developing a non-speciesist green criminology. In a related vein, Sollund (2020) asserts that green criminology has not explored the gendered aspects of crimes and harms against nonhuman animals. Invoking eco-feminist theories to better understand the gendered dynamics involved in wildlife trafficking and the theriocide of large carnivores, Sollund makes a case for greater collaboration between eco-feminism and green criminology. Finally, Lynch (2018) adopts an intersectional approach to examining green victimisation, arguing that green criminology neglects female victims of green crimes. Drawing upon medical and epidemiological literature, Lynch suggests that eco-feminist inquiry into green victimisation of economically marginalised women can inform eco-feminist studies by more fully elaborating a position of the environmental/green victimisation.

While this summary of the cross-fertilisation of eco-feminist and green criminological research is not exhaustive, it does provide an idea of how these literatures offer new opportunities for analyses. Clearly, eco-feminism and green criminology offer intellectual fervour and activist energy for defending people and the environment from harm. In the following section, I contribute to the literature, providing an analysis of the parallels between DFN swaps and underage marriage in Indonesia through an eco-feminist lens.

Indonesia's debt-for-nature swap

The origins of DFN transactions can be traced to debt-for-equity exchange campaigns during the Latin American debt crisis in the early 1980s. While primary markets are where companies sell new stocks and bonds to the public for the first time, the secondary market is where both equity and debt can be purchased internationally. Chile was one of the first countries to explore the parameters of debt-for-equity exchanges, allowing investors to redeem its external debt at a discount on the secondary market. This, in turn, allowed Chile (the debtor) to reinvest local currency as equity in national companies (Moye, 2001, 2003). First proposed in 1984 by Dr Thomas Lovejoy, then vice-president of the World Wildlife Foundation, these swaps were also linked to the privatisation of public assets, valued up to a combined swap of US$27 billion in 1990 (Kaiser and Lambert, 1996). Debt-for-equity exchanges started to wane in the mid-1990s onwards because the value of debt in developing countries appreciated within the secondary market as a result of the improved stability and solvency of major economies such as Argentina, Brazil, Chile and Mexico (Ruiz, 2007).

The 2009 United Nations Climate Change Conference, commonly known as the Copenhagen Summit, renewed interest in DFN swaps by claiming that such transactions reduced debt burdens, saved scarce hard currency, and freed up budgetary resources for environmental (or other) spending (Gugler, 1997). Specifically, a developing country's debt was sold on the secondary market in exchange for a domestic currency instrument used to finance environmental and conservation initiatives. Today, different variations of this swap procedure exist. For example, some transactions are conducted directly between a creditor and a debtor government, while other transactions include an international NGO assisting a creditor government. These arrangements are perceived as 'win-win-win' because environmental NGOs can leverage their funds by raising their profile and expanding their network through the positive difference between the redemption value and the secondary market value of the debt purchased or received creditors – namely, developed country governments or private banks – avail themselves of the value of debt claims and improve their environmental credentials. Finally, developing countries can seek debt relief (Occhiolini, 1990; Arslanalp and Henry, 2005).

Indonesia's decision to receive funding during the United Nations Convention on Climate Change set in motion an 'external debt swap/ relief' deal between Conservation International, the United States and an Indonesian environmental foundation. This US$30 million agreement allowed the Indonesian government to 'swap' its debt over the next eight years in exchange for conservation efforts in Sumatra, a large Indonesian island west of Java and south of the Malay Peninsula. This was one of the largest debt swaps to date, conducted under the Tropical Forest Conservation Act (TFCA) (Buckley, 2009).

Pursuant to the TFCA, the United States agreed to forgive six debt claims owed by Indonesia to the United States Agency for International Development (USAID). As a part of the debt forgiveness measures, Indonesia would have to spend an equivalent amount on grants to support local NGOs involved in tropical forest conservation projects in Sumatra. USAID received US$20 million from the US Treasury, and US$1 million each from Conservation International and an Indonesian environmental foundation – Yayasan Keanekaragaman Hayati Indonesia. For this arrangement, the Nature Conservancy and the World Wildlife Fund for Nature joined the agreement as well, with each organisation contributing US$2 million to reduce Indonesia's debt payments to the US government over the next eight years by nearly US$28.5 million (Knicley, 2012).

Cassimon et al (2011), however, question the efficacy of Indonesia's DFN swap. The authors suggest that the swaps generally fail to deliver additional resources for sectoral or public goods in debtor countries, exacerbating poverty and inequality. While the Indonesian government prepared to avail itself of the increased net financial transfers, opening up the 'fiscal space' to divert public resources to domestic spending on environmental concerns, this 'fiscal space' was never created because the entire previous outstanding principal and interest sums were still due, ending up into the Debt Service Account instead of being repaid to USAID. In other words, the original debt service schedule did not provide sufficient opportunity for Indonesia to tackle its debt. Rather, the government of Indonesia was still liable for the entire previous outstanding principle and interest sums, enjoying little to no positive discount (Mold, 2009).

Additionally, Indonesia's government did not witness hard currency relief. In fact, when analysing the Indonesian case, despite being the largest DFN agreement yet conducted under the TFCA, the US$30 million earmarked for debt relief was not properly allocated to Indonesia's total outstanding debt of US$149.7 billion and the swap only concerned 0.02 per cent of Indonesian debt (in nominal terms). Indonesian environmental activists questioned if the US–Indonesian swap deal even allowed Indonesia to increase its resources by US$30 million, or if the extra US$30 million was really spent on environmental purposes. While the deal benefited the global community through global public goods such as the conservation of forests, carbon sequestration and reduced carbon emissions, it did very little for the Indonesian people (Knicley, 2012). The controversy surrounding DFN swaps, then, is that these swaps result in the debtor country relinquishing aspects of its ecological sovereignty to international environmental groups.

According to Cassimon et al (2011: 21), 'although typical debt-for-nature contracts required NGO projects to be consistent with government policies, host government involvement was generally limited to veto rights on the projects proposed. Often there was no room for governments to actively define programs and/or projects according to their own national development or sectoral priorities'. Greener (1991) and Gugler (1997) refer to these swaps as examples of 'eco-imperialism' or 'eco-colonialism', respectively, because the control over various conservation efforts is taken out of the hands of the debtor country and transferred into the hands of global actors. From an environmental perspective, Indonesia's DFN swap raised concerns about the key principles of conservation commitments: *additionality*, *permanence*, *monitoring* and *enforceability*. Knicley (2012) explains that *additionality* refers to the extent to which conservation efforts, either organised or de facto, would have occurred without the DFN exchange, while *permanence* signifies the longevity and sustainability of the conservation commitment hold over time. Lastly, problems of *monitoring* and *enforceability* of the conservation commitment include how the debtor country's Indigenous peoples will be treated amid interference with the debtor country's sovereignty. Knicley (2012: 29) concludes that Indonesia's DFN swap remained 'shrouded in mystery' when considering the aforementioned principles of conservation commitments and their overall efficacy.

Hassoun (2012) explores DFN agreements from a human rights perspective, revealing that loans are conditional – that is, creditors make loans on the condition that debtors will pay them back via economic conditionalities which force developing countries to implement economic reforms designed to promote macroeconomic stability and growth. In the case of Indonesia, Hassoun argues that the conditionalities of the conservation programmes prevented Indonesia from maintaining the economic growth, infrastructure and redistributive mechanisms necessary to protect current and future generations of Indonesians from extreme poverty. As such, the DFN transaction was morally questionable, if not impermissible from human rights perspective. The author elaborates: 'the original loan could be *odious*. Odious loans are either (1) accepted by those without legitimate authority or (2) used for purposes that do not benefit the debtor country or its inhabitants' (Hassoun, 2012: 12; emphasis in original). Indonesia's loans were impermissible because they were exploitative; international creditors used their disproportionate bargaining power to coerce the government into accepting conditions that would not otherwise be reasonable to accept. This was precisely the observation made by Human Rights Watch (2019), an international NGO, which revealed that Northern actors developed conservation projects in Indonesia on the basis of market-based instruments, asymmetrical relationships and weak laws – all of which were exacerbated by poor government oversight. These factors merely accelerated the commodification of nature and its subjugation to Northern powers which disproportionately affected Indigenous peoples' rights to their forests, livelihood, food, water and culture.

The 'logic of domination' was invoked by non-Indonesian actors when implementing the DFN swap. Illuminating a power struggle between different competing interest groups, the DFN exchange pitted Western-based approaches to conservation and environmental governance against Indonesian approaches, with the former seen as 'superior' on the grounds of the (perceived) inferiority of the latter. Indonesia's rainforests were situated in an oppressive conceptual framework which justified Western, hegemonic values and assumptions about conservation over local and Indigenous approaches to conservation. Leading to the creation of a value dualism and the social construction of nature as the 'Other', this served as an impetus for geographical expansion, economic restructuring, wage reduction, cuts to social programmes and privatisation of public goods at the expense of nature. Through an eco-feminist lens, we can examine the symbolic connections between (disordered) nature and anthropocentric attempts to commodify, control and bring order to nature through neoliberal, market-oriented instruments. The environmental security of Sumatra's rainforests was violated, as its lush and fertile grounds, species and ecosystems were perceived as the 'Other' – to be dominated and controlled for business reasons under the guise of debt repayment and conservation. This value dualism was clearly evident in the implementation of Indonesia's DFN arrangement because NGO contracts flagrantly limited the country's veto rights on the projects proposed, subordinating national development and sectoral priorities (Knicley, 2012). Can parallels be drawn between the 'logic of domination' behind Indonesia's DFN agreement and other forms of domination in the country? It is this issue to which I turn in the following section.

Underage marriage in Indonesia

South of Sumatra, in the province of Bali, another type of debt is repaid in Indonesia – this one, with young girls and women. With a population of 276 million people, and home to the world's largest Muslim population, Indonesia boasts geographic and cultural diversity, emerging as a significant economic and political power. However, despite these advances, children in Indonesia face a number of serious challenges – chief among these is poverty and infant mortality, which as of 2018, was recorded at 40 deaths per 1,000 live births. Maternal mortality, on the other hand, was recorded at 359 deaths per 100,000 live births and has been on the increase, giving rise to GBV (Rumble et al, 2018: 411).

Indonesia is the eighth-highest ranked country in the world for recorded child marriages, with one in nine girls marrying before the age of 18 (Arshad, 2020). Himawan (2019: 225) reveals that child prostitution has also become an epidemic for Indonesia where sex tourism runs rampant: approximately 100,000 children and women are trafficked each year in Indonesia, and 30 per cent are below the age of 18. The malign intensity of this trade is captured in the data collected by the Women's Institute, based in West Java, which reports that nearly 43.5 per cent of trafficking victims are as young as 14 years old (Himawan, 2019). Indonesia has one of the worst records for underage marriage, according to the NGO Girls Not Brides (Lih Yi, 2019). The imperfect implementation of the Child Protection Act, social acceptance of child labour and the lack of birth registration serve as the drivers of child prostitution, sex trafficking and underage marriage (Lih Yi, 2019). Additionally, abject poverty leads families to force their children into underage marriage; on average, over 3,500 Indonesian girls are married every day and in some cases religious courts have endorsed the marriages of girls younger than 16 (Lih Yi, 2019). What is more tragic, however, is the complicity of families and close friends in underage 'money marriages', the purpose of which is to pay off familial debt. Young women and girls are sold and trafficked as far as China, Thailand and Eastern Europe, while in West Kalimantan, girls are forced to prostitute themselves in jungle brothels (Arshad, 2020).

In the early 2000s, the government of Indonesia introduced its National Plan of Action on human rights and combating trafficking, sexual exploitation and underage marriages. However, Hastanto (2020) suggests that structural determinants and the sociocultural dimensions of unregistered marriages, or what locals refer to as '*nikah siri*', pervade the rural areas, condoning the practice of selling young girls to repay debts and 'favours'. Hastanto explains:

> Despite being illegal, child marriage remains rampant in Indonesia, with some parents forcing their children into unions at a young age to pay off debts. Just last month, for instance, another adoptive father named Sappa, also known as Father Alif, married his 12-year-old daughter off to a 44-year-old man named Baharuddin. The marriage was performed according to traditional Muslim customs, despite the local religious affairs office having already rejected the request because the girl was underage. (Hastanto, 2020: 3)

According to Dharma Negara, head of the Pinrang District Police Criminal Unit, data from the Central Statistics Agency's National Socio-Economic Survey suggest that there were 1.2 million cases of child marriage in 2018, despite the Legal Aid Foundation of the Indonesian Women's Association for Justice demanding amendments to the Marriage Law (Arshad, 2020: 2). The Indonesian government has even set a target to reduce the ratio of child brides to 8.7 per cent of all marriages by 2023, from the 11.2 per cent ratio of 2020 (Rahiem, 2021). Erica Hall, World Vision's child marriage expert, suggests that the COVID-19 pandemic will severely affect the livelihood of girls from poor families burdened by added economic shocks in Indonesia. Hall explains: 'four million girls are at risk of child marriage in the next two years as a result of COVID-19. When you have any crisis like a conflict, disaster or pandemic – rates of child marriage go up' (ASEAN, 2020: 7). Relatedly, the UN predicts that the pandemic could place another 130 million people into extreme poverty by 2030, forcing families to explore child prostitution, trafficking and underage marriage as debt repayment measures. From a human security perspective, the commodification of young girls' and women's bodies as currency for debt repayment is a violation of Indonesia's Child Protection Act, the National Plan of Action and a suite of international laws – namely, the United Nations Committee on the Convention on the Rights of the Child, the Beijing Declaration and Platform for Action, and the Convention on the Elimination of All Forms of Discrimination against Women. Moreover, young girls sold to repay familial debt are susceptible to domestic violence and poor health; young brides also have a higher risk of dying during childbirth because their reproductive organs have not fully developed and are not ready for sexual intercourse and pregnancy.

Child prostitution, sex trafficking and underage marriage in Indonesia stem from a 'logic of domination'. Whereas nature is commodified under DFN swaps, young girls' bodies, too, are commodified in the name of familial debt relief and, in some cases, are viewed as simply disposable. Considering the disproportionality of underage marriage of girls, an eco-feminist analysis of this phenomenon warrants extended reflection: women and girls are viewed as inferior within historically and traditionally male gender-identified beliefs, values and attitudes. In the Indonesian context, a patriarchal conceptual framework assigns greater value, status or prestige to that which traditionally has been identified as 'male' than to that which traditionally has been identified as 'female', maintaining value dualisms and the perpetuation of GBV through the domination and subordination of girls and women.

DFN swaps and underage marriage, I contend, are two pressing issues that have the potential to hasten the hitherto slow-paced application of eco-feminist thought to human and environmental security. Theoretically minded readers may notice the common denominator in both examples is debt. Feminists Lucí Cavallero and Verónica Gago (2021) develop a feminist understanding of debt, revealing its impact on women and members of the 2SLGBTQ+ community. The authors assert that debt 'tries to present itself as a true black box, in which decisions are made in a mathematical, algorithmic way about what has value and what does not. … A feminist reading of debt involves detecting how debt is linked to violence against feminized bodies' (Cavallero and Gago, 2021: 6). Simply put, debt is intimately linked to gendered violence and patriarchal notions of the family. Massaging the conceptual parameters of debt, we can begin to see how DFN transactions and underage marriage are the result of oppressive conceptual frameworks. In the case of the former, neoliberal policies and national debt justify the domination and commodification of Sumatra, perpetuating an economy of ecoviolence in which top-down, Western-based conservation efforts enclose land and resources. As the authors suggest, real-life impacts of debt fall mainly on the shoulders of women. I argue that the burden of national debt, too, negatively affects nature and this was clearly evident in the manner in which Sumatra was 'opened up' to geopolitical and eco-colonial forces – some of which introduced new thresholds of debt and the concomitant neocolonial dispossession of resources.

While Cavallero and Gago demonstrate how neoliberal policies constitute a specific economy of violence via femicide and travesticide, I argue that underage marriage must, per force, be theorised as an extension of this economy of violence, and linked to the very debt regimes necessitating

DFN swaps. It is debt, and debt regimes, which inform decision-making power over women and young girls' bodies, but also nature, land and resources in Indonesia, extracting value from various forms of life. It is high time green criminologists adopt eco-feminism as a theoretical framework to unpack how debt operates for both nature and young girls and women, interrogating the violence experienced by the indebted. Gender inequality and anthropocentrism supposes a form of moralisation directed towards feminised bodies and nature, maintaining value dualisms and a differential of exploitation due to the corresponding relations of subordination within a 'logic of domination'.

Conclusion

As I compose the concluding section of this chapter, in 2022, Coordinating Minister for Human Development and Cultural Affairs, Muhadjir Effendy, is calling for collaboration among organisations such as the National Population and Family Planning Agency to eliminate underage marriage in the next five years (Bhwana, 2022). Indonesia's officials have taken heed of the negative effects of forced marriages: domestic violence, divorce and the emergence of new poor families. Effendy hosts online seminars entitled 'It's time for the youth to make a voice' every Saturday, in the hope that education and awareness campaigns will empower communities and prevent GBV.

The Indonesian case study demonstrates how the 'logic of domination' and feminist understandings of debt can be invoked to explain threats to human security and environmental security. In both cases, nature and young women's bodies are commodified, subordinated and oppressed. The 'logic of domination' explains, justifies and maintains the subordination of the 'inferior' lands and human bodies by a 'superior' group on the grounds of their (perceived) superiority. The belief that superiority justifies subordination perpetuates beliefs, values, attitudes and assumptions about Western-led conservation efforts, but also patriarchal customs in Indonesia. Human security and environmental security, then, serve as links between the environment and human activities, with eco-feminist inquiry playing a crucial role in designing ecological and security policies aimed at environmental, social and gender justice.

To summarise, this chapter offered a brief analysis of human and environmental security through the prism of GBV. This was followed by an examination of eco-feminist theory and Indonesia's DFN swap and prevalence of underage marriage, analysing how eco-feminist inquiry and the 'logic of domination' can expand the conceptual terrain of debt. Finally, I offered some concluding thoughts on the interconnections of DFN transactions and various forms of violence against women in the context of human and environmental security. However, much more work needs to be done. The IUCN is attempting to advance better legal protections for women, drawing undeniable links with environmental issues and related socioeconomic inequalities in policy and legal frameworks. Drawing from existing expertise, the IUCN is calling upon governments and environmental organisations to partner with humanitarian and health organisations that have knowledge and expertise on GBV to develop environmental policies and projects which prevent violence and enable women to safely engage in activism. Systematic efforts to invest in knowledge and solutions are also underway. The IUCN publishes extensively on how GBV is exacerbated by environmental degradation but many gaps in the organisation's understanding remain. Eco-feminism research and evidence-based action can fill these gaps, giving a voice to the forgotten 'daughters of dust'.

References

Akerman, M. (2003) What does 'natural capital' do? The role of metaphor in economic understanding of the environment. *Environmental Values*, 12, 431–48.

Arshad, A. (2020) Child marriages on the rise in Indonesia amid COVID-19 outbreak. *The Straits Times*, 3 October. Available at: www.straitstimes. com/asia/se-asia/child-marriages-on-the-rise-in-indonesia-amid-covid-19-outbreak (accessed 10 July 2022).

Arslanalp, E. and Henry, P.B. (2005) Is debt relief efficient? *Journal of Finance*, 60(2), 1017–51.

ASEAN (2020) Poverty and underage marriage in Indonesia. *The ASEAN Post Team*, 7 November. Available at: https://theaseanpost.com/article/poverty-and-underage-marriage-indonesia (accessed 27 July 2022).

Bakker, K. (2003) *An Uncooperative Commodity: Privatizing Water in England and Wales*. Oxford: Oxford University Press.

Bakker, K. (2005) Neoliberalizing nature? Market environmentalism in water supply in England and Wales. *Annals of the Association of American Geographers*, 95(3), 542–65.

Bakker, K. (2007) The 'commons' versus the 'commodity': Alter-globalization, anti-privatization and the human right to water in the global south. *Antipode*, 39(3), 430–55.

Barnett, J. (2009) Environmental security. In R. Kitchin and N. Thrift (eds) *International Encyclopedia of Human Geography*. Amsterdam: Elsevier, pp 553–7.

Belluck, D., Hull, R., Benjamin, S., Alcorn, J. and Linkov, I. (2006) Environmental security, critical infrastructure and risk assessment: Definitions and current trends. In B. Morel and I. Linkov (eds) *Environmental Security and Environmental Management: The Role of Risk Assessment*. New York: Springer, pp 3–16.

Benjaminsen, T.A. and Bryceson, I. (2012) Conservation, green/blue grabbing and accumulation by dispossession in Tanzania. *The Journal of Peasant Studies*, 39(2), 335–55.

Bhwana, P. (2022) Joint efforts essential to prevent child marriage, says Minister. *TEMP.CO*. Available at: https://en.tempo.co/read/1605802/joint-efforts-essential-to-prevent-child-marriage-says-minister (accessed 10 July 2022).

Bigger, P., Dempsey, J., Adeniyi, P., Asiyanbi, K., Lave, R., Mansfield, B., Osborne, T. and Robertson, M. (2018) Reflecting on neoliberal natures: An exchange. *Environment and Planning*, 1(1–2), 25–75.

Blomley, N. (2006) Uncritical critical geography? *Progress in Human Geography*, 30(1), 87–94.

Bruggeman, W. (2008) Failing global justice and human security. In M. den Boer and J. de Wilde (eds) *The Viability of Human Security*. Amsterdam: University of Amsterdam Press, pp 47–69.

Buckingham, S. (2015) Ecofeminism. In J. Wright (ed) *International Encyclopedia of the Social & Behavioral Sciences*, 2nd edn. Amsterdam: Elsevier, pp 845–50.

Buckley, R. (2009) Debt-for-development exchanges: The origins of a financial technique. *Law and Development Review*, 2(1), 24–49.

Camey, C.I., Sabater, L., Owren, C., Boyer, A.E., Wen, J., Itza, C.C., Sabater, L., Owren, C. and Boyer, A. (2020) *Gender-based Violence and Environment Linkages: The Violence of Inequality*. Gland: International Union for Conservation of Nature.

Cassimon, D., Prowse, M. and Essers, D. (2011) The pitfalls and potential of debt-for-nature swaps: A US-Indonesian case study. *Global Environmental Change*, 21(1), 93–102.

Castree, N. (2003) Commodifying what nature? *Progress in Human Geography*, 27(3), 273–97.

Cavallero, L. and Gago, V. (2021) *A Feminist Reading of Debt*. London: Pluto Press.

Commission on Human Security (2003) *Human Security Now*. Geneva: Commission on Human Security.

Davies, P. (2014) Green crime and victimization: Tensions between social and environmental justice. *Theoretical Criminology*, 18, 300–16.

Erickson, P. (2010) Knowing nature through markets: Trade, populations, and the history of ecology. *Science as Culture*, 19(4), 529–51.

Fairhead, J., Leach, M. and Scoones, I. (2012) Green grabbing: A new appropriation of nature? *The Journal of Peasant Studies*, 39(2), 237–61.

Gaarder, E. (2013) Evading responsibility for green harm: State-corporate exploitation of race, class and gender inequality. In N. South and A. Brisman (eds) *Routledge International Handbook of Green Criminology*. Abingdon: Routledge, pp 272–81.

Galtung, J. (1969) Violence, peace and peace research. *Journal of Peace Research*, 6(3), 167–91.

Galtung, J. (1990) Cultural violence. *Journal of Peace Research*, 27(3), 291–305.

Garrity-Bond, C. (2018) Ecofeminist epistemology in Vandana Shiva's the feminine principle of Parkriti and Ivone Gebara's trinitarian cosmology. *Feminist Theology*, 26(2), 185–94.

Green, K.E. and Adams, W. (2015) Green grabbing and the dynamics of local-level engagement with neoliberalization in Tanzania's wildlife management areas. *The Journal of Peasant Studies*, 42(1), 97–117.

Greener, L.P. (1991) Debt-for-nature swaps in Latin American countries: The enforcement dilemma. *Connecticut Journal of International Law*, 7, 123–80.

Gugler, A. (1997) The win–win–win-scenario: Conversion of debt. Paper presented at the ECDPM Anniversary Seminar on Alternatives to Cooperation: Sustainable Finance as an End to Development. Maastricht, 12–13 May.

Hassoun, N. (2012) The problem of debt-for-nature swaps from a human rights perspective. *Philosophy Faculty Scholarship*, 15, 1–38.

Hastanto, I. (2020) Indonesian family forces 12-year-old girl to marry 45-year-old man to 'repay a favor.' *VICE*, 7 July.

Heynen, N., McCarthy, J., Prudham, S. and Robbins, P. (eds) (2007) *Neoliberal Environments: False Promises and Unnatural Consequences*. Abingdon: Routledge.

Himawan, K. (2019) Either I do or I must: An exploration of the marriage attitudes of Indonesian singles. *The Social Science Journal*, 56(2), 220–7.

Human Rights Watch (2019) Indonesia: Indigenous peoples losing their forests. Available at: www.hrw.org/news/2019/09/22/indonesia-indigenous-peoples-losing-their-forests (accessed 26 June 2022).

Jakobsen, H. (2014) What's gendered about gender-based violence? An empirically grounded theoretical exploration from Tanzania. *Gender and Society*, 28(4), 537–61.

Jones, N. (2019) Dying to eat? Black food geographies of slow violence and resilience. *ACME: An International Journal for Critical Geographies*, 18(5), 1076–99.

Kaiser, J. and Lambert, A. (1996) *Debt Swaps for Sustainable Development: A Practical Guide for NGOs*. Gland: IUCN.

Knicley, J. (2012) Debt, nature and indigenous rights: Twenty-five years of debt-for-nature evolution. *Harvard Environmental Law Review*, 36, 80–122.

Kwan, M. (2002) Feminist visualization: Re-envisioning GIS as a method in feminist geographic research. *Annals of the Association of American Geographers*, 92(4), 645–61.

Lahiri-Dutt, K. (2017) Thinking 'differently' about a feminist critical geography of development. *Geographical Research*, 55(3), 326–31.

Lane, P. (1998) Ecofeminism meets criminology. *Theoretical Criminology*, 2(2), 235–48.

Leahy, T. (2003) Ecofeminism in theory and practice: Women's responses to environmental issues. *Journal of Interdisciplinary Gender Studies*, 7(1), 106–25.

Lih Yi, B. (2019) End child marriage 'hell' for Indonesian girls, lawmaker urges. *Reuters*, 19 August.

Lonergan, S. (1999) *Environmental Change, Adaptation and Security*. New York: Springer.

Lynch, M. (2017) Conceptualizing green victimization, green criminology and political economy: A reply. *Critical Sociology*, 43(3), 473–8.

Lynch, M. (2018) Acknowledging female victims of green crimes: Environmental exposure of women to industrial pollutants. *Feminist Criminology*, 13(4), 404–27.

Lynch, M. and Stretsky, P. (2003) The meaning of green: Contrasting criminological perspectives. *Theoretical Criminology*, 7(2), 217–38.

Mansfield, B. (2004) Rules of privatization: Contradiction in neoliberal regulation of North Pacific fisheries. *Annals of the Association of American Geographers*, 94(3), 565–84.

McCarthy, J. and Prudham, S. (2004) Neoliberal nature and the nature of neoliberalism. *Geoforum*, 35, 275–83.

Merchant, C. (1980) The death of nature: A retrospective. *Organization & Environment*, 11(2), 198–206.

Missemer, A. (2018) Natural capital as an economic concept, history and contemporary issues. *Ecological Economics*, 143, 90–6.

Mold, A. (2009) *Policy Ownership and Aid Conditionality in the Light of the Financial Crisis: A Critical Review*. New York: Development Centre Studies.

Mondal, C.C. and Majumder, P. (2019) Ecofeminism: Encouraging interconnectedness with our environment in modern society. *American Journal of Educational Research*, 7(7), 482–4.

Moye, M. (2001) *Overview of Debt Conversion. Publication No. 4*. London: Debt Relief International.

Moye, M. (2003) *Bilateral Debt-for-Environment Swaps by Creditor*. London: WWF Center for Conservation Finance.

Occhiolini, M. (1990) *Debt-for-Nature Swaps*. PRE Working Paper. Washington, DC: World Bank.

Omrow, D. (2017) *Abyssal Ideology and the Amerindians of Guyana: An Eco-crimes Analysis of Power, Discourse and Cognitive Injustice* (unpublished doctoral dissertation). York University.

Plumwood, V. (1993) *Feminism and the Mastery of Nature*. Abingdon: Routledge.

Rahiem, M. (2021) COVID-19 and the surge of child marriages: A phenomenon in Nusa Tenggara Barat, Indonesia. *Child Abuse & Neglect*, 118, 1–36.

Redclift, M. (2001) Environmental security and the recombinant human: Sustainability in the twenty-first century. *Environmental Values*, 10(3), 289–99.

Redclift, M. (2003) Sustainability and human security: Challenges for the twenty-first century. *Journal of Interdisciplinary Economics*, 14(4), 403–14.

Robertson, M. (2002) No net loss: Wetland restoration and the incomplete capitalization of nature. *Antipode*, 32(4), 463–93.

Robertson, M. (2004) The neoliberalization of ecosystem services: Wetland mitigation banking and problems in environmental governance. *Geoforum*, 35(3), 361–73.

Robertson, M. (2006) The nature that capital can see: Science, state, and market in the commodification of ecosystem services. *Environment and Planning*, 24(3), 367–87.

Ruiz, M. (2007) *Debt Swaps for Development: Creative Solution or Smoke Screen?* Paris: EURODAD.

Rumble, L., Peterman, A. and Irdiana, N. (2018) An empirical exploration of female child marriage determinants in Indonesia. *BMC Public Health*, 18, 407–22.

Sollund, R. (2020) Wildlife crime: A crime of hegemonic masculinity? *Social Sciences*, 9(6), 1–16.

Taylor, N., and Fitzgerald, A. (2018) Understanding animal (ab)use: Green criminological contributions, missed opportunities and a way forward. *Theoretical Criminology*, 22(3), 402–25.

Thorpe, H. and Chawansky, M. (2021) Gender, embodiment and reflexivity in everyday spaces of development in Afghanistan. *A Journal of Feminist Geography*, 28(3), 1–27.

United Nations (1993) *Declaration on the Elimination of Violence against Women.* United Nations.

United Nations Development Programme (1994) *Human Development Report 1994.* Oxford University Press.

Varona, G. (2020) The contribution of critical ecofeminism to the criminological debate in Spain: Debating all rules of all tribes. In S. Walklate, K. Fitz-Gibbon, J. Maher and J. McCulloch (eds) *The Emerald Handbook of Feminism, Criminology and Social Change.* Bradford: Emerald Publishing, pp 119–36.

Warren, K. (1988a) Toward an ecofeminist ethic. *Studies in the Humanities: Special Issue on Feminism, Ecology, and the Future of the Humanities*, 15(2), 140–56.

Warren, K. (1988b) Critical thinking and feminism. *Informal Logic*, 10(1), 31–44.

Warren, K. (1990) The power and promise of ecological feminism. *Environmental Ethics*, 12(1), 125–44.

Warren, K. (1994) *Ecological Feminism.* Abingdon: Routledge.

Watts, M. (2013) Toward a political ecology of environmental security. In R. Floyd and R. Matthew (eds) *Environmental Security*. Abingdon: Routledge, pp 82–102.

Watts, M. (2014) Building resilience, making secure: Biopolitical security, catastrophism and the food–climate question. In N. Chen and L. Sharp (eds) *Biosecurity*. School of American Research, pp 145–72.

WHO (2021) *Violence Against Women Prevalence Estimates, 2018*. Gland, Switzerland: World Health Organization.

Wynne-Jones, S. (2012) Negotiating neoliberalism: Conservationists' role in the development of payments for ecosystem services. *Geoforum*, 43(6), 1035–44.

Zurlini, G. and Müller, F. (2008) Environmental security. In S.E. Jørgensen and B.D. Fath (eds) *Encyclopedia of Ecology*. Cambridge, MA: Academic Press, pp 1350–6.

PART III

Resistance

12

Women's Experiences of Environmental Harm in Colombia: Learning from Black, Decolonial and Indigenous Communitarian Feminisms

Daniela Suárez Vargas and Rachel Killean

Introduction

This chapter interrogates the gendered impacts of environmental harm in Colombia. Following intersectional theory as developed by Crenshaw (1989), we focus on Colombia's Indigenous and Afro-descendent communities, who are placed in a particularly disadvantaged position due to both gendered and ethnoracialised systems of oppression (Tovar-Restrepo and Irazábal, 2014). To do so, we bring green criminology into conversation with the decolonial, Black and Indigenous feminism that has emerged from across the Americas as a way of exposing the connections between gendered harms and structural violence. In doing so, we wish to highlight how platforming and learning from this feminist scholarship and activism can lead to a more nuanced understanding of environmental harm and its legacies in Colombia.

The chapter proceeds as follows. The next section provides the research context for this chapter. We highlight the range of theoretical perspectives which have been adopted to interrogate the continuum of environmental harms perpetrated in Colombia and outline how we feel engagement with feminist scholarship can enhance green criminological knowledge in this area. In the third section we outline the continuum of environmental harm in Colombia, focusing on its impacts on Indigenous and Afro-descendent populations.[1] We detail how Colombia's colonisation established harmful and exclusionary practices of land dispossession, which in turn lay the groundwork for one of the longest running conflicts in the Western world. We then turn to the environmental legacies of that conflict, before considering how demobilisation and 'peace' have created contemporary challenges from illegal and state-sanctioned extractivism. We highlight the sustained resistance to environmental harm in Colombia, and the increasing risks that environmental defenders face. In the fourth sections we turn to the feminist literature, drawing out four key insights regarding gendered environmental harm: first, the theorisation of the connections between territorial and female bodily autonomy; second, the exposure of the gendered vulnerabilities associated with dispossession and forced displacement; third, the 'everyday' gendered nature of environmental harms; and fourth, the important role of women in resisting environmental injustice and protecting their communities and territories.

Theorising gendered experiences of environmental harm in Colombia

Colombia is one of the most biodiverse countries in the world. Over half of Colombia's territory is covered in natural forest, including part of the Amazon rainforest (Bickerton, 2017), and 50 per cent of the world's moorlands are in Colombia (San Pedro, 2019). The country is home to around 63,303 reported species and an endless number of endemic species, making it the world's number one country for birds, orchids and butterflies; the country with the second highest number of plants, amphibians and freshwater fish; the country with the third highest number of palms and reptiles; and the country with the sixth highest number of mammals (SiB Colombia, 2020).

Colombia is also a country that has faced significant threats to its biodiversity, and one in which Indigenous and Afro-descendent communities' access to and control over natural resources has been shaped by legacies of colonialism, sustained armed conflict, and illegal and state-sanctioned extractivism (understood as both the extraction of minerals and agribusiness). The range of historic and contemporary environmental harms and their impacts on marginalised populations in Colombia have been explored in a vast body of multidisciplinary work spanning political geography (for example, Vélez-Torres, 2014), political ecology (see, for example, Gudynas, 2021), transitional justice (for example, Gomez-Betancur, 2020) and environmental peacebuilding (for example, Baptiste et al, 2017), among many others.

Yet, historically, green criminologist engagement with Colombia has been relatively sparse. This situation has changed over the last seven years, and a growing green criminological literature has considered conflicts over land use (Rodríguez Goyes, 2015), biopiracy (Rodríguez Goyes and

South, 2016), the wildlife trafficking trade (Rodríguez Goyes and Sollund, 2016; Sollund, 2017), the harms of palm oil production (Mol, 2017), and the risks of ecotourism (McClanahan et al, 2019). Some of this literature has also explicitly engaged with the harms, experiences and perspectives of various Indigenous communities living in Colombia, including research on the ecocide–genocide nexus (Rodríguez Goyes et al, 2021a), communities' resistance to extractivism (Rojas-Páez, 2017) and representations of nature within four Indigenous communities (Rodríguez Goyes et al, 2021b). However, to date there is little green criminological literature that explicitly engages with feminist perspectives or the gendered impacts of environmental harm in Colombia (although Rodríguez Goyes notes the influence of decolonial feminism on his work, see Rodríguez Goyes, 2017, 2019). This chapter offers a modest contribution towards addressing this gap.

There is an expansive feminist literature documenting gender-based violence in Colombia. However, this has tended to centre physical and sexual violence, particularly those harms experienced during the country's armed conflict (Céspedes-Báez, 2019). Acknowledging and responding to such harms are important feminist endeavours. Yet, this approach has often overlooked the broader range of structural harms experienced by women in Colombia, including those linked to environmental injustice (Paarlbeg-Kvam, 2021), and particularly those harms experienced by Indigenous and Afro-descendent women. This blind spot has been attributed to a tendency within Global North and urban Latin American feminists to create 'universal' feminisms grounded in individual rights and development (Rodriguez Castro, 2021c). This approach is critiqued for 'genderwashing' women's experiences, that is, homogenising women's experiences under a single gender lens, while ignoring their racial and ethnic realities (Rodríguez Castro, 2021b; see also Bastian Duarte, 2012; Hernández Castillo, 2014).

In response, other forms of feminist thought have emerged across Latin America and in solidarity with feminists from across the Americas, including decolonial, Black and Indigenous communitarian approaches. Black feminism has emerged as a means of revealing the overlapping oppressions of race, gender, class and sexuality (Roth, 2003), while decolonial and Indigenous feminisms have sought to highlight the ways in which the coloniality of power and the coloniality of gender have shaped women's lives (Lugones, 2008, 2014; Federeci, 2010). These bodies of work demonstrate that Indigenous and Black women's 'struggles cannot be understood through fragmented viewpoints that delink their past from their present' (Hernández Reyes, 2019: 224), or which separate patriarchal heterosexualism from processes of racialisation (Mack and Na'puti, 2019). They also share a commitment to highlighting the agency and resistance of marginalised women, challenging assumptions made within white feminist scholarship regarding their passivity in the face of oppression (Carby, 2007;

Cusicanqui, 2012). In doing so, they seek to 'open up spaces to think the new – from the social spaces of activism and research' (Millán, 2014, cited in Rodríguez Castro, 2021c: 49).

As an urban Colombian woman and a white woman from the Global North, it is not our place to speak for Indigenous and Afro-descendent communities and we do not seek to do so. Rather, we aim to engage in the 'politics of citation' (Kim, 2020), platforming and highlighting some of the wealth of feminist knowledge and scholarship that can and should inform green criminological work on the gendered impacts of environmental harm in Colombia. In doing so, we are influenced by the call from Rodríguez Goyes and others for a Southern green criminology, one '[w]here the distinctive traces of colonially informed environmental harms are acknowledged; research is guided by the knowledge of its victims; Western theory and concepts are affected by the knowledge, realities and understanding from the South; and Western theory is used too to better understand such phenomena' (Rodríguez Goyes, 2017: 339; see also Brisman et al, 2015; Rodríguez Goyes, 2019; Rodríguez Goyes et al, 2021b).

Continuums of environmental harm in Colombia

Colonial land dispossession

The 'continuum of harms' (Santamaria et al, 2019) experienced by Indigenous and Afro-descendent women in Colombia must be understood in the context of the country's colonial history. Colonialism is a dynamic of political and economic power which shapes the way the world is structured and understood (Wright et al, 2022). The effects of colonisation do not end with the formal cessation of colonial governance (Balint et al, 2014). On the contrary, the legacies of colonial projects continue to organise and define the relationships, experiences and aspirations of individuals in formerly colonised societies (Park, 2015). This colonial 'debris' is often the foundations of present-day injustices, with access to resources and wellbeing built on categories such as race, ethnicity, gender and religion, among others (Stoler, 2008, cited in Wright et al, 2022). This is particularly true for Indigenous peoples. As Aymara scholar Cuscicanqui notes: 'The Indigenous world does not conceive of history as linear; the past-future is contained in the present' (Cusicanqui, 2012: 96).

In the country now known as Colombia, colonisation involved the expeditions of 'conquistadors', who were granted the 'right' to conquer territory by the Spanish Crown and to extract 'rewards' from that territory and its inhabitants (Batchelder and Sanchez, 2013). Todorov argues that the conquest of America involved the hierarchisation of individuals into certain categories for their control. At the top of the hierarchy, there were the full individuals – the owner, the coloniser – who were seen as rational beings, with rights and autonomy; at the bottom, there were the non-subjects – enslaved people – who were seen as 'beasts', nonhuman; and between these categories were the intermediate subjects – women, Indigenous people – who were considered as producers of objects or services, with no capacity for self-definition or rebellion (Todorov, 2007: 189–90).

One prominent colonial mechanism was the 'encomienda' system, whereby colonisers were granted Crown permission to dispossess Indigenous land, exploit their resources and forcibly extract labour from Indigenous communities (Cabnal, 2010; Batchelder and Sanchez, 2013). These practices constituted a fundamental shift from pre-colonial relationships to land. Andean Indigenous cosmologies, beliefs and ways of living are often premised on the interrelationship between individual physical wellbeing, community life and land, with 'land' extending beyond a plot of land to include the concept of 'Pachamama',[2] or 'the territory' (Rivera Zea, 1999).

The labour extracted from Indigenous communities was often connected to the exploitation of the land, for example, involving working in mines, factories and transport. This extractive and challenging work directly contributed to the death of thousands of Indigenous people (Batchelder and Sanchez, 2013). From as early as the 1500s until the 1800s, enslaved Afro-Caribbean people were also brought to the country, particularly along the Pacific coast. Enslaved peopled were forced to engage in mining (Leal and Van Ausdal, 2013), one of the most important economic activities in the colonial system (Hernández Reyes, 2019), as well as on plantations and as labourers (Gruner and Mina Rojas, 2018).

After Colombia's independence, the ruling classes continued to implement colonial logics of segregation, exploitation, assimilation and annihilation against Indigenous peoples (Gargallo Celentani, 2014). Dispossession of land continued to be prevalent (Rodríguez Castro, 2021a), justified through proclamations as to the economic inferiority and underdevelopment of the traditional community work structures and cultures of Indigenous peoples. Such proclamations justified the leading classes in 'aggressive recolonization', allocating fertile land to the tenure of the agrarian bourgeoisie and engaging in 'speculative investment in land and mining concessions' (Cusicanqui, 2012: 100). As Tuck and Yang note, this focus on the land's productivity is characteristic of settler colonialism, as the settler 'can only make his identity as a settler by making the land produce, and produce excessively, because "civilization" is defined as production in excess of the natural world' (Tuck and Yang, 2012: 6; see also Cooper, 2005).

The seizure of land, uneven distribution of resources, and political and territorial marginalisation led to bipartisan violence in the early 20th century. The discovery of significant oil deposits alongside increases in drug trafficking and administrative decentralisation fanned the flames of violence in the 1980s and 1990s. The result was prolonged multi-party conflicts over

Colombian territory and resources (Céspedes-Báez, 2019), and the longest running conflict in the Western hemisphere (ICTJ, 2009).

Environmental legacies of conflict

While in some instances the presence of armed groups prevented environmentally destructive development (Sánchez-Cuervo and Aide, 2013), Colombia's conflict also had profound impacts on nature. Land-grabbing, illegal mining, forced displacement of farmers into forest areas, the establishment of coca and poppy crops, and glyphosate spraying by the state all played a role (Rodríguez et al, 2020). Coca and poppy crops in particular played a fundamental financing role for non-state armed groups, leading to deforestation and degradation of extensive areas of natural forest (Torres Rodríguez et al, 2019). An estimated 1 million hectares of forest was lost over the course of the conflict, while unstable and informal economies further depleted forests' natural resources (Baptiste et al, 2017). Several million hectares of land have also been abandoned due to violence in the area (Tovar-Restrepo and Irazábal, 2014).

The exploitation of Colombia's natural resources was accompanied by severe human suffering: hundreds of thousands of individuals lost their lives over the course of the conflict, while millions experienced internal displacement. Civilians faced impossible choices between abandoning their land, engaging in illegal extractive activities or joining armed groups (Ibañez, 2009). Indigenous and Afro-descendent peoples were at particular risk of violence and forced displacement. Their territories were the 'site of combat, massacres, kidnappings, bombardments, forced disappearances, forced drafts, antipersonnel land mines, and displacement' (Tovar-Restrepo and Irazábal, 2014: 46), while war was used as a means of forcing capitalist development onto their resource-rich territories (Escobar, 2004). Experiences of physical violence were accompanied by cultural harms, through isolation from traditional customs and community structures (Sousa Santos and Sena Martins, 2021).

In 2016, the Colombian government and the Revolutionary Armed Forces of Colombia guerrilla group (FARC-EP) signed a historic peace agreement establishing a legal framework promoting, among other aspects, the return of victims to areas previously occupied by the FARC-EP (Government of Colombia and FARC-EP, 2016). However, the laws allow eligible claimants to be denied the return of their land if it is being used for agro-industrial projects, unless they sign contracts enabling ongoing exploitation of the land (Congress of the Republic of Colombia, Law 1448, 2011, art. 99). Furthermore, the prospect of peace has raised new environmental challenges for the country, as natural areas that were inaccessible due to the conflict are becoming increasingly open for exploitation (Torres Rodríguez et al, 2019).

Contemporary challenges

Colombia has entered a time of ever-increasing socioenvironmental conflict (Camargo, 2019), through the 'reproduction of colonial patterns of landownership' (Motta, 2011: 4) and the 'increasing concentration of resources in the hands of the few' (Maldonado-Torres, 2016: 1). The country has one of the highest rates of inequality in terms of land distribution and rural populations continue to face forced displacement. This sometimes occurs at the hands of armed groups, who repurpose land for extractive practices (Sánchez-Cuervo and Aide, 2013). Conflicts are arising everyday over the creation of new mines, plantations and dams, including on environmental reserves (San Pedro, 2019), with 131 environmental conflicts recorded in the Environmental Justice Atlas as of June 2022 (Temper et al, 2022). The government has provided significant incentives for extractive companies and foreign investors (San Pedro, 2019), and has even deployed its armed forces for the protection of foreign investment. Violence has been particularly common in regions experiencing tensions between the interests of extractive companies, communities and illegal armed groups (San Pedro, 2019).

Indigenous and Afro-descendent communities continue to be vulnerable to the development of extractive projects on their territories, as well as the ongoing presence of legal and illegal armed groups (IWGIA, 2021). It is important to note the role of conservation projects in this process: as Ojeda and others have noted, investors have also used 'green pretexts' to garner state support for ecotourism and conservation projects at the expense of dispossessed local populations (Ojeda, 2012). Since 2020, Indigenous communities have also seen the erosion of their right to free, prior and informed consultation and consent in decisions impacting their lives and territories (IWGIA, 2021). This erosion has emerged through the government's temporary introduction of virtual consultations, the introduction of a 'proportionality test' deemed capable of replacing consultation in cases where consent was not achieved, and disregard for the law from powerful actors (Hernández Reyes, 2019; IWGIA, 2021). This continued expropriation for extractive purposes can be understood as a continuation of the same colonial doctrines which 'conceptualised the New World as "terra nullis"' (Picq, 2014; see also Grosfoguel, 2016).

Resistance and risk

Indigenous and Afro-descendent communities have been at the forefront of environmental activism in Colombia, which have been committed to denouncing the risks and damage caused to the environment by extractivism and conflict (Tobasura Acuña, 2007; Dixon, 2016; Open Democracy, 2021; Paz Cardona, 2021). They have been termed 'troublemakers' and 'bad-faith disruptors' due to their environmental activism and critical stance towards the extractive sector and armed actors (San Pedro, 2019). In some cases, Indigenous communities have been associated with guerrilla groups, which has justified the excessive use of force against them by the army and police (San Pedro, 2019; IWGIA, 2021). Similarly, Indigenous protesters have been prosecuted for rebellion and terrorism following their social mobilisation (San Pedro, 2019).

This situation reflects broader trends: the Special Rapporteur on the rights of Indigenous peoples and the Special Rapporteur on situation of human rights defenders have noted the dramatic escalation of threats and violent attacks aimed at Indigenous peoples (UNGA, 2018, 2020), while the Special Rapporteur on the situation of human rights defenders and the Special Rapporteur on human rights and the environment have each raised concerns over a 'global crisis' of attacks against environmental human rights defenders (UNGA, 2018). Yet, the situation is particularly serious in Colombia. In 2019, Colombia and the Philippines recorded half of all killings of environmental rights defenders (Global Witness, 2020; Somos Defensores reported 124 murders in 2019). The militarisation of the country has added to the threats faced by environmental defenders: killings are most common in areas where the FARC-EP were historically present, while the continued presence of the National Liberation Army (ELN) aggravates the situation (UNGA, 2018). Most targeted activists are Afro-descendent and Indigenous, with women representing an ever-increasing ratio of victims (Paarlberg-Kvam, 2021).

In summary, Colombia has a history of land dispossession and the unequal distribution of natural resources. Extractive practices and violent displacement have characterised its history and have only increased since the 2016 peace agreement. These environmental harms have been acutely felt and have been resisted by Indigenous and Afro-descendent groups, leading to additional forms of state and armed violence being perpetrated against them. In the following section, we turn to considering the ways in which Black, decolonial and Indigenous feminist literature and activism can inform our understandings of the gendered impacts of these harms.

Gendering environmental harm in Colombia: learning from Black, decolonial and Indigenous communitarian feminisms

Theorising gendered violence

Decolonial and Indigenous communitarian feminists have drawn attention to the 'coloniality of gender', meaning the interlinked patriarchal and racialised systems that govern who has control over land, production and authority. They have argued that the ongoing coloniality of gender has played a fundamental role in facilitating dispossession and extraction on their territories (Rodriguez Castro, 2021c). In turn, this dispossession and extraction has had gendered impacts, with feminists exposing and theorising the relationship between harms against territory and violence against women (Paredes, 2012; Zaragocin and Caretta, 2021).

Lorena Cabnal (2010, 2016a, 2016b) and others have used the concept of *Cuerpo-Territorio* ('body-territory' or 'body-land') to illuminate interconnected legacies of colonial violence in several ways. First, the concept frames 'territory' as 'concrete territorial space, where the life of bodies is manifested' (Cabnal, 2010: 22), but also as 'resembling the human body and the collective spirit' (Tovar-Restrepo and Irazábal, 2014: 51), making clear that territory is something beyond a plot of land. Second, it understands women's bodies as the continuation of 'territory' (Sweet and Escalante, 2017), recognising that women's bodies, like 'territory' are both 'the givers of life' (Santamaria et al, 2019: 237) and frequent sites of violence (Cabnal, 2016b). Such an approach recognises territory as being at the centre of Indigenous women's agency (Gargallo Celentani, 2014; Zaragocin and Caretta, 2021), with 'agency' meaning the ability to create a 'satisfactory life through access to various resources, including decision making and the satisfaction of one's needs and interests' (El-Bushar, 2000: 67, cited in Tovar-Restrepo and Irazábal, 2014: 43).

The concept *Cuerpo-Territorio* is visible in the ways in which Indigenous and Afro-descendent women describe the violence they experience (Santamaria et al, 2019). For example, Hernández Reyes cites the following statement made by a Black women's group: 'we are threatened with both *bodily and cultural death … we are threatened by illegal mining and its backhoes that are destroying the environment and territory* that we have cared for centuries' (Hernández Reyes, 2019: 227; emphasis added). Similarly, Blanca Chancoso, a Quichua leader, has noted that for Indigenous women, violence is not something understood only interpersonally, but as something perpetrated through the seizing and destruction of territory (Rivera Zea, 1999: 19). This correlation between 'degraded bodies and environments' (Pulido, 2017: 525) has been expressed by drawing parallels between colonialism and rape since both involve an unwanted invasion. On one hand, the violation of a woman's body, which no woman desires. On the other hand, colonial invasion, which no community desires (Cabnal, 2010; Paredes, 2012).

The gendered impacts of attacks on territory are also expressed in terms of spiritual and communal harm. Yusmidia Solano, an organiser with the Red de Mujeres del Caribe in San Andrés, has explained that 'the effect [of extractivism] is grave, because women lose their territory, and with the loss of the territory comes the loss of identity, and the loss of family relationships, and everything else' (in interview with Paarlberg-Kvam, 2021: 13). Similarly, during fieldwork with Arhuaco women, Tovar-Restrepo and Irazábal were told that acts of violence towards their land caused 'negative imbalances to the whole universe' (2014: 51).

Dispossession, displacement and gendered violence

Black, decolonial and Indigenous feminists have also drawn attention to the ways in which harms perpetrated against territory, including dispossession and displacement, place women at increased risk of gendered forms of violence. These risks are compounded for those encountering patriarchal structures. For example, some communities do not allow for women to exercise land ownership rights (Meertens, 2012), or participate in political spaces (Santamaria et al, 2019). Further vulnerabilities exist for those living in isolated areas, for those who have relied on the land for survival, and for those who lack access to a 'formal' education and the majority-spoken Spanish language (Tovar-Restrepo and Irazábal, 2014). The power asymmetries that exist between marginalised women and armed actors (whether illegal or state) places them at risk of 'great losses, both human and health-related and with regard to land and other material resources' (Tovar-Restrepo and Irazábal, 2014).

This compounding of vulnerability can be traced through Colombia's history. For example, Paredes has noted how the imposition of colonial control over territory was accompanied by the imposition of gendered norms around 'femininity', which served to disenfranchise women and attacked their understanding of their own bodies, their place in their community and their traditional territorial knowledge (Paredes, 2012). This disenfranchisement had physically violent implications: framings of Indigenous and Afro-descendent women as sexually perverse or immoral facilitated (and continues to facilitate) a narrative that sexual violence against them was not a 'crime' (Gargallo Celentani, 2014). These colonial gender norms have had long-term implications, creating legacies of power over Indigenous women's bodies and power through the solidification of heterosexual patriarchy (Zaragocin, 2018).

Displacement from territory and environmental degradation, whether through the presence of armed groups, the imposition of extractive industries by the state, or illegal land-grabbing, continues to increase women's vulnerability to physical violence. For example, processes of forced displacement through land-grabbing have been shown to often be accompanied by sexual violence perpetrated by armed groups (Céspedes-Báez, 2019; Santamaria et al, 2019). As Gruner and Mina Rojas argue, women's bodies are used 'to infuse terror throughout their communities' (2018: 218). Displacement and the associated pressures of poverty have also been linked to increased risks of intrafamilial violence and intimate partner violence, including physical and sexual violence and reproductive control (Wirtz et al, 2014).

Displacement and the imposition of extractive industries on territory place women at greater risk of exploitation as well as violence (Collins, 2000). For example, in their ethnographic work with Afro-descendent women in María La Baja, Berman-Aréval and Ojeda (2020) trace the ways in which legacies of land-grabbing, the appropriation of water and paramilitary violence continue to shape the lived realities of women in the region. As they note, the paramilitary imposition of oil palm agribusiness marginalised women in the workforce, reducing them to 'activities that include domestic and sexual work in near urban centres' and increasing their dependency on men. This has in turn led to an increase in exposure to sexual and gender-based violence, within and outside the home (Berman-Aréval and Ojeda, 2020). Santamaria et al's (2019) interviews with Indigenous women reveal similar continuums of violence: their participants shared stories of having been denied access to an education after leaving their territories to engage in domestic work, where they encountered labour exploitation, abuse from employers and intimate partner abuse when they were deemed to have failed to provide adequate domestic labour.

Similarly, Rodríguez Castro's research in Toca and Sierra Nevada de Santa Marta highlights how the arrival of large businesses (flower companies and ecotourist initiatives) led to women taking on low-paid work, a shift which was often accompanied by their families selling land to large landowners and moving to urban areas. The work sometimes involved 'gendered' tasks, such as hosting tourists, or creating 'crafts', which women found difficult without sufficient training or time, and which was sometimes unpaid. At the same time, the dominance of patriarchal gendered roles meant women continued to take on caring and domestic roles, 'despite the physical and mental stress of working long hours' (Rodríguez Castro, 2021c: 53–4, see also Rodríguez Castro, 2021a). A decolonial feminist lens is important in understanding these series of events as violence. While they may not be constructed as 'criminality' in mainstream development literature, the dispossession through the introduction of agribusiness and exploitative tourism has 'clear elements of economic and epistemic violence embedded in dispossession and extractive logics of the body-land' (Rodríguez Castro, 2021c: 55).

Extractivism and 'everyday' violence

Black, decolonial and Indigenous feminist approaches stress the importance of centring the 'everyday' experiences of women in the spaces in which they operate. This approach acknowledges that women's positionality in informal economies and the private sphere, particularly when accompanied by their exclusion from public life, can lead to their experiences being overlooked (Rodríguez Castro, 2021c). Centring the 'everyday' ensures women's visibility and recognises the power structures that shape women's lives (Zaragocin, 2018; Paarlberg-Kvam, 2021). While women have certainly suffered from more 'conspicuous' forms of violence as noted previously, gendered violence may also be 'subtle and often taken-for-granted' (Berman-Arévalo and Ojeda, 2020: 1584), becoming so embedded that it is no longer recognised as violence (Menjívar, 2011). A feminist approach draws attention to these less visible forms of violence, which in Colombia often occur where the legacies of colonialism, conflict and extractive practices meet.

For example, drawing on ethnographic research with Epera Indigenous women on the Colombian–Ecuadorian border, Zaragocin demonstrates the ways in which the destruction of native forest and pollution of soil and riverways due to mining impact every aspect of life, including by making the community sick. As one of her respondents notes: '[W]hy do I have to worry that [my son] may die from diarrhoea? And I ask all of us women here – which one of us hasn't been hospitalised?' (Zaragocin, 2018: 382). Similar concerns have been noted in other contexts, with women who carry familial caring duties facing challenges in the face of limited resources, poverty and sickness (Carrillo, 2009). In the face of these challenges, women can find themselves expected to take on extra labour, whether as unpaid work on plantations under the tutelage of their male relatives, or by travelling further from home to seek clean water and uncontaminated food for their families (Berman-Arévalo and Ojeda, 2020). As noted previously, this increased their vulnerability to other forms of violence and exploitation, while risks of familial separation have also been exacerbated by socioeconomic deprivation (Carrillo, 2009).

In addition to polluting land and water, the imposition of extractive industries can change the physical shape of the community, introducing 'fences, security guards and the physical occupation of spaces of sociality and mobility' by industries (Berman-Arévalo and Ojeda, 2020). Such restrictions have 'everyday' impacts on women, reducing the space available to engage in family and community activities and increasingly confining their lives to their homes and domestic labours. Thus, women can find their lives simultaneously reduced in terms of their autonomy and access to space and expanded in terms of the labour that is expected of them.

Agency and resistance

Black, decolonial and Indigenous feminist approaches often challenge the 'inevitable fate of female victimization' when analysing women's experiences of violence (Santamaria et al, 2019). In common with environmental movements across Latin America and the world (UNGA, 2018), grassroots women have been central in resisting resource extraction and associated environmental destruction on their territories (Rodríguez-Garavito and Arenas, 2005; Seppälä, 2016). Indigenous and Afro-descendent women have repeatedly challenged the commercial development model and the actions of armed actors, calling out their disregard for the interplay between the natural environment, Indigenous territories, self-determination and Indigenous cultural survival (Dixon, 2016; UNGA, 2018; IWGIA, 2021).

Cabnal (2010) draws attention to the connections between defending land and feminist activism (see also Gargallo Celentani, 2014). As she argues, the *Cuerpo-Territorio* approach recognises that for these women, defending territory is not solely about assuring access to the natural resources needed to live and pass on a dignified life, but about reclaiming bodies that have been expropriated by colonisation. Indigenous communitarian feminism argues that as the dispossession of territory also involves the dispossession of women's bodies, so must the territorial struggles of Indigenous peoples go hand in hand with the defence of women's bodies. In this context, defending the body-territory means acquiring vitality, enjoying pleasures previously forbidden by the coloniser, and building liberating knowledge for autonomous decision-making. As Cabnal highlights, Indigenous women defend their territory because they 'cannot conceive of this woman's body, without a space on earth that dignifies [their] existence and promotes [their] life in plenitude' (2010: 23). Similarly, Hernández Reyes has argued that 'Black people's activism and their collective resistance … constitute a response to the enduring historical, political, and economic project of capitalism in which structural racism and the devaluation of black bodies interlock' (Hernández Reyes, 2019: 219). She quotes a Black woman activist, who states that Black people 'are concerned about defending the territory, because without the territory they will be nobody', reasserting the social costs of the racialised commodification of land (Hernández Reyes, 2019: 228–9).

In practice, Indigenous and Afro-descendent women have stood out for their commitment to peaceful resistance against activities that put their livelihood, territory, culture and communities at risk (San Pedro, 2019). This has involved a range of activities, including mobilising their communities (sometimes in the face of patriarchal resistance within their communities), making legal and public complaints (San Pedro, 2019), and fighting for space at the table during Colombia's peace process (Gruner and Mina Rojas, 2018). This has also involved stepping onto the international stage. For example, at the World People's Conference on Climate Change and Mother Earth in 2020, the Assembly of Communitarian Feminism made a statement demanding the end to 'patriarchal property and male control over land, territory, seas, lakes, and sky'. The group further asserted that 'by abolishing neoliberal and masculine property, the war that preys on territory and makes women its booty may be ended', highlighting the interconnected nature of environmental harm and gender-based violence in their lived experiences (Asambleas del Feminismo Comunitario, 2010).

This activism has come at a cost: Indigenous and Afro-descendent women environmental defenders in Colombia have become targets of assassinations, attacks on their integrity, harassment, violence, threats and surveillance (San Pedro, 2019). Women represent 'collectivity and ways of being on land' which are deemed particularly threatening (Gruner and Mina Rojas, 2018), and the attacks on women have several gendered qualities. According to the Colombian Ombudsman's Office, unlike threats directed against men, the language used in threatening messages against women human and environmental rights defenders includes sexist content, allusions to women's bodies, sexual slurs and explicit threats of sexual violence (Defensoria del Pueblo Colombia, 2018). It is also more likely to be accompanied by threats against women's families, a tactic which reflects patriarchal assumptions about the role of women in the family (Defensoria del Pueblo Colombia, 2018; San Pedro, 2019). For the Ombudsman's Office, this violence against women defenders seeks to reprimand and inhibit their public leadership, which is considered contrary to the stereotypical gender roles imposed through Colombia's colonial history (Defensoria del Pueblo Colombia, 2018). Women can also face risks from within their own communities, where these stereotypical gender roles are reproduced and enforced through disapproval and in some cases physical violence (Santamaria et al, 2019). Nevertheless, women continue to defy traditional macho values, and calls to restrict themselves to domestic activities (San Pedro, 2019).

Conclusion

This chapter has sought to highlight the continuum of environmental harms perpetrated against Indigenous and Afro-descendent women due to colonial practices of land distribution maintained over centuries, a long-standing armed conflict of several decades, and the ever-more extractive economy that has emerged in the aftermath of that conflict. By platforming Black, decolonial and Indigenous feminist scholarship and activism, it has aimed to highlight how much can be learned about the relationship between environmental and gendered harm by engagement with these bodies of work and praxis. These lessons include:

- a deeper understanding of the connections between territorial and bodily autonomy for Indigenous and Afro-descendent women;
- an awareness of the gendered vulnerabilities that can result from displacement due to armed conflict and extractivism;
- greater visibility for the 'everyday' forms of violence women experience; and

- an appreciation for the agency and resistance of Indigenous and Afro-descendent women in the face of attacks on their territory and communities.

The Global North academy (where both authors are currently based) has been fairly critiqued for failing to 'interact with the Andean social sciences in any meaningful way', or for selecting certain voices and arguments while leaving other, less 'fashionable' voices in the shadows (Cuiscanqui, 2012). As we have explored here, recognising and learning from Black, decolonial and Indigenous feminism provides a counter to 'universalising' narratives, enhancing nuanced understandings of gendered victimhood and agency (Hernández Reyes, 2019), making visible analyses that are grounded in *Cuerpo-Territorio* and 'the anti-extractive resistance of subaltern women' (Paarlberg-Kvam, 2021: 309). Such an approach is particularly crucial in the context of green criminology, due to the profound connections between feminist and eco-centric scholarship and activism. As argued by Svampa, 'one cannot be a feminist, struggle against the patriarchy, and foment the destruction of territory and the advance of extractivism. Neither the planet in flames nor the country groaning under the weight of unsustainable development models will tolerate it' (Svampa, 2020, translated by Paarlberg-Kvam, 2021). Similarly, green criminology's commitment to scrutinising 'the harms and crimes of the powerful' (Sollund, 2021) and relationship with political activism (Rodríguez Goyes, 2016) necessitate an awareness of the gendered nature of power and the intersectional impacts of environmental harm. Black, decolonial and Indigenous feminisms play a crucial role in exposing these interconnected issues and calling attention to the social, economic, racial and gendered injustices that racialised women continue to face in the neoliberal extractivist system.

Notes

[1] It is important to note that there is not one history of Indigenous and Afro-descendent populations in Colombia. While we aim to highlight some dominant themes in the literature, we acknowledge that histories and experiences vary significantly across physical place and time. See, for example, LeGrand (2016).

[2] It is often translated as 'Mother Earth'. Pachamama is one of the deities of the Indigenous communities of the Andes. Pachamama has the power to sustain life on this earth.

References

Asambleas del Feminismo Comunitario (2010) Pronunciamiento del Feminismo Comunitario latinoamericano en la Conferencia de los pueblos sobre Cambio Climático. *Biodiversidad en América Latina*. Available at: www. biodiversidadla.org/Documentos/Pronunciamiento_del_Feminismo_ Comunitario_latinoamericano_en_la_Conferencia_de_los_pueblos_sob re_Cambio_Climatico (accessed 13 June 2022).

Balint, J., Evans, J. and McMillan, N. (2014) Rethinking transitional justice, redressing indigenous harm: A new conceptual approach. *International Journal of Transitional Justice*, 8(2), 194–216.

Baptiste, B., Pinedo-Vasquez, M., Gutierrez-Velez, V.H., Andrade, G.I., Vieira, P., Estupiñán-Suárez, L.M., Londoño, M.C., Laurance, W. and Ming Lee, T. (2017) Greening peace in Colombia. *Nature Ecology & Evolution*, 1(4), 1–3.

Bastian Duarte, Á.I. (2012) From the margins of Latin American feminism: Indigenous and lesbian feminisms. *Signs*, 38(1), 153–78.

Batchelder, R.W. and Sanchez, N. (2013) The encomienda and the optimizing imperialist: An interpretation of Spanish imperialism in the Americas. *Public Choice*, 156(1/2), 45–60.

Berman-Arévalo, E. and Ojeda, D. (2020) Ordinary geographies: Care, violence, and agrarian extractivism in 'post-conflict' Colombia. *Antipode*, 52(6), 1583–602.

Bickerton, P. (2017) *Keeping Colombia Megadiverse*. Available at: www. earlham.ac.uk/articles/keeping-colombia-megadiverse (accessed 26 February 2022).

Brisman, A., South, N. and White, R. (2015) *Environmental Crime and Social Conflict: Contemporary and Emerging Issues*. Farnham, Surrey: Ashgate.

Cabnal, L. (2010) *Feminismos Diversos: El Feminismo Comunitario*. Spain: ACSUR-Las Segovias.

Cabnal, L. (2016a) De las opresiones a las emancipaciones: Mujeres indígenas en defensa del territorio cuerpo-tierra. *Biodiversidad en América Latina*. Available at: www.biodiversidadla.org/Documentos/De_las_opresiones_ a_las_emancipaciones_Mujeres_indigenas_en_defensa_del_territorio_cue rpo-tierra (accessed 13 June 2022).

Cabnal, L. (2016b) Comunidad, cuerpo y territorio: feminismos y ambiente. Available at: www.youtube.com/watch?v¼gyjSxocXVf4 (accessed 26 July 2022).

Camargo, X.S. (2019) The ecocentric turn of environmental justice in Colombia. *King's Law Journal*, 30(2), 224–233.

Carby, H. (2007) White woman listen! Black feminism and the boundaries of sisterhood. In A. Gray, J. Campbell, M. Erickson, S. Hanson and H. Wood (eds) *CCCS Selected Working Papers*. London: Routledge, pp 753–74.

Carrillo, A.C. (2009) Internal displacement in Colombia: Humanitarian, economic and social consequences in urban settings and current challenges. *International Review of the Red Cross*, 91(875), 527–46.

Céspedes-Báez, L.M. (2019) A (feminist) farewell to arms: The impact of the peace process with the FARC-EP on Colombian feminism. *Cornell International Law Journal*, 52, 39–63.

Collins, P. (2000) Gender, black feminism, and black political economy. *Annals of the American Academy of Political and Social Science*, 568(1), 41–53.

Congress of the Republic of Colombia (2011) *Law 1448*.

Cooper, F. (2005) *Colonialism in Question: Theory, Knowledge, History*. Berkeley: University of California Press.

Crenshaw, K. (1989) Demarginalizing the intersection of race and sex: A black feminist critique of antidiscrimination doctrine, feminist theory and antiracist politics. *The University of Chicago Legal Forum*, 140, 139–68.

Cusicanqui, S.R. (2012) Ch'ixinakax utxiwa: A reflection on the practices and discourses of decolonization. *The South Atlantic Quarterly*, 111(1), 95–109 .

Defensoria del Pueblo Colombia (2018) *Alerta Temprana No. 026–18*. Colombia: Defensoria del Pueblo Colombia.

Dixon, L. (2016) *Comunidades indígenas asumen el mando de la conservación en Colombia*. Available at: https://es.mongabay.com/2016/11/colombia-indigenas-conservacion-medioambiente/ (accessed 20 February 2022).

El-Bushar, J. (2000) Transforming conflict: Some thoughts on a gendered understanding of conflict cesses. In S. Jacobs, R. Jacobson and J. Marchbank (eds) *States of Conflict: Violence, and Resistance*, London: Zed Books, pp 66–86.

Escobar, A. (2004) Desplazamientos, desarrollo y modernidad en el Pacífico colombiano. In E. Restrepo and A. Rojas (eds) *Conflicto e (in) visibilidad: Retos en los estudios de la gente negra en Colombia*. Cali: Editorial Universidad del Cauca, pp 53–72.

Federici, S. (2010) *Calibán y la bruja: Mujeres, cuerpo y acumulación primitiva*. Madrid: Traficantes de Sueños.

Gargallo Celentani, F. (2014) *Feminismos desde Abya Yala. Ideas y proposiciones de las mujeres de 607 pueblos en nuestra América*. Ciudad de Mexico: Editorial Corte y Confección.

Global Witness (2020) *Defending Tomorrow: The Climate Crisis and Threats against Land and Environmental Defenders*. Available at: www.globalwitn ess.org/en/campaigns/environmental-activists/defending-tomorrow/ (accessed 12 May 2023).

Gomez-Betancur, L. (2020) The rights of nature in the Colombian Amazon: examining challenges and opportunities in transitional justice setting. *UCLA Journal of International Law and Foreign Affairs*, 25(1), 41–84.

Government of Colombia and FARC-EP (2016) *Acuerdo Final para la Terminación del Conflicto y la Construcción de una Paz Estable y Duradera*.

Grosfoguel, R. (2016) Del 'extractivismo económico' al 'extractivismo epistémico' y al 'extractivismo ontológico': Una forma destructiva de conocer, ser y estar en el mundo. *Tabula Rasa*, 24, 123–43.

Gruner, S. and Mina Rojas, C. (2018) Black and indigenous territorial movements: Women striving for peace in Colombia. *Canadian Woman Studies*, 33(1–2), 211–21.

Gudynas, E. (2021) *Extractivisms: Politics, Economy and Ecology*. Black Point, Nova Scotia: Fernwood Publishing.

Hernández Castillo, R.A. (2014) Entre el etnocentrismo feminista y el esencialismo étnico. Las mujeres indígenas y sus demandas de género. In Y. Espinosa Miñoso, D.M. Gómez Correal and K. Ochoa Muñoz (eds) *Tejiendo de otro modo: feminismo, epistemología y apuestas descoloniales en Abya Yala*. Popayán: Editorial Universidad del Cauca, pp 279–94.

Hernández Reyes, C.E. (2019) Black women's struggles against extractivism, land dispossession, and marginalization in Colombia. *Latin American Perspectives*, 46(2), 217–34.

Ibañez, A. (2009) Forced displacement in Colombia: Magnitude and causes. *The Economics of Peace and Security Journal*, 4(1), 48–54.

ICTJ (2009) *An Overview of Conflict in Colombia*. International Center for Transitional Justice. Available at: www.ictj.org/publication/overview-confl ict-colombia (accessed 12 May 2023).

IWGIA (2021) *Indigenous World 2021*. International Work Group for Indigenous Affairs. Available at: www.iwgia.org/en/resources/publicati ons/4513-the-indigenous-world-2021.html (accessed 12 May 2023).

Kim, A. (2020) The politics of citation. *Diacritics*, 48(3), 4–9.

Leal, C. and Van Ausdal, S. (2013) *Landscapes of Freedom and Inequality: Environmental Histories of the Pacific and Caribbean Coasts of Colombia*. Working Paper No. 58, Research Network on Interdependent Inequalities in Latin America.

LeGrand, C. (2016) *Colonización y Protesta Campesina en Colombia (1850–1950)*. Bogotá: Universidad de los Andes, Ediciones Uniandes and Universidad Nacional de Colombia, CINEP.

Lugones, M. (2008) Colonialidad y género: hacia un feminismo descolonial. In W. Mignolo (ed) *Género y descolonialidad*. Buenos Aires: Ediciones del Signo., pp 13–54.

Lugones, M. (2014) Rumo a um feminismo descolonial. *Estudios Feministas, Florianópolis*, 22(3), 935–52.

Mack, A. and Na'puti, T. (2019) 'Our bodies are not Terra Nullius': Building a decolonial feminist resistance to gendered violence. *Women's Studies in Communication*, 42(3), 347–70.

Maldonado-Torres, N. (2016) *Outline of Ten Theses on Coloniality and Decoloniality*. Fondation Frantz Fanon. Available at: http://fondation-fran tzfanon.com/outline-of-ten-theses-on-coloniality-and-decoloniality/ (accessed 12 May 2023).

McClanahan, B., Sanchez Parra, T. and Brisman, A. (2019) Conflict, environment and transition: Colombia, ecology and tourism after demobilisation. *International Journal for Crime, Justice and Social Democracy*, 8(3), 74–88.

Meertens, D. (2012) *Forced Displacement and Gender Justice in Colombia: Between Disproportional Effects of Violence and Historical Injustice*. International Center for Transitional Justice. Available at: www.ictj.org/es/node/15653 (accessed 12 May 2023).

Menjívar, C. (2011) *Enduring Violence: Ladina Women's Lives in Guatemala*. Berkeley: University of California Press.

Millán, M. (ed) (2014) *Mas Allá del Feminismo: Caminos Para Andar*. México, D.F.: Red de Feminismos Descoloniales.

Mol, H. (2017) *The Politics of Palm Oil Harm: A Green Criminological Perspective*. Cham: Palgrave Macmillan.

Motta, S.C. (2011) Social movements and/in the postcolonial: Dispossession, development and resistance in the Global South. In S.C. Motta and A.G. Nilsen (eds) *Social Movements in the Global South*. London: Palgrave Macmillan, pp 1–31.

Ojeda, D. (2012) Green pretexts: Ecotourism, neoliberal conservation and land grabbing in Tayrona National Natural Park, Colombia. *Journal of Peasant Studies*, 39(2), 357–75.

Open Democracy (2021) *La gigantesca mina de carbón de Cerrejón, denunciada por atentar contra derechos humanos y ambientales*. Available at: www.opende mocracy.net/es/gigantesca-mina-de-carb%C3%B3n-colombiana-denunci ada-por-atentar-contra-derechos-humanos-y-ambientales/ (accessed 27 February 2022).

Paarlberg-Kvam, K. (2021) Open-pit peace: The power of extractive industries in post-conflict transitions. *Peacebuilding*, 9(3), 289–310.

Paredes, J. (2012) Las trampas del patriarcado. In *Pensando los feminismos en Bolivia*. La Paz: Conexión Fondo de Emancipación, 89–112.

Park, A.S.J. (2015) Settler colonialism and the politics of grief: Theorising a decolonising transitional justice for Indian residential schools. *Human Rights Review*, 16(3), 273–93.

Paz Cardona, A.J. (2021) *Las deudas ambientales de Colombia en 2020: defensores asesinados, más deforestación y la polémica sobre el glifosato*. Available at: https:// es.mongabay.com/2021/01/balance-deforestacion-asesianto-lideres-colom bia-2020/ (accessed 27 February 2022).

Picq, M. (2014) Self-determination as anti-extractivism: How indigenous resistance challenges world politics. In M. Woons (ed) *Restoring Indigenous Self-Determination: Theoretical and Practical Approaches*. Bristol: E-International Relations, pp 26–33.

Pulido, L. (2017) Geographies of race and ethnicity II: Environmental racism, racial capitalism and state sanctioned violence. *Progress in Human Geography*, 41, 524–33.

Rivera Zea, T. (ed) (1999) *El andar de las mujeres indígenas*. Lima: Chirapaq, Centro de Culturas Indias.

Rodríguez, A., Binda, E., Ochoa Quintero, J.M., Garcia, H., Gómez, B., Soto, C., Martínez, S. and Clerici, N. (2020) Answering the right questions: Addressing biodiversity conservation in postconflict Colombia. *Environmental Science and Policy*, 104, 82–7.

Rodríguez Castro, L. (2021a) *Decolonial Feminisms, Power and Place: Sentipensando with Rural Women in Colombia*. Cham: Palgrave Macmillan.

Rodríguez Castro, L. (2021b) 'We are not poor things': Territorio cuerpo-tierra and Colombian women's organised struggles. *Feminist Theory*, 22(3), 339–59.

Rodríguez Castro, L. (2021c) Extractivism and territorial dispossession in rural Colombia: A decolonial commitment to *campesinas*' politics of place. *Feminist Review*, 128, 44–61.

Rodríguez-Garavito, C.A. and Arenas, L. (2005) Indigenous rights, transnational activism, and legal mobilization: The struggle of the U'Wa people in Colombia. In B. de Sousa Santos and C.A. Rodríguez-Garavito, *Law and Globalization from Below: Towards a Cosmopolitan Legality*. Cambridge: Cambridge University Press, pp 241–66.

Rodríguez Goyes, D.R. (2015) Land uses and conflict in Colombia. *Journal of Comparative and Applied Criminal Justice*, 37(1), 1–15.

Rodríguez Goyes, D.R. (2016) Green activist criminology and the epistemologies of the south. *Critical Criminology*, 24(4), 503–18.

Rodríguez Goyes, D.R. (2017) Green criminology as decolonial tool: A stereoscope of environmental harm. In K. Carrington, R. Hogg, J. Scott and M. Sozzo (eds), *The Palgrave Handbook of Criminology and the Global South*. Cham: Palgrave Macmillan, 323–46.

Rodríguez Goyes, D.R. (2019) *Southern Green Criminology: A Science to End Ecological Discrimination*. Bingley: Emerald Publishing.

Rodríguez Goyes, D.R. and Sollund, R. (2016) Contesting and contextualising CITES: Wildlife trafficking in Colombia and Brazil. *International Journal for Crime, Justice and Social Democracy*, 5(4), 87–102.

Rodríguez Goyes, D.R. and South, N. (2016) Land-grabs, biopiracy and the inversion of justice in Colombia. *The British Journal of Criminology*, 56(3), 550–77.

Rodríguez Goyes, D.R., South, N., Astroina Abaibira, M., Baicué, P., Cuchimba, A., Tatiana, D. and Ñeñetofe, R. (2021a) Genocide and ecocide in four Colombian indigenous communities: The erosion of a way of life and memory. *The British Journal of Criminology*, 61(4), 965–84.

Rodríguez Goyes, D.R., Astroina Abailbira, M., Baicué, P., Cuchimba, A., Tatiana, D., Ñeñetofe, R., Sollund, R., South, N. and Wyatt, T. (2021b) Southern green cultural criminology and environmental crime prevention: Representations of nature within four Colombian indigenous communities. *Critical Criminology*, 29, 469–85.

Rojas-Páez, G. (2017) Understanding environmental harm and justice claims in the global south: Crimes of the powerful and people's resistance. In D.R. Rodríguez Goyes, H. Mol, A. Brisman and N. South (eds) *Environmental Crime in Latin America: The Theft of Nature and the Poisoning of the Land*. London: Palgrave Macmillan, pp 57–83.

Roth, B. (2003) Second-wave black feminism in the African diaspora: News from new scholarship. *Agenda: Empowering Women for Gender Equity*, 58, 46–58.

San Pedro, P. (2019) *Women Defenders of Agricultural, Territorial, and Environmental Rights in Colombia: Risking their Lives for Peace*. Oxfam. Available at: https://policy-practice.oxfam.org/resources/women-defend ers-of-agricultural-territorial-and-environmental-rights-in-colombi-620 872/ (accessed 12 May 2023).

Sánchez-Cuervo, A. and Aide, T. (2013) Consequences of the armed conflict, forced human displacement, and land abandonment on forest cover change in Colombia: A multiscaled analysis. *Ecosystems*, 16(6), 1052–70.

Santamaria, A., García, D., Hernández, F. and Pardo, A. (2019) Kaleidoscopes of violence against indigenous women (VAIW) in Colombia: The experiences of Pan-Amazonian women. *Gender, Place & Culture*, 26(2), 227–50.

Seppälä, T. (2016) Feminizing resistance, decolonizing solidarity: Contesting neoliberal development in the Global South. *Journal of Resistance Studies*, 2(1), 12–47.

SiB Colombia (2020) *Biodiversidad en Cifras*. Available at: https://cifras.biodiv ersidad.co/ (accessed 26 February 2022).

Sollund, R. (2017) The use and abuse of animals in wildlife trafficking in Colombia: Practices and injustice. In D.R. Rodríguez Goyes, H. Mol, A. Brisman and N. South (eds) *Environmental Crime in Latin America: The Theft of Nature and the Poisoning of the Land*. London: Palgrave Macmillan, pp 215–43.

Sollund, R. (2021) Green criminology: Its foundation in critical criminology and the way forward. *The Howard Journal of Crime and Justice*, 60(3), 304–22.

Somos Defensores (2020) *Informe Annual 2019. La Ceguera*. Bogotá: Programa Somos Defensores and Asociación Minga.

Sousa Santos, B. and Sena Martins, B. (2021) *The Pluriverse of Human Rights: The Diversity of Struggles for Dignity*. New York: Routledge.

Stoler, A.L. (2008) Imperial debris: Reflections on ruins and ruination. *Cultural Anthropology*, 23(2), 191–219.

Svampa, M. (2020) Svampa y su 'mensajería' para los 'progres' que integran el gobierno de Alberto. *El Extremo Sur de la Patagonia*, 12 January. Available at: www.elextremosur.com (accessed 23 July 2022).

Sweet, E.L. and Escalante, S.O. (2017) Engaging Territorio Cuerpo-Tierra through body and community mapping: A methodology for making communities safer. *Gender, Place & Culture*, 24, 1–13.

Temper, L., del Bene, D. and Martinez-Allier, J. (2022) Colombia. *Environmental Justice Atlas*. Available at: https://ejatlas.org/country/colom bia (accessed 2 June 2022).

Tobasura Acuña, I. (2007) Ambientalismos y ambientalistas: una expresión del ambientalismo en Colombia. *Ambiente & Sociedade*, 10, 45–60.

Todorov, C. (2007) *La conquista de América: el problema del otro*. Mexico: Siglo Veintiuno Ed.

Torres Rodríguez, A., Binda, E., Ochoa Quintero, J.M., García, H., Gómez, B., Soto, C., Martínez, S. and Clerici, N. (2019) Answering the right questions: Addressing biodiversity conservation in post-conflict Colombia. *Environmental Science & Policy*, 104, 82–7.

Tovar-Restrepo, M. and Irazábal, C. (2014) Indigenous women and violence in Colombia: Agency, autonomy, and territoriality. *Latin American Perspectives*, 41(1), 39–58.

Tuck, E. and Yang, K.W. (2012) Decolonization is not a metaphor. *Decolonization: Indigeneity, Education, and Society*, 1(1), 1–40.

UNGA (2018) *Report of the Special Rapporteur on the Rights of Indigenous Peoples*. UNGA Human Rights Council, 10 August, A/HRC/39/17.

UNGA (2020) Final warning: Death threats and killing of human rights defenders. *Report of the Special Rapporteur on the Situation of Human Rights Defenders*, 24 December 2020, A/HRC/46/35.

Vélez-Torres, I. (2014) Governmental extractivism in Colombia: Legislation, securitization and the local settings of mining control. *Political Geography*, 38, 68–78.

Wirtz, A., Pham, K., Glass, N., Loochkartt, S., Kidane, T., Cuspoca, D., Rubenstein, L.S., Singh, S. and Vu, A. (2014) Gender-based violence in conflict and displacement: Qualitative findings from displaced women in Colombia. *Conflict and Health*, 8(10), 1–14.

Wright, C., Rolston, B. and Ní Aoláin, F. (2022) Navigating colonial debris: Structural challenges for Colombia's peace accord. *Peacebuilding*, 11(1), 1–16.

Zaragocin, S. (2018 Gendered geographies of elimination: Decolonial feminist geographies in Latin American settler contexts. *Antipode*, 51(1), 373–92.

Zaragocin, S. and Caretta, M.A. (2021) Cuerpo-Territorio: A decolonial feminist geographical method for the study of embodiment. *Annals of the American Association of Geographers*, 111(5), 1503–18.

13

Vegan Feminism Then and Now: Women's Resistance to Legalised Speciesism across Three Waves of Activism

Corey Lee Wrenn and Lynda M. Korimboccus

Introduction

There is an implicit deviance in women's activism, exemplified in the popular adage: 'Well-behaved women seldom make history' (Ulrich, 2007). This is certainly the case in anti-speciesism work beginning in late-18th and early 19th-century Britain and Europe. Advocacy on behalf of other animals was stealthily adopted by women who played on gender stereotypes such as 'angels of the home' and 'nature's caretakers' in order to enter the patriarchal public sphere and resist anthroparchal oppression (Unger, 2012). Originally, these campaigns focused on especially male pursuits, such as 'hunting' (a euphemism for male violence against other animals) and vivisection. Women's contributions to campaigns, funding and public support were invaluable to early efforts. Yet, as was typical of the time, women were also frequently prevented from leadership positions due to Victorian mores and concerns that the cause might face delegitimisation via feminisation (a fear that persists today) (Groves, 2001).

A small but significant body of research has documented the efforts of women in the nonhuman animal rights movement as founders of leading charities, authors of seminal texts, and community organising and education (Ferguson, 1998; Kean, 1998; Gaarder, 2011; Donald, 2020), but this chapter is interested in the extra-institutional ingenuity of women who eschewed prevailing laws, tactfully adopted the deviant mantle, and advanced anti-speciesist practice and theory through their actions. To achieve this, we highlight the efforts of one notable woman in each of the three waves of Western anti-speciesist activism. Charlotte Despard is offered as a representative of the first wave. This initial wave transpired over the Victorian and early Edwardian eras, emphasising humane education and challenging vivisection head on. Next, Patty Mark is chosen as a representative of the second wave which rose in the mid-20th century, inspired by goals and strategies of the civil rights movement and characterised by increased attention to 'farmed' animal welfare. Lastly, we examine Sarah Kistle as a representative of the current wave of nonhuman animal rights activism. This third wave is distinguished by a commitment to veganism, conscious attention to intersectionality, and access to new social media technologies. These three women are not only interesting in their explicit challenge to the legal system to advance nonhuman animal interests; they also demonstrate the deeply entangled nature of oppression in their resistance to speciesist subjugation through explicitly gendered (and sometimes racialised) lenses. The intersectional lens that has become popularised in current green criminology and eco-feminist discourses is an extension of foundational connections made by Victorian-era activists more than a century ago (Lahar, 1991). This survey of three women defying speciesism across the centuries interweaves generations of activism and demonstrates the centrality of intersectional, feminist thought to critical vegan, anti-speciesist and environmental work.

As this chapter furthers, the *activist* element of eco-feminist theory is also particularly relevant (see Chapter 2 on eco-feminism for further exploration). That is, collective consciousness and theoretical awareness of social injustices and their intersectional nature are important, but practitioners are expected to participate, disrupt and bring theory into action (Sturgeon, 1997). Said theory combines various approaches to support the victims of injustice. Feminist criminology, for instance, highlights a wider system of domination that is maintained, at least in part, by the stigmatisation or criminalisation of various aspects related to identities of 'otherness' (Crenshaw, 1991). Feminist criminology has been unforthcoming with regard to the otherness of nonhuman animals, however. Green criminologists have been better representatives in this regard, although they are primarily interested in the injustices faced by 'wild' nonhuman animals. More recently, some of these scholars have started to address this gap in turning their attention to the millions of other nonhuman animals imprisoned in agribusiness (Taylor and Fitzgerald, 2018; Sollund, 2021). This 'critical animal turn' in green criminology has roots in the more established environmental ethic of eco-feminism. Eco-feminism understands environmental and nonhuman oppression as a consequence of complex power relations manufactured by capitalist, patriarchal and white supremacist systems. By the 1980s, a vegan eco-feminist (or, more broadly, vegan feminist) branch had emerged, explicitly acknowledging the plight of other animals, both domesticated and free-living (Adams, 1991; Gaard, 2002; Foster, 2021).

As the overlapping emphases of these theoretical traditions suggest, an intersectionality in injustices faced by various marginalised groups is worth examining, and the result of these examinations must be acted upon. This chapter argues that many anti-speciesist activists, particularly *feminist* anti-speciesists, have acknowledged that intervention will be necessary to disrupt unjust power structures, and that this disruption will need to consider the entanglements of species, gender, 'race' and other identities. The ways in which this gendered green philosophy has transpired in real-world efforts are diverse and, of course, shaped by their historical contexts. The following examination of three waves of nonhuman animal rights activism as exemplified by three extraordinary women, however, identifies one predominant commonality: some element of extra-institutional participation will be vital to achieving both short- and long-term social change for the benefit of nonhuman animals and other marginalised groups.

Charlotte Despard and Edwardian anti-speciesist activism

The history of nonhuman animal rights activism, we would argue, is relatively unknown to the average activist in the West. The dominant narrative usually begins with the modern incarnation of the movement in the 1970s, ushered in with the hugely popular work of Peter Singer (1975). Some might be familiar with the work of British activist and author Henry Salt (1851–1939) or founder of the American Society for the Prevention of Cruelty to Animals, Henry Bergh (1813–88). However, few may be familiar with the extraordinary efforts of so many powerful women of the era. Charlotte Despard is one such titan who deserves to be a household name in the vegan home.

Despard's obscurity in nonhuman animal rights history may be due to the fact that she is more popularly historicised as a leading suffragette. Like many feminists of the time, the various strains of her activism often merged (Leneman, 1997). She served as president to the Women's Freedom League in Britain, which regularly served vegetarian food at events, offered demonstrations in vegetarian cookery, and even opened up wartime vegetarian restaurants. She was also on the Council of the London Vegetarian Society and a plethora of other social justice organisations. Although born in Edinburgh and living for most of her life in the south of England, Despard (whose father was Irish) was a key supporter of Irish nationalism and co-founded the Irish Women's Franchise League in 1908. At one point, Despard operated a jam-making enterprise on her property as a means of providing employment independent of Britain, food independent of colonial control, and vegetarian produce free of nonhuman animal suffering (Farr, 2019) (she would likely be horrified that a modern London pub in her name and neighbourhood sells all manner of nonhuman animal bodies on its snacking menu).

Although her tactical and organisational accomplishments in the field of social justice are many, for the purposes of this chapter, we focus on her contributions to nonhuman animal liberation. In this area, she dedicated particular attention to anti-vivisection work in London. Despard took the lead on the 1906 erection of a statuesque fountain in Battersea to commemorate some of the victims of the exploding and controversial vivisection industry. The statue depicted a small, stray dog who had been publicly vivisected upon in 1903 at University College London. The subsequent campaign and its associated legal case are probably best remembered in the nonhuman animal rights annals as the 'Brown Dog Affair' (less graciously known as the 'Brown Dog Riots') (Lansbury, 1985). Battersea was strategically chosen as the statue's location as it was home to England's largest dog shelter. Though the borough was a poor one, it also enjoyed a rather democratic (and anti-vivisectionist) council and offered housing estates that were a point of pride. The small brown dog, just one of the hundreds vivisected at University College London that year, stood atop the statue in one of these estates. More than a symbolic protest against vivisection, the little dog came to represent the injustices experienced by the residents of humble Battersea and beyond, all those who had been languishing under a rigid class system. Despard was far from working class herself, but she was nonetheless accepted as a local resident in the industrial Nine Elms area of London (Farr, 2019).

The statue quickly became a lightning rod in the clash between University College London, its students, and middle- and upper-class medical practitioners, and the disenfranchised women, Irish immigrants, working-class labourers and slum-dwellers. As one historian observes: 'the brown dog stared across to the neat ranks of council houses, and if ever a riot had been deliberately instigated, this was it' (Lansbury, 1985: 15). University students, boisterous, male, and intent on preserving their threatened power and entitlement, descended on the borough with sledgehammers and crowbars to (unsuccessfully) dismantle the statue, and were met with retaliatory bonfires on campus. Participants on both sides of the row were arrested, including Despard. In a later attempt, pro-vivisection medical students created trouble in Trafalgar Square, engaging in fights with the working-class men there and necessitating police intervention. They then attempted an attack on the flagship National Anti-Vivisection Hospital in Battersea (later Battersea General Hospital), an institution designed to defy prevailing medical authority by providing treatment without the need for vivisection or state support (Bates, 2017). Anti-vivisection meetings were also targeted by pro-vivisection students, leading to considerable violent disruption.

In this remarkable campaign, activists not only recognised the terrible injustices enacted on nonhuman animals, but also the deeply entangled nature of human and nonhuman oppression. The blatant torture of nonhuman animals in scientific and medical establishments for the privilege of the elite, and rationalised by the elite as being in the so-called 'greater good', represented similar sufferings enacted on vulnerable human groups. In some cases, the sufferings were identical – humans were frequently forced or coerced into vivisection, especially women, children, poor persons and enslaved persons (Beecher, 1959; Savitt, 1982; Kenny, 2015). With many suffragettes on the forefront of anti-vivisection campaigning, the women's movement soon became conflated with the nonhuman animal rights movement (Donald, 2019). Yet, as the working classes continued to step up to protect the statue (and what it represented), this conflation shifted. When council leadership in Battersea moved away from the political left in 1908, for instance, the brown dog came to symbolise the threat of socialism (Lansbury, 1985). The working-class character of the borough became a point of vulnerability when the citizens were ordered to pay for the heavy policing required to deal with constant rioting the statue provoked. Although anti-vivisectionist leaders moved quickly for an injunction, the council moved quicker, and in 1910, under cover of darkness, the brown dog was removed with the protective assistance of 120 policemen. Just over a week later, a crowd of 3,000 assembled in Trafalgar Square to hear stirring anti-vivisection orations and calls for the dog to be reinstalled, but the original statue was never seen again.

Almost simultaneously, women in Scotland were battling similar heavy-handedness and what they experienced as oppressive London-centric sentiment. In 1908, the Research Defence Society in Britain was formed in an attempt to counter the increase in anti-vivisection narratives and activities. In 1911, a dispute between the Scottish Branch of the National Anti-Vivisection Society (NAVS) and its national representatives from England saw its female activists at the helm of the breakaway Scottish Co-operative Anti-Vivisection Society, looking to achieve 'financial independence' but 'friendly cooperation' with the UK-wide organisation. Like Despard, these (mostly middle-class) women were supporters of the women's suffrage and other social movements of the time, often outspoken and frequently adorned with sandwich board campaign slogans. Included in an address to their 1911 meeting by activist Louise Lumsden, was an important sentiment: 'to bring before them … the danger to human beings involved in this practise of vivisection' (Kean and Pakeman, 2013: 1).

More than a century on, vivisection continues practically unabated. The medical institution, now greatly industrialised, has monetised the practice.

Millions continue to suffer and die in the name of research, usually for unnecessary studies and product development (Hermann and Jayne, 2019). Vivisection is now protected by secrecy and a veil of welfare laws that make civil the very barbaric reality of nonhuman animals' experiences. However, the public's attitude towards vivisection is considerably more sceptical today, at least with regard to familiar species such as dogs, cats, monkeys and apes. More welfare laws exist to at least alleviate some of the pain, stress and poor living conditions endured by individuals in some experiments (Hall and Favre, 2004). In 1985, the National Anti-Vivisection Society (originally formed as the 'Victoria Street Society' by Despard's contemporary Frances Power Cobbe a century before) reinstated a new statue in Battersea Park, and more recently, author of the fictionalised *Little Brown Dog* (2021), Paula Owen, has created a replica based on original photographs, which she is campaigning to install in the original Latchmere location (Thorpe, 2021).

Despard's campaign ultimately failed when the justice system sided with the powerful medical establishment, but her effort was nonetheless ground-breaking as the largest protest against speciesism of the time. Her efforts to mobilise change simultaneously across a number of causes in a concerted effort to resist oppression also illustrates her awareness of the intersectional nature of multispecies advocacy. Even the single-issue brown dog campaign was not isolated in its reach. Despard intentionally chose the location and framing of the protest to draw on parallel campaigns and the momentum of communities who recognised a shared enemy. It is exactly this intersectional awareness that green criminologists such as Beirne and South (2007: xx) believe to be key to overthrowing multigenerational harms that have manifested in 'gender inequalities, racism, dominionism and speciesism, classism, the north/south divide, the [lack of] accountability of science, and the [lacking] ethics of global capitalist expansion'. The success of social justice and ecological movements, in other words, will depend on solidarity across campaigns as Despard's efforts helped to initiate.

Patty Mark and modern anti-speciesist activism

As the nonhuman animal rights movement entered its second wave in the 1970s, activists were heavily influenced by the grassroots, direct action of the American civil rights movement and tended to operate outside of the large non-profit institutions that had grown increasingly conservative since their establishment in the Victorian era (Ryder, 1989; Wrenn, 2019). The Animal Liberation Front (ALF), frustrated with peaceful protest and slow-moving non-profits, began to physically disrupt laboratories, 'fur' farms and other speciesist industries. In doing so, they filmed and photographed spaces that were largely unknown to the public. This material was vital for galvanising the movement, with many images gathered by ALF activists featuring in the campaigning materials for grassroot groups and charities. However, the ALF also pursued property damage as a strategy of economic disruption, a tactic that backfired by decreasing public support for the cause and heightening legal restrictions on anti-speciesist activism. The United States' Animal Enterprise Protection Act of 1992, for instance, was designed in response to ALF interferences (in 2006 it was strengthened and renamed the Animal Enterprise Terrorism Act) (Fiber-Ostrow and Lovell, 2016). Similar legislative (and many argue, disproportionate) restrictions were placed upon activism in the UK in response to the increase in nonhuman animal rights activism throughout the late 1990s and at the turn of the 21st century, for example, Stop Huntingdon Animal Cruelty and other anti-vivisection campaigns (see Ellefsen and Busher, 2020) prompted similar amendments to several UK Bills, though open rescue was not a much-utilised British tactic.

Also in the 1990s, the Western nonhuman animal rights movement entered another phase of transition (Wrenn, 2019). The grassroots, direct action approach that had characterised mobilisation in the 1970s and 1980s was giving way to the professional non-profit pathway. The ALF lost crucial movement support and became a far more marginal player. Yet, the materials the ALF were able to source through illegal entries were essential for campaigning, and the direct interference with speciesism was still seen as admirable and motivating.

In 1978, Australian Patty Mark was organising what would become one of her country's leading anti-speciesist charities, Animal Liberation Victoria. At the time, no other organisation in that country with its vast expanse of agricultural land was campaigning for farmed nonhuman animals (ALV, nd). By the 1990s, Mark, vegetarian since 1974 and vegan since 1991, had developed open rescue as a middle-ground tactic that was *not* intentionally clandestine as was typical of the ALF but *was* intentionally and directly disruptive – atypical among competing charities. Starting in 1993, Mark would regularly enter facilities such as chicken barns or piggeries, alert the media to her plans, and remove injured nonhuman animals in broad daylight. Following her first open rescue, Mark mused of its utility in demonstrating the reality of speciesism to the world: 'How could we expect the general public to comprehend the situation without visual proof?' (OpenRescue.org, 2003). In some cases, Mark and fellow activists would liberate nonhuman animals, then chain themselves to the farms (what they termed as a 'lock-on'), again, without secrecy or property damage. The impact was effective, not only in documenting the horrors within, but also in gaining media coverage to heighten public awareness of the brutal reality of nonhuman animal 'farming', despite the risk to their own liberty. Open rescue as a tactic has since become a valuable staple in the repertoire of many within the nonhuman animal rights movement.

Open rescue, as Mark developed it, is an intentional and strategic challenge to the law. First, open rescue rejects the validity of laws that commodify sentient beings as owned property. The act of rescue is a protest against the legalised, widespread killing of other animals. Open rescue is distinct from 'intervention purchases' in which activists buy nonhuman animals (usually 'livestock') to spare them from slaughter or other nefarious ends. Other activists often frown on this approach as it does little to disrupt the commodity status of nonhuman animals, maintains the profitability of the system, and does not prevent the replacement of those rescued nonhuman animals by the sellers. By way of an example, the Friends of Philip Fish Sanctuary[1] monitor 'pet' shops and Craigslist (a classified advertisement website) for vulnerable nonhuman animals, offering them a safe home, but not compensating the 'owners'. In the case of 'pet' shops, it is usually the fishes who are near death and unlikely to sell that are released to the sanctuary by management. The sanctuary then documents the fishes' recovery and improved quality of life on Facebook and Instagram, where the experiences of Philip and his comrades become educational resources.

In most cases, however, activists enter industrial farming facilities which house hundreds if not thousands of inmates. Unlike 'pet' shops, security measures are far more foreboding in nonhuman animal agriculture. It is not possible to rescue all nonhuman animals, nor is it even possible to rescue those who appear to be experiencing greater than normal levels of pain and stress. This tactic comes at considerable psychological cost to activists (Gorski et al, 2019) as they come face to face with speciesism at its worst. It is fair to say that most activists rarely witness extreme speciesist violence against living nonhumans, such as 'slaughter', given that most speciesist industries are located in rural, isolated and protected areas. Worse still, liberators must make the wrenching decision of who to rescue from a sea of suffering, crying and desperate prisoners. Activists will often negotiate adoptive homes and sanctuaries for survivors before and after rescues, but it is generally the case that a small number of activists (between one and five) will enter a facility and abscond with as many nonhuman animals as possible, though this may not be many. Although limited in their capacity, liberation for even a tiny percentage is a monumental achievement, and an unquantifiable delight for the birds, pigs, rabbits, rats and others removed from farms and laboratories, able thereafter to live their lives naturally with dignity, care, medical attention and rehabilitative therapy. Shelters, sanctuaries and activist collectives frequently document these journeys. For example, vegan-run Tribe Animal Sanctuary Scotland is home to dozens of nonhuman animals, including turkeys, pigs, cows, donkeys and goats. Narratives of their lives and liberation are publicly detailed for educational purposes (TASS, 2017).

The second premise of open rescue is to gather information. Activists enter speciesist facilities not only to directly rescue victims, but also to document the reality of living and dying conditions. Advertisements for some restaurants, for instance, regularly depict nonhumans as happy captives who look forward to their death for human consumption (for example, cartoons of smiling pigs relaxing atop barbeques, or dressed as chefs). Nonhuman animal agriculture operates in relative secrecy compared with the rest of the food system. By way of example, exceedingly few farmers open the doors to their layer hen housing systems for public tours, not least because the noxious levels of ammonia and chicken litter would make doing so unsafe (farmers themselves typically must suit up in protective gear to enter) (HSE, 2017). More importantly, though, the conditions required to mass produce affordable nonhuman animal products will never match the advertised (more palatable) ideal that is presented to consumers. The nonhuman animal agriculture industry is rightfully concerned with the negative potential of exposure, leading to a series of 'ag gag' laws across the United States and Canada (Sorenson, 2016). These legislative manoeuvres criminalise whistle-blowing from within the industry's ranks, rendering possible allies silent.

Third, open rescue is designed to invite media coverage and sometimes even police involvement. Direct Action Everywhere, for instance, regularly enters farming establishments to remove suffering nonhuman animals and welcomes arrest, hoping to utilise the opportunity within the legal system to challenge speciesist laws. Direct Action Everywhere operates an Open Rescue Network, formed in 2015 by activist Wayne Hsiung, who was convicted of felony larceny in late 2021 for liberating a sick six-week-old baby goat from a North Carolina farm. Hsiung (2021) cites Ganz's sociological concept of 'strategic capacity' in defence of what he believes to be morally obligatory in such circumstances, and states: 'by going in with our faces proudly uncovered, we dare the industry to try our actions in the court of public opinion' (2021: np), a rationale that reflects that which Mark initiated almost 30 years prior. Also aligned with Mark's open rescue strategy is the direct activism of various environmental campaigners. This might include interference with any proposed ecologically damaging development projects, by way of camps and vigils or lock-ons to bulldozers, trees or corporate offices. So-called 'ecoterrorists' join so-called 'animal rights terrorists' on government watch lists, despite their goals of preserving life and liberty by disrupting the 'corporate colonisation of nature' (South, 2007: 230). The comparison with the well-documented political activism of women's liberation and civil rights across time is also clearly evident (Sturgeon, 1997). Consider, for instance, suffragette Emily Davidson's fatal trespass onto Epsom racecourse in 1913, the Miss America beauty pageant demonstrations of 1968, or more recently, 21st-century Russian feminist performance protestors, Pussy Riot, objecting to state oppression.

Sarah Kistle and intersectional anti-speciesism

Like activists of the second wave, third wave anti-speciesists continue to be influenced by civil rights efforts. The language of anti-slavery abolition, for instance, persists among radicals who hope to liberate other animals entirely from speciesism, while campaigns draw on modern civil rights themes including intersectionality, food justice and challenges to carceral logics (Sturgeon, 1997). This contemporary style of activism resists the criminal justice system and its oppression of vulnerable humans and other animals, but does so in developing alternatives to speciesism and by building alliances across social justice movements (Phillips and Rumens, 2016). The efforts of Sarah Kistle[2] exemplify this approach. Kistle is an adopted Korean-American activist who pioneered intersectional social justice work in the Western nonhuman animal rights movement in the early 2010s. A long-time vegan, she began her activist career volunteering in the abolitionist faction of the movement through extensive social media campaigning and the founding of the short-lived Abolitionist Vegan Society. Abolitionism pushes for the total liberation of other animals and is explicitly critical of the dominant welfarist paradigm in the movement, a paradigm which primarily seeks to improve living and working conditions of commodified nonhuman animals through legal initiatives (Wrenn, 2016). Abolitionist theory, by contrast, is deeply critical of the legal system, positing that the property status of nonhuman animals renders impotent any liberatory ambitions, offering instead merely ameliorative results, successful only where they prove profitable to speciesist industries. Abolitionists are divided on whether the legal system is at all useful for achieving liberation. 'Animal rights' itself is a contested term, believed by some to be a project intent on realising actual legal rights for other animals, while for others it is shorthand for a more general approach to opposing speciesism.

Abolitionism in nonhuman animal rights was developed by philosopher and environmental ethicist Tom Regan (1984). Regan often spoke of nonhuman animals in terms of their 'inherent value'. This concept aligns with that of individual rights: 'the formal principle of justice stipulates that each individual is to be given his or her due' (Regan, 1984: 263). He furthered that 'the rights view will not be satisfied with anything less than the total dissolution of the animal industry as we know it' (1984: 395), but just how to achieve such abolition using the legal system is questionable. The ending of many practices falls within welfarism rather than abolitionism (for example, killing without stunning first: stunning was eventually supported by the industry as it made the process more efficient) (Welty, 2007), or simply pushes the practise elsewhere (which has been the case with outlawing horse slaughter in the United States) (*Meat Trades Journal*, 2012).

A century prior, however, recall that abolitionists became disillusioned with the capabilities of the law following failed efforts to curb vivisection. Frances Power Cobbe, instrumental in bringing to fruition Britain's Cruelty to Animals Act in 1876, hoped to make vivisection legally difficult to engage in, and to restrict the use of nonhuman animals in science. Instead, the Act had the opposite effect, creating a veneer of legitimacy to, and state approval of, the blossoming industry. Cobbe was incensed, thereafter committing herself and her organisation to an abolitionist approach, hoping to end rather than regulate the practise (Bates, 2017). Since then, the legal and criminal justice systems have merely strengthened support for animal-based industries. Rather than reduce society's reliance on speciesism, industrialisation has rationalised and expanded it. Today, the number of nonhuman animals who are harmed and killed for food, clothing, entertainment, product development, medicine, companionship, labour and land acquisition is so vast it is not possible to estimate with any accuracy. Subsequently, abolitionists today remain deeply critical of the legal system's ability to acknowledge fairness for all of earth's sentient inhabitants, but increasingly they draw on the language of social justice to increase recognition of nonhuman animals as stakeholders and individuals, worthy of rights and recognition.

Furthermore, vegan abolitionism of the nonhuman animal rights movement's third wave insists that resistance to the oppression of nonhuman animals is inherently linked to that of marginalised human groups. Kistle (2015) suggests that any notion that people of colour, or women, are less worthy of anti-speciesists' attention is a reflection of the white-supremacy and androcentrism that continues to plague movement structures (Kistle, 2015). The growing cultural presence of Black Lives Matter proved to be a turning point in the vegan abolitionist faction, with some activists embracing intersectional values and others rejecting the relevance of other movements. By the mid-2010s, Kistle was disinvested in social media activism and brought her charity, the Abolitionist Vegan Society, to a close. To some extent, this was a response to the intense racism she experienced in the nonhuman animal rights movement (Wrenn, 2019), but Kistle was also drawn to the possibilities of collaborating more directly with Black Lives Matter, particularly given its heightened activity where she lived, in Minneapolis.

Among others, Kistle assisted activist and father-of-four Louis Hunter during and after criminal trial where he had been wrongfully arrested and falsely accused of throwing missiles at police while protesting the fatal police shooting of his cousin. Following extensive and relentless campaigning by numerous activists (organised in part by Kistle), the case was dismissed. Following the ordeal, Hunter went on to collaborate in community food justice with Kistle and her partner, launching a food truck to serve, heal and connect the surrounding communities with healthy vegan comfort food. Successful vegan pop-up restaurants followed until finally, funded by the community, the three opened a permanent vegan restaurant, Trio Plant-Based (2020). Understanding the importance of community activism against injustice and oppression, Kistle's approach was unique at a time when the emerging third wave of nonhuman animal activism was just coming to remember and reimagine the importance of intersectional activism. It is an approach that, at the time of writing, is widely acknowledged as important in anti-speciesism campaigning. But this taken-for-grantedness was hard won by the likes of determined women of colour such as Kistle.

Conclusion

There is something to be said about the goals and motivations of these women, and the millions of other women like them, who risk personal safety and liberty to advocate for other animals in a society that has legalised speciesism and criminalised anti-speciesism. Although environmental wellbeing is of concern to anti-speciesist activists, it can in some cases be secondary in importance. The effort to gender green criminology is commendable in drawing attention to the explicitly patriarchal motivation behind men's war on nature, but anti-speciesism can sometimes make for an awkward disciplinary inclusion. These women are not just eco-feminists, they are social justice activists. The growing attention to nonhuman animals in green criminology is an important development in the recognition of nonhuman personhood, but to truly recognise their personhood, scholars will need to retrieve nonhuman animals from the wilderness of 'other'. Most nonhuman animals that humans interact with exist *within* human society, languishing in aquariums, zoos, pet shops, laboratories, feedlots, dairies, battery cages, live export cargo ships, transport trucks and slaughterhouses. Many others interact with us in more categorically ambiguous spaces such as parks, cities and homes. Indeed, millions of species inhabit areas we would typically categorise as 'green', but just as we would not subsume feminist criminology to green criminology because of the stereotypical association between women and nature, we should be likewise hesitant to do the same with other animals. Anti-speciesism is not just a matter of environmental justice, it is a matter of *social* justice.

Given the speciesist, sexist, racist and classist nature of many social institutions, some activists have opted to circumvent institutional means for social change, opting instead for a variety of physically and ideologically disruptive tactics. While various waves have been characterised by different tactical styles, many strategies have persisted across time (Sturgeon, 1997). Open rescue today remains a staple of anti-speciesist resistance, for instance, while intersectional campaigning is perhaps even stronger in 21st-century campaigning than it was in the 19th century. Modern vegan feminists, for instance, regularly employ food justice as a means to liberate human and nonhuman animals alike. Across all activist generations, the recognition that the criminal justice system must be interrogated if not outright provoked is evident. The criminalisation of certain ecological and environmental harms, but not others, reflects the anthropocentric nature of the system (White, 2007) as well as its tendency to protect the interests of upper-middle-class white men. Achieving social justice for so many nonhuman animals, women, people of colour, working-class individuals, and other marginalised groups, can and will happen with intersectional discourse and an interdisciplinary approach (Cudworth, 2005; South, 2014). Social equity and environmental justice *is* possible (Lynch and Stretsky, 2003) if scholars and activists are willing to engage the moral logic of anti-speciesism within the wider eco-feminist movement and green criminological perspectives.

Notes

[1] This rescue group is named in honour of Philip, the organisation's first rescue beta.

[2] For many years Sarah went by the name of Sarah K. Woodcock, particularly during the years when the Abolitionist Vegan Society was active.

References

Adams, C.J. (1991) Ecofeminism and the eating of animals. *Hypatia*, 6(1), 125–45.

ALV (Animal Liberation Victoria) (nd) About ALV. Available at: https://www.alv.org.au/the-facts/about-alv/ (accessed 20 May 2022).

Bates, A. (2017) *Anti-Vivisection and the Profession of Medicine in Britain: A Social History*. London: Palgrave Macmillan.

Beecher, H.K. (1959) Experimentation in man. *Journal of the American Medical Association*, 169(1), 461–78.

Beirne, P. and South, N. (eds) (2007) *Issues in Green Criminology: Confronting Harms against Environments, Humanity and Other Animals*. Devon: Willan Publishing.

Crenshaw, K. (1991) Mapping the margins: Intersectionality, identity politics, and violence against women of color. *Stanford Law Review*, 43(6), 1241–99.

Cudworth, E. (2005) *Developing Ecofeminist Theory: The Complexity of Difference*. Basingstoke: Palgrave Macmillan.

Donald, D. (2020) *Women against Cruelty: Protection of Animals in Nineteenth-Century Britain*. Manchester: Manchester University Press.

Ellefsen, R. and Busher, J. (2020) The dynamics of restraint in the Stop Huntingdon Animal Cruelty campaign. *Perspectives on Terrorism Special Issue: Restraint in Terrorist Groups and Radical Milieus*, 14(6), 165–79.

Farr, M. (2019) Charlotte Despard: Suffragist, vegetarian, radical. Available at: www.exploringsurreyspast.org.uk/wp-content/uploads/2019/09/Charlotte-Despard-full-text-for-PDF2.pdf (accessed 28 January 2022).

Ferguson, M. (1998) *Animal Advocacy and Englishwomen, 1780–1900*. Ann Arbor: University of Michigan Press.

Fiber-Ostrow, P. and Lovell, J.S. (2016) Behind a veil of secrecy: Animal abuse, factory farms, and Ag-Gag legislation. *Contemporary Justice Review*, 19(2), 230–49.

Foster, E. (2021) Ecofeminism revisited: Critical insights on contemporary environmental governance. *Feminist Theory*, 22(2), 190–205.

Gaard, G. (2002) Vegetarian ecofeminism. *Frontiers: A Journal of Women Studies*, 23(3), 117–46.

Gaarder, E. (2011) *Women and the Animal Rights Movement*. New Brunswick: Rutgers University Press.

Gorski, P., Lopresti-Goodman, S. and Rising, D. (2019) 'Nobody's paying me to cry': The causes of activist burnout in United States animal rights activists. *Social Movement Studies*, 18(3), 364–80.

Groves, J. (2001) Animal rights and the politics of emotion. In J. Goodwin, J. Jasper, and F. Polletta (eds) *Passionate Politics: Emotions and Social Movements*. Chicago: University of Chicago Press, pp 212–32.

Hall, C.F. and Favre, D.S. (2004) Comparative animal welfare laws. Available at: www.animallaw.info/article/comparative-national-animal-welfare-laws-0 (accessed 30 April 2022).

Hermann, K. and Jayne, K. (eds) (2019) Animal experimentation: Working towards a paradigm change.

HSE (Health & Safety Executive) (2017) *Farmwise: Your Essential Guide to Health and Safety in Agriculture*, 3rd edn. London: The Stationery Office.

Hsiung, W. (2021) The trial of Open Rescue. Available at: https://simpleheart.substack.com/p/the-trial-of-open-rescue (accessed 27 January 2022).

Kean, H. (1998) *Animal Rights: Political and Social Change in Britain Since 1800*. London: Reaktion Books.

Kean, H. and Pakeman, D. (2013) Report on origins of SSPV. Available at: www.onekind.scot/wp-content/uploads/Origins-of-SSPV-by-Hilda-Kean.pdf (accessed 28 January 2022).

Kenny, S.C. (2015) Power, opportunism, racism: Human experiments under American racism. *Endeavour*, 39(1), 10–20.

Kistle, S. (2015) 'It is anything but': Sarah K. Woodcock comments on equality in abolitionist spaces. [Blog]. Available at: http://academicactivstvegan.blogspot.com/ (accessed 14 February 2022).

Lahar, S. (1991) Ecofeminist theory and grassroots politics. *Hypatia*, 6(1), 28–45.

Lansbury, C. (1985) *The Old Brown Dog: Women, Workers, and Vivisection Edwardian England*. Madison: The University of Wisconsin Press.

Leneman, L. (1997) The awakened instinct: Vegetarianism and the women's suffrage movement in Britain. *Women's History Review*, 6(2), 271–87.

Lynch, M.J. and Stretsky, P.B. (2003) The meaning of green: Contrasting criminological perspectives. *Theoretical Criminology*, 7(2), 217–38.

Meat Trades Journal (2012) Horse slaughter ban has lessons, says US speaker. *Meat Trades Journal*, 22 June, p 13.

OpenRescue.org (2003) Latest news: Interview with Open Rescue pioneer Patty Mark. Available at: www.openrescue.org/news/20030327.html (accessed 14 February 2022).

Owen, P. (2021) *Little Brown Dog*. Aberystwyth: Honno Welsh Women's Press.

Phillips, M. and Rumens, N. (2016) *Contemporary Perspectives on Ecofeminism*. Abingdon: Routledge.

Regan, T. (1984) *The Case for Animal Rights*. Berkeley: University of California Press.

Ryder, R. (1989) *Animal Revolution: Changing Attitudes Toward Speciesism*. Oxford: Blackwell.

Savitt, T.L. (1982) The use of blacks for medical experimentation and demonstration in the old south. *The Journal of Southern History*, 48(3), 331–48.

Singer, P. (1975) *Animal Liberation: A New Ethics for Our Treatment of Animals*. New York: New York Review.

Sollund, R. (2021) Green criminology: Its foundation in critical criminology and the way forward. *The Howard Journal of Crime and Justice*, 60(3), 304–22.

Sorenson, J. (2016) *Constructing Ecoterrorism: Capitalism, Speciesism and Animal Rights*. Nova Scotia: Fernwood Publishing.

South, N. (2007) The 'corporate colonization of nature': Bio-prospecting, bio-piracy and the development of green criminology. In P. Beirne and N. South (eds) *Issues in Green Criminology: Confronting Harms against Environments, Humanity and Other Animals*. Devon: Willan Publishing, pp 230–47.

South, N. (2014) Green criminology: Reflections, corrections, horizons. *International Journal for Crime, Justice and Social Democracy*, 3(2), 6–21.

Sturgeon, N. (1997) *Ecofeminist Natures: Race, Gender, Feminist Theory and Political Action*. Abingdon: Routledge.

TASS (Tribe Animal Sanctuary Scotland) (2017) Our animals. *TASS*. Available at: https://tribesanctuary.co.uk/our-animals/ (accessed 12 February 2022).

Taylor, N. and Fitzgerald, A. (2018) Understanding animal ab(use): Green criminological contributions, missed opportunities and a way forward. *Theoretical Criminology*, 22(3), 402–25.

Thorpe, V. (2021) How the cruel death of a little stray dog led to riots in 1900s Britain. *The Guardian*, 12 September. Available at: www.theguardian.com/artanddesign/2021/sep/12/how-the-cruel-death-of-a-little-stray-dog-led-to-riots-in-1900s-britain (accessed 6 January 2022).

Trio Plant-Based (2020) Our story. *Trio Plant-Based*. Available at: www.trioplant-based.com/our-story (accessed 20 May 2022).

Ulrich, L.T. (2007) *Why Well-Behaved Women Seldom Make History*. New York: Alfred Knopf.

Unger, N. (2012) *Beyond Nature's Housekeepers: American Women in Environmental History*. London: Oxford University Press.

Welty, J. (2007) Humane slaughter laws. *Law & Contemporary Problems*, 70(1), 175–206.

White, R. (2007) Green criminology and the pursuit of social and ecological justice. In P. Beirne and N. South (eds) *Issues in Green Criminology: Confronting Harms against Environments, Humanity and Other Animals*. Devon: Willan Publishing.

Wrenn, C. (2016) *A Rational Approach to Animal Rights: Extensions in Abolitionist Theory*. London: Palgrave.

Wrenn, C. (2019) *Piecemeal Protest: Animal Rights in the Age of Nonprofits*. Ann Arbor, MI: University of Michigan Press.

14

'To Preserve and Promote': Gendering Harm in Green Cultural Criminology

Angeline Marie Letourneau

Introduction

Green criminology is an invitation to rethink how we conceptualise categories of harm and who the offenders or victims are (for example, Agnew, 2012; Sollund, 2017; Brisman and South, 2018). Accepting this invitation is no small feat. Feminist scholars have an established history of outlining the social and environmental repercussions of traditional, Western expressions of masculinities (Carson, 1962; Merchant, 1980; Plumwood, 1993; MacGregor, 2009). In this chapter I will take a cultural criminology approach to examine how masculinity serves to shape cultural understandings of harm while simultaneously justifying harmful activities associated with resource development in the oil and gas industry of the Canadian province of Alberta. I will illustrate how gendered cultural discourses 'turn elite beliefs and values into common sense perceptions' (Seiler and Seiler, 2004: 173–4) and foster public support for industrial development that is harmful both socially and environmentally.

Cultural green criminology

I begin this inquiry by considering what differentiates those harms which are criminalised under the law and those which scientific evidence would caution do have the potential to cause significant harm but are not yet criminalised. Cultural dynamics like the distribution of power within society determine the meaning of crime and, therefore, what harms will be criminalised (Ferrell et al, 2015). The distinction between political protest or civil disobedience, for example, is often a fine line. In many cases, scientific research recognises environmental harms that the law fails to capture (Lynch and Stretesky, 2001). In other situations, the law determines what counts as scientific evidence, often to the benefit of state interests (for example, Whitt, 2009). But as with the beginning of every apocalyptic film, the warnings of scientists often go unheeded, with environmental laws reflecting the economic interests that benefit from ecological withdrawals and additions (Stretesky et al, 2014). For example, ecological additions, like pollution, or the withdrawal of resources, are often framed as unfortunate but necessary costs of progress (Brulle and Pellow, 2006; Gould et al, 2008). Of course, this does not mean that environmental harms are never criminalised. After all, significant portions of both bureaucratic and corporate resources are devoted to navigating the immense regulatory processes meant to prevent unrestricted harm to the environment. Yet regulations rarely capture the full extent of harms, making trade-offs that require further scrutiny. To accept the legal definition of crime is to accept the subjectivity of the law's interpretation of crime, the interests it reflects, and ultimately limits the focus of the study of crime to street crimes (Stretesky et al, 2014). Non-criminalised ecological harm continues to exist because governments and industry share the same goal of increased production: industry wants to increase their profits, and government wants to increase their tax revenue generated by industry. These shared goals create a subjective distinction between crime and harm that illustrate the political power and influence of industry and elites. Environmental enforcement, or the lack thereof, is an organisational outcome intimately related to power struggles between civil society, state authorities, producers and labour (Gould et al, 1996).

These subjective distinctions are not confined to c-suite boardrooms either. Individuals are socialised within organisations to act in ways that damage the environment – and the communities dependent on it – in ways that meet organisational goals and increase or secure production (Pearce and Tombs, 1998). Activities that may have had an alarming ecological impact at first, slowly become normalised and justified. To understand crime and harm systemically, we must attend to the discourses that emerge around crime within its cultural context and intersectionally, both at the elite level and through the everyday organisational life of working- and middle-class people.

Culture within a cultural criminological lens refers to collective meanings and collective identity – the process through which meaning and importance is negotiated among different groups of people (Ferrell et al, 2015). Culture is not a social structure in itself, but it is deeply entangled with structures of power (Naegler and Salman, 2016). Within a cultural lens, crime is a social product emerging from specific social conditions (Hayward and Young, 2004). The subjective understandings of crime and harm shift across social contexts, requiring critical attention to cultural norms, who these norms serve, and how they are reproduced.

As gender is one of the central organising categories within Western social culture (Ahmed, 2017), attending to it is particularly important in a cultural criminological approach. The lack of analytical gender lenses has been a main critique of cultural criminology (Naegler and Salman, 2016), and criminology more generally (Leonard, 1982). Green criminology has also suffered from this same lack of attention to gender. Despite considering who is committing harm and why they might be doing so (Brisman and South, 2018; see also Gibbs et al, 2010; White, 2011; Agnew, 2013; Kramer, 2013; Brisman, 2014), green criminology has yet to outline the influence that gender plays in shaping cultural understandings of harm and maintaining such framings to the benefit of powerful actors.

The absence of a gender lens risks the unintentional adoption of what O'Brien (1981) refers to as malestream norms, a foundational cause of sexism in many contexts and one which I will explore further in the next section. Much scholarly attention to gender and environmental harms focuses on the impacts disproportionally experienced by women and gender minorities, especially those who experience intersecting identities (for example, racialised, abilities, working class). While this work is important for mitigating impacts against these groups, they leave the gendered dynamics and cultural norms that created such systems of inequality in the first place uninterrogated (Bell et al, 2020). Efforts within green criminology to apply an intersectional lens, including gender, often focus on those who are victimised, namely women, children, racialised populations and working-class communities (Williams, 1996; Wachholz, 2007; Sollund, 2013; Rodríguez Goyes and South, 2016). In doing so, they treat gender and other identifiers as a demographic category, ignoring the ways that identity can provide a foothold for cultural narratives that perpetuate harm. In short, gendering green criminology requires us to go beyond gender as an 'added on' line of inquiry and integrate it more holistically (Naegler and Salman, 2016, p 355).

In the next section, I will explore hegemonic masculinity as a key driver of existing norms in Western culture. Due to the self-reinforcing nature of malestream norms, whereby the narratives that establish gender and racial hierarchies are informed and maintained by male as neutral, there is significant incentive for those that benefit from the current gender dynamics to ensure they remain in positions of influence or hegemony.

Hegemonic masculinity and malestream norms

Masculinities as identities arise from the everyday performances and social interactions of people (Connell, 2005; Connell and Messerschmidt,

2005). They are socially situated within the everyday organisational life of a given social context, iteratively constructed in the social (Connell, 2001; Connell and Messerschmidt, 2005; Hearn et al, 2012). Expressions of masculine identities are multidimensional, mobile (Brekhus, 2020) and as unique as the contexts which they are a part of (Connell, 2005; Messerschmidt, 2018). Masculinities are arranged within a hierarchy which shifts with changing social contexts. What might be popular expressions of masculinity in one context may be rejected in another, though the hierarchies, particularly in Western culture, typically situate femininity at the bottom and those most distinct from femininity at the top (Connell, 2001; Connell and Messerschmidt, 2005). In any context, the dominant masculinity is referred to as hegemonic masculinity, which refers not to the most popular or common expression of masculinity, but the most privileged in gender relations.

Hegemony is described as the process through which those in power establish and maintain their influence within social groups (Donaldson, 1993). The ability of hegemony to impose meaning and set social terms allows the influencing of social organisation to occur in a way that appears natural or normal. Hegemony must be enforced to maintain influence, and non-conformity can be punished in a number of ways, though these often occur in passive ways, such as through social exclusion (Connell, 2005. This process of hegemonic masculinity working to maintain power inspired Hultman and Pulé (2018) to refer to masculinity as a politics of domination.

In *The Politics of Reproduction*, Mary O'Brien deconstructs the normalisation of male domination through a critique of what she terms 'malestream norms' (1981: 62). Her reflections on cultural androcentrism are echoed by Bem (1993), who points out the centrality of male-ness in everything as neutral, with women or feminine being a gendered deviation from that norm. If gender is not specified, it is assumed to be male. Malestream norms extend through everything, from discourses, to social institutions, to individual psyches. Examining how these norms shape social processes requires shifting our focus away from assigned biological sex towards the sociocultural constructions of gender (Keener and Mehta, 2017).

Within Western culture, malestream norms are most often characterised by the prioritisation of data, so-called logic, and economic rationality in decision-making (Mellström, 1995). Both competition and the individualism necessary to thrive in a cut-throat world are highly valued, and the range of emotions allowed is limited to anger, as most other emotions are regarded as a sign of weakness. There is also a lot of faith put into technology as a solution to any and all problems, as it combines the merit of both logical scientific processes and the free market. It is important to acknowledge the racialised dimensions of malestream norms. Historically, only white men have inhabited this position of uncriticised neutrality, and therefore have assumed the majority of the privileges that accompany this position; however, intersectional scholars have highlighted that racialised men and white women have benefited from malestream norms in various and complex ways.

The gender influence of men and masculinities in social and environmental issues have gone unquestioned for so long because of these malestream norms. To begin untangling the influence of malestream norms, a new reference point must be established – one which deviates away from male as normative and acknowledges the intersectional and gendered narratives that create such hierarchies in the first place (Hultman and Pulé, 2018).

Feminist scholars have noted the entanglement of social and environmental impacts through masculinities (for example, Carson, 1962; Merchant, 1980; Plumwood, 1993; MacGregor, 2009). Eco-feminists have focused on the ways patriarchy has subjugated both woman and the natural world through the Western cultural products of colonialism, capitalism and androcentric epistemologies (for example, Haraway, 1988, 2016; Gaard, 2001; Cudworth, 2005; Mies et al, 2014). One of the most foundational conclusions to arise from this scholarship is the belief that the social construction of nature is inseparable from the social construction of gender, whereby the exploitation and destruction of both non-men and the natural world are connected (Griffin, 1978).

Feminist critics of science emphasise the dominance of malestream norms throughout the Western scientific tradition (Haraway, 1988; Harding, 1991). Though, ecologically speaking there is no discernible separation between humans and nature, Western culture conceives of Nature and Reason as separate (Plumwood, 2018). This duality of Nature and Reason is informed by malestream norms, which, historically, considered humanity – capable of rational thought – to be more narrowly conceptualised as mankind, excluding women and racialised bodies and relegating them into the far broader category of nature. Nature and mankind are conceptualised within a hierarchy, with mankind occupying a small but privileged status, similar to Connell's theory of hegemonic masculinity (Connell, 2005; Connell and Messerschmidt, 2005). This hierarchy served to justify the mistreatment and exploitation of nature, including the biophysical world and most of humanity. Racism, colonialism, sexism and heteronormativity all draw their strength from this hierarchy as evidence that subordination of most peoples is simply the natural order of things (Plumwood, 1993).

There are substantiated gendered differences in concern for the environment. Women are more likely to express higher concerns regarding risk in general (Davidson and Freudenburg, 1996). They are also more likely to express concern for the environment (Briscoe et al, 2019) and more likely to act on it (Brown and Ferguson, 1995; Bell and Braun, 2010; Shriver et al,

2013). In terms of environmental impacts or harms, women are more likely to express concern for the safety of their children or grandchildren (Krauss, 1993; Brown and Ferguson, 1995; Culley and Angelique, 2003; Peeples and DeLuca, 2006; Bell and Braun, 2010). When scientific evidence indicates that there is environmental harm occurring, or, if there is insufficient evidence to draw conclusions about the degree to which damage may be occurring, women are more likely to advocate for caution.

In contrast, men are less likely to be involved in environmental activism, which Bell and Braun (2010) and Bell and York (2010) attribute to the role identities many men have as economic providers working in environmentally damaging industries. This is especially true for working-class and racialised men, who are more likely to be victims of economic blackmail, where they are coerced into accepting environmental harm in their communities and insufficient wages in exchange for a pay cheque (Bullard, 1993). Generally, men express significantly higher risk tolerances than women do (Davidson and Freudenberg, 1996). This is particularly true when examining risk acceptance intersectionally: gender, 'race' and power all culminate in what is referred to as the 'white male effect', where white men demonstrate a significantly higher risk acceptance in social, economic and environmental circumstances than any other demographic (Finucane et al, 2000: 160; Slovic et al, 2005). For example, they are the most likely demographic to deny the science of climate change (McCright and Dunlap, 2011; Vowles and Hultman, 2021).

The connections between white male domination and capitalism have been well established by Marxist scholars (Keith, 2017: 2–3). It is the entanglements between the two that secure their parallel reproduction. So long as an activity fits within the characteristics of malestream norms – as capitalism and resource extraction undoubtedly do – it is unlikely to be criminalised. Those gendered narratives responsible for upholding malestream norms shape the decisions that are made about the environment and what constitutes harm sufficient enough for criminalisation. These narratives are harnessed by industry and industry-supporting governments to maintain power and perpetuate the resource-dependent treadmill of production (Stretesky et al, 2014).

Like hegemonic masculinity, the strength or influence of these gendered narratives differs across social contexts. In some places, malestream norms may create a broader acceptance of environmental and social harms, while in others there may be higher expectations to perform some consideration of these other components (for example, Hultman and Anshelm, 2017). I now turn my attention to the Canadian province of Alberta – the largest producer of oil and gas in Canada and home to the third largest deposit of oil in the world (Government of Alberta, 2015) – to demonstrate the impact of gendered politics in defining harm and shaping criminalisation.

Alberta

Gendered cultural narratives

Even before the drilling of the first oil well in Turner Valley, Albertan identity has been deeply entangled with traditional representations of 'rough and tough' masculinity. Images of the frontier cowboy and pioneer explorer define much of the iconography of Albertan identity (Wright, 2001; Miller, 2004; Williams, 2021). The annual Calgary Stampede, self-proclaimed as the 'Greatest Outdoor Show on Earth', celebrates this so-called Western heritage (Calgary Stampede, 2022a) and provides an illustrative place to begin this inquiry into the relationship between culture, industry and criminalisation in the province (Figure 14.1).

Bringing in over 4 million people to the city over a two-week period, the Stampede is a combination of rodeo events, an agricultural trade fair, music shows, midway attractions, and a range of parties held throughout the city (Calgary Stampede, 2022a). These events are all performative, reproducing that which they simultaneously define (Williams, 2021). It is unquestionably the largest event held annually within the province and as described by Hanvelt, it is 'a place where the major tenets of masculinity are represented, enacted, and thereby reinforced. This includes ideologically divided gender roles, a privileging of whiteness, and a compulsory heterosexuality' (Hanvelt, 2004: ii).

The Stampede takes place in Calgary, the city within Alberta that houses the vast majority of oil and gas company head offices. The Stampede's pedagogical intent rests in 'its mission to "preserve and promote" the "heritage, cultures, and community spirit" of the Canadian prairie West' (Williams, 2021: 8). But what is this heritage exactly? Like most of Canada's legacy, the Stampede is built upon deeply rooted ties with colonialism. The Stampede was originally launched in 1912 – just seven years after Alberta joined Canada as a province – to attract white settlers and tourists to Southern Alberta to experience the wildness of the frontier (Seiler and Seiler, 2004).

Miller (2004) describes the dominant, hegemonic masculinity in the extractive industries of Alberta as frontier masculinity. This is because the same romanticisation that once drove settlement of the West continues to inform everyday interactions and decisions within Alberta's oil and gas sector. This frontier masculinity is portrayed front and centre in the Stampede experience and plays a crucial role in justifying ongoing extractive activities in the province (Lowman and Barker, 2015). Because of the importance of the Stampede in upholding these frontier narratives, exceptions to the rules are made for the event to continue. The Stampede itself is a non-profit subsidiary of the City of Calgary that pays no property taxes on its prime, downtown land (Campbell, 1984). The leaders elected to head the Stampede organisation every year are almost exclusively drawn from Calgary's elite community of oil

Figure 14.1: A rider and his horse take a tumble in the saddle bronc event at Calgary Stampede

Source: Sean Robertson, Unsplash, 15 September 2020 (https://unsplash.com/photos/HmiC m1iWewQ)

and gas executives and reflect that demographic by being largely comprised of white men (Williams, 2021). Women have been excluded from Stampede leadership roles and events in all but tangential and symbolic ways, while being significantly over-represented among the low-wage service staff and volunteers that the event depends on to operate each year.

Also erased in all but symbolic ways have been Indigenous peoples. Elbow River Camp (formerly Indian Village) serves as one of the 'attractions' of the Stampede where one can 'experience the cultures of the Kainai, Piikani, Siksika, Stoney Nakoda, and Tsuut'ina First Nations of Treaty 7 in Elbow River Camp, presented by Enbridge. Here you will find 26 tipis to visit, local artisans selling jewelry and art, bannock and saskatoonberry jam, traditional dancing and much more' (Calgary Stampede, 2022b). This 'attraction' persists, sponsored by energy companies residing in the city, despite the dispossession of Indigenous peoples through oil extraction and pipeline development from these same companies (Richards, 2019). This erases the historical and ongoing violence Indigenous communities have experienced and replaces it with a whitewashed version of history, one which promotes the opportunities that resource extraction provides while ignoring the processes of colonisation that have made traditional ways of life unfeasible and resource dependency unavoidable (see Parlee, 2015). The Elbow River Camp, and the experience it provides to visitors, serves to sell and celebrate a settler colonial narrative that situates Indigenous cultures as historical, points of interest on the frontier landscape, but benign and non-threatening.

Finally, there is the role of the frontier hero, embodied not only in the modern-day cowboys who compete via invitation only in the Stampede rodeo, but also in those white men who are employed in the oil industry. The frontier myth and its animation throughout the two weeks of the Stampede 'enables white-collar workers to see themselves as cowboys and rugged individualists ... rather than as corporate bureaucrats subject to government regulation' (Richards, 2019: 153). This oil economy boosterism that manifests in everything from the makeup of the Stampede's decision-making board to the sponsorship of virtually every element of the event, reproduces a colonial, misogynistic culture rooted in frontier and petromasculinities (Williams, 2021). After all, what is the purpose of preserving a tradition if not to celebrate it into the future? The frontier myth fits well into Alberta's extraction-dependent economy by overlaying the mythical subject of the frontier cowboy on Alberta's contemporary oilmen, constructing their work as a willingness to go to battle on the people's behalf.

The manifestation of gendered cultural narratives

The frontier masculinity narrative – that expression of masculinity that is most readily observed at Stampede events and throughout Alberta political and economic discourses – can be summed up by four characteristics (Wright, 2001):

- A vast, empty landscape, full of resources and waiting to be conquered.
- A heroic cowboy – the conqueror of this new land.

- The subjugation of white settler women, who are coerced through petrosexual violence to provide domestic labour.
- Clearly defined villains in local Indigenous groups and land defenders, whom the cowboy must eliminate in order to access the bounty of the land.

The first characteristic is the vast, empty landscape, full of resources and waiting to be conquered. Alberta's economy is deeply entangled with resource extraction, with over 26 per cent of the provincial gross domestic product dependent on extractive activities (Government of Alberta, 2020). The province has also resisted any federal efforts to restrict or impede development of resources within that landscape, including taking the federal government all the way to the Supreme Court over the carbon tax and its impact on resource development (Dryden, 2021).

The heroic cowboy lives on in the modern oil worker, who has been portrayed both by provincial government officials and on social media as a selfless hard worker, singlehandedly responsible for all modern comforts. Variations of the following quote can often be found circulating through social media networks whenever there is an extreme cold snap: 'Thank you oilfield workers for working so hard, for giving up nights, weekends, and holidays. For risking your lives to provide for our families and our society. We appreciate you' (OilfieldWifeSorority, 2014). This quote has circulated so extensively that the original author is difficult to trace; however, it is noteworthy in that this self-sacrificing, heroic narrative resonates so strongly in relation to the oil and gas industry, particularly within the province.

Women are also systemically subjugated and undervalued in Alberta's frontier culture. The oil and gas industry, like the Stampede, is heavily dependent on the gendered division of labour to supplement production and profitability (Dorow, 2015). Work associated with masculinity, such as operating machinery and equipment, engineering, and business operations, are all well-paid positions, and are dominated by men. On the other hand, domestic labour such as cooking, cleaning, and administrative labour, which is necessary to support a workforce in extraction, are feminised and relatively undervalued (Dorow and Mandizadza, 2018). These positions tend to be dominated by women and racialised individuals and are paid less than tasks directly involved in extraction. This disproportionate compensation plays a key role in the high gender wage gap in Alberta, which is second highest across the Canadian provinces, second only to Newfoundland, where much of the labour force is also employed in the oil industry (Conference Board of Canada, 2020). Despite both types of labour being critical to work in the oil and gas sector in the province – especially in the work camps that many projects rely on for their operations – that work which has culturally been associated with masculinity and manliness is paid significantly more compared with its feminised counterpart (Dorow, 2015). This remains true at the household level as well. The second shift refers to the labour performed in the 'domestic sphere' which is additional to the labour performed in the sphere of paid work (Hochschild and Machung, 1989). It is a phenomenon which impacts women across many social contexts, yet the long hours and weeks away from home that are required for much of the work in the oil and gas sector place a significantly higher responsibility on women for social reproduction activities, such as childcare and homemaking due to the extended absences of their male partners (Dorow and Mandizadza, 2018).

Financial compensation, or the lack thereof, presents an ample opportunity to view how, culturally, men and women, and the labour that has come to be associated with each group in the Western gender binary, are valued differently. Yet more evidence of this differential emerges when we examine the high rates of violence against women within Alberta, particularly in areas with high economic dependency on resource extraction (National Inquiry into Missing and Murdered Indigenous Women and Girls, 2019. Anti-violence experts consider the gendered violence in the province to be a significant issue that appears to be getting worse (Trembath, 2020). Yet despite this interconnection between resource extraction and gender violence, the premier of the province, Jason Kenney, expresses a lot more concern for maintaining the heroic frontier hero's image than he does about addressing the violence experienced by women. In 2018, when the Canadian Prime Minister Justin Trudeau announced new federal legislation that would require gender-based analysis in all major development projects going forward to address these impacts (Dawson, 2018), Premier Kenney tweeted the following in response: 'Justin Trudeau says pipelines must go through a "gender-based analysis" because male construction workers have "impacts." Darned right they do. They build things, create wealth, pay taxes, take care of their families. But this trust fund millionaire thinks they can't be trusted' (Kenney, 2018).

This tweet is one of many comments made by the premier, members of his party and supporters who place more value in the perpetuation of the oil and gas sector and its related narratives than in the quantifiable impacts this industry has on women. This is particularly true in the case of violence against Indigenous women. Alberta has the highest rate of missing and murdered Indigenous women and girls in Canada (National Inquiry into Missing and Murdered Indigenous Women and Girls, 2019), yet the province has taken no substantive action to address this tragic loss of life.

Given this relative disregard for the loss of human life it should come as no surprise that environmental protection falls even lower on the list of priorities for provincial elites. The ruling United Conservative Party, a group that has been in power since 1971 (Graney, 2019) – excluding a brief term by the New Democratic Party – has never been known for its green-friendly policies; however, this antagonism has become particularly evident in recent years, as has the connection between government policy and the perpetuation of fossil fuel development (for example, see Kenney, 2022).

The first action taken by the United Conservative government upon re-election in 2019, was to pass Bill 1: The Carbon Tax Repeal Act, legislation which repealed the Climate Leadership Act and eliminated the provincial carbon tax. This sent a clear message regarding the priorities of the new administration: they were here to encourage development and eliminate as much red tape that might slow down that process (Government of Alberta, nd). This process has included providing over 1.32 billion Canadian dollars in subsidies to the fossil fuel industry in the 2020/1 fiscal year alone, as the province promotes fossil fuel development as a COVID-19 recovery strategy (McKenzie et al, 2022).

Many of the examples provided up to this point may leave the reader questioning whether this is really a cultural issue within Alberta, or if the province is simply dealing with a particularly extreme government administration. An example which demonstrates the cultural tolerance of some sacrifices in the name of oil and gas development and the outright rejection of other activities that have no connection with the province's key industries, is the fate of the provincial coal mining policy. Early in 2020, without any public consultation and at the height of the COVID-19 pandemic, the United Conservative government repealed the 1971 coal policy that banned open pit coal mining on the eastern slopes of many parts of the Rocky Mountains (Fletcher and Omstead, 2020). The original ban had been implemented due to the extremely destructive nature of mountain top removal coal mining, as well as the sensitive ecological nature of these slopes and the role they play as the headwaters for many of Canada's major river systems. Unlike when the government suspended environmental monitoring for oil and gas operations during the early months of the COVID-19 pandemic (Weber, 2020), the elimination of legislation protecting the eastern slopes from coal mining resulted in significant public outrage (Nikiforuk, 2020). While one source of people's frustration stemmed from the lack of consultation prior to the policy's elimination, another sentiment seemed to be emerging, one which united people across the full political spectrum within the province: you don't mess with the mountains. While practical reasons related to the ecology of the area certainly emerged in these discussions, there was also a consistent theme related to Albertan identity and the imagery that is deeply entangled with the Rocky Mountains and surrounding foothills (Figure 14.2) that many people drew upon in articulating their lack of support for the decision. It is this imagery that is notably central to the frontier masculine narrative. The dissatisfaction expressed by people across the province eventually led to the reinstatement of the 1976 Coal Policy in February of 2021 (Government of Alberta, 2022).

Figure 14.2: Cattle grazing on the foothills of Chief Mountain in southern Alberta

Note: Ranching imagery plays a significant role in the Albertan identity.
Source: David Thielen, Unsplash, 15 July 2021

Coal mining plays a minimal role in Alberta's economy and threatens natural areas that are central to the frontier masculine aesthetic that is at the forefront in Albertans' collective sense of self. As a result, yielding to citizen concerns about water quality and ecological integrity was an obvious political decision for the United Conservatives. Environmental activists that emphasise equally damaging activities within the oil and gas sector, however, are not treated with the same regard. Instead, within the provincial narrative they are often cast as public enemy number one. While tensions between fossil fuel companies, government officials and environmentalists have been high for a long time, this tension was uncharacteristically formalised by the creation of the 30 million Canadian dollars energy 'war room' by the United Conservative government in 2019, formally named the Canadian Energy Center (Anderson, 2019; Canadian Energy Centre, nd). Created as a quasi-public relations entity meant to fight the misinformation on Alberta's energy industry, the Canadian Energy Center was launched alongside a public inquiry into the supposed foreign funding of anti-energy campaigns within the province (Allan, 2021). Both of these initiatives frame environmentalists, particularly Indigenous land defenders, as naïve, anti-Canadian and irrelevant (for example, Jaremko, 2022; Pappano, 2022)

The government's struggle with environmental groups came to a head when the United Conservative Party passed the Critical Infrastructure Defence Act, legislation that criminalises protestors who block access to or trespass on infrastructure deemed critical in any way, including pipelines, refineries, utility corridors, roads and railways (Kenney, 2020). This broad piece of legislation was introduced during a number of high-profile railroad blockades from Indigenous land defenders and their supporters in protest of fossil fuel development projects, specifically the Coastal Gas Link pipeline on the unceded territory of the Wet'suwet'en (for example, BBC, 2020; Leavitt, 2020; Turnbull and Aiello, 2020). This restriction on the ability to peacefully protest is yet another tactic employed to ensure that the Albertan frontier remains open and that the extraction of its resources does not become inhibited.

Conclusion

A cultural criminological approach invites us to consider the role that culture plays in the criminalisation of harmful activities. Efforts within green criminology have not yet considered the influence that gender plays in shaping cultural norms and narratives around environmental harm, a gap which this volume attempts to address. As the Alberta case demonstrates, industries which are intimately tied not only economically to a place but also culturally, tend to be subject to far less regulatory oversight and criminalisation than industries which are not. The oil industry is entangled with the culture in the province through the frontier masculine narrative and this serves to support the oil and gas industry even when it may not be the most economic choice for the province. This gendered narrative often escapes uncriticised, as it is obscured by the male-as-normative epistemology within Western culture. This is despite the role this narrative plays in shaping how harms are both conceptualised and whether they are criminalised or enforced at all.

Generally speaking, criminal law tends to exclude those harms which are exacted by elites (Stretesky et al, 2014). This is not the case universally, particularly in situations where there is significant citizen action (Jarrell and Ozymy, 2010), as occurred with the Coal Policy Repeal. What is worth noting is that despite the far more significant impact that the oil and gas industries have on natural areas within the province, the oil and gas sector has not been subject to the same public scrutiny that coal mining was, highlighting the infiltration of oil and gas into the provincial identity and culture through processes like frontier masculinity.

Frontier masculinity in Alberta captures the narratives and discourses that sustain colonialism, hidden within the obscurity of male-as-normative ways of knowing. Within a colonial nation state, everyone is racialised and gendered, sustaining a relationship to colonialism whether they wish to acknowledge it or not (Arvin et al, 2013). Frontier masculinity furthers the goal of colonialism, opening up the frontier as a new territory, full of resources for exploitation while simultaneously erasing the claim that Indigenous peoples have to that land and mobilising Western conceptualisations of femininity and masculinity to maintain fossil fuel hegemony. In this way, frontier masculinity serves as 'the master narrative which explains the culture to itself and seems to express its overriding purpose' (Francis, 1997: 10), a self-fulfilling prophecy that is difficult to address. Shifting towards criminalisation on the basis of scientific standards for harm, rather than subjective interpretations of what may be too detrimental to the economy (Lynch and Stretesky, 2001), is one possible measure for dealing with frontier masculinity in the short term, though such efforts must integrate both feminist (Haraway, 1988; Martin et al, 2015; D'Ignazio and Klein, 2020) and anticolonial perspectives (Liboiron, 2021; see also Daston, 2006; Donald, 2012) to avoid reproducing existing systems of harm. So long as the colonial legacies of frontier masculinity are actively celebrated within the provincial identity, true change will remain as fictional as the frontier narrative standing in its way.

References

Agnew, R. (2013) The ordinary acts that contribute to ecocide: A criminological analysis. In N. South and A. Brisman (eds) *Routledge International Handbook of Green Criminology*. London: Routledge, pp 58–72.

Ahmed, S. (2017) *Living a Feminist Life*. Durham, NC: Duke University Press.

Allan, J.S. (2021) *Alberta Public Inquiry into Anti-Alberta Energy Campaigns*. Public Inquiry, p 650. Available at: https://open.alberta.ca/dataset/f4f39 b9e-48cb-4f6a-b491-25ee6f9c281e/resource/8db15e6c-5826-4ac5-b804-675e95867e9e/download/lbr-alberta-mining-and-oil-and-gas-extract ion-industry-profile-2020.pdf (accessed 20 June 2022).

Anderson, D. (2019) Alberta's energy 'war room' launches in Calgary. *CBC News*. Available at: www.cbc.ca/news/canada/calgary/alberta-war-room-launch-calgary-1.5392371 (accessed 16 June 2022).

Arvin, M., Tuck, E. and Morrill, A. (2013) Decolonizing feminism: Challenging connections between settler colonialism and heteropatriarchy. *Feminist Formations*, 25(1), 8–34.

BBC (2020) The Wet'suwet'en conflict disrupting Canada's rail system. *BBC News*, 20 February. Available at: www.bbc.com/news/world-us-canada-51550821 (accessed 16 June 2022).

Bell, S.E. and Braun, Y.A. (2010) Coal, identity, and the gendering of environmental justice activism in central Appalachia. *Gender and Society*, 24(6), 794–813.

Bell, S.E. and York, R. (2010) Community economic identity: The coal industry and ideology construction in West Virginia. *Rural Sociology*, 75(1), 111–43.

Bell, S.E., Daggett, C. and Labuski, C. (2020) Toward feminist energy systems: Why adding women and solar panels is not enough. *Energy Research and Social Science*, 68 (101557), 1–13.

Bem, S.L. (1993) *The Lenses of Gender: Transforming the Debate on Sexual Inequality*. New Haven: Yale University Press.

Brekhus, W. (2020) *The Sociology of Identity: Authenticity, Multidimensionality, and Mobility*. Medford: Polity Press.

Briscoe, M.D., Givens, J.E., Hazboun, S.O. and Krannich, R.S. (2019) At home, in public, and in between: Gender differences in public, private and transportation pro-environmental behaviors in the US Intermountain West. *Environmental Sociology*, 5(4), 374–92.

Brisman, A. (2014) Environmental and human rights. In G. Bruinsma and D. Weisburd (eds) *Encyclopedia of Criminology and Criminal Justice*, vol. 3. New York: Springer Verlag, pp 1344–53.

Brisman, A. and South, N. (2018) Green criminology and environmental crimes and harms. *Sociology Compass*, 13(1), e12650.

Brown, P. and Ferguson, F.I.T. (1995) 'Making a big stink': Women's work, women's relationships, and toxic waste activism. *Gender & Society*, 9(2), 145–72.

Brulle, R. and Pellow, D.N. (2006) Environmental justice: Human health and environmental inequalities. *Annual Review of Public Health*, 27(3), 3.1–3.22.

Bullard, R.D. (ed) (1993) *Confronting Environmental Racism: Voices from the Grassroots*. Boston, MA: South End Press.

Calgary Stampede (2022a) *About Us*. Available at: www.calgarystampede.com/stampede/about (accessed 6 June 2022).

Calgary Stampede (2022b) *Elbow River Camp*. Available at: www.calgarystampede.com/stampede/attractions/elbow-river-camp (accessed 15 June 2022).

Campbell, C.S. (1984) The stampede: Cowtown's sacred cow. In C. Reasons (ed) *Stampede City: Power and Politics in the West*. Toronto: Between the Lines, pp 103–22.

Canadian Energy Centre (nd) *Canadian Energy Centre*. Available at: www.canadianenergycentre.ca/ (accessed 21 June 2022).

Carson, R. (1962) *Silent Spring*. Boston: Houghton Mifflin.

Conference Board of Canada (2020) *Gender Wage Gap – Society Provincial Rankings – How Canada Performs*. Available at: www.conferenceboard.ca/hcp/provincial/society/gender-gap.aspx (accessed 2 June 2022).

Connell, R.W. (2001) Understanding men: Gender sociology and the new international research on masculinities. *Social Thought & Research*, 24(1/2), 13–31.

Connell, R.W. (2005) *Masculinities*, 2nd edn. London: Routledge.

Connell, R.W. and Messerschmidt, J.W. (2005) Hegemonic masculinity: Rethinking the concept. *Gender and Society*, 19(6), 829–59.

Cudworth, E. (2005) Introduction. In E. Cudworth (ed) *Developing Ecofeminist Theory: The Complexity of Difference*. London: Palgrave Macmillan, pp 1–15.

Culley, M.R. and Angelique, H.L. (2003) Women's gendered experiences as long-term Three Mile Island activists. *Gender & Society*, 17(3), 445–61.

D'Ignazio, C. and Klein, L.F. (2020) *Data Feminism*. Cambridge, MA: The MIT Press.

Daston, L. (2006) The history of science as European self-portraiture. *European Review*, 14(4), 523–36.

Davidson, D.J. and Freudenburg, W.R. (1996) Gender and environmental risk concerns: A review and analysis of available research. *Environment and Behavior*, 28(3), 302–39.

Dawson, T. (2018, December 5) What you need to know about Trudeau's comments on 'gender impacts' and construction workers. *National Post*. Available at: https://nationalpost.com/news/canada/heres-what-you-need-to-know-about-the-controversy-over-trudeaus-comments-about-gender-impacts-and-construction-workers (accessed 24 May 2023).

Donaldson, M. (1993) What is hegemonic masculinity? *Theory and Society*, 22(5), 643–57.

Dorow, S. (2015) Gendering energy extraction in Fort McMurray. In L. Stefanick and M. Shrivastava (eds) *Alberta Oil and the Decline of Democracy in Canada*. Edmonton: Athabasca University Press, pp 275–92.

Dorow, S. and Mandizadza, S. (2018) Gendered circuits of care in the mobility regime of Alberta's oil sands. *Gender, Place & Culture*, 25(8), 1241–56.

Dryden, J. (2021) The Supreme Court has ruled in favour of the carbon tax—here's what might happen next. *CBC News*, 28 March. Available at: www.cbc.ca/news/canada/calgary/kenney-jonathan-wilkinson-west-centre-kathleen-petty-1.5967214 (accessed 24 May 2023).

Ferrell, J., Hayward, K.J. and Young, J. (2015) *Cultural Criminology: An Invitation*, 2nd edn. Abingdon: Sage.

Finucane, M.L., Slovic, P., Mertz, C.K., Flynn, J. and Satterfield, T.A. (2000) Gender, race, and perceived risk: The 'white male' effect. *Health, Risk & Society*, 2(2), 159–72.

Fletcher, R. and Omstead, J. (2020) Alberta rescinds decades-old policy that banned open-pit coal mines in Rockies and Foothills. *CBC News*. Available at: www.cbc.ca/news/canada/calgary/alberta-coal-policy-rescinded-mine-development-environmental-concern-1.5578902 (accessed 16 June 2022).

Francis, D. (1997) *National Dreams: Myth, Memory, and Canadian History*. Vancouver: Arsenal Pulp Press.

Gaard, G. (2001) Women, water, energy: An ecofeminist approach. *Organization & Environment*, 14(2), 157–72.

Gibbs, C., Gore, M.L., McGarrell, E.F. and Louie Rivers, I.I.I. (2010) Introducing conservation criminology: Towards interdisciplinary scholarship on environmental crimes and risks. *The British Journal of Criminology*, 50(1), 124–44.

Gould, K.A., Schnaiberg, A. and Weinberg, A. (1996) *Local Environmental Struggles: Citizen Activism in the Treadmill of Production*. Cambridge: Cambridge University Press.

Gould, K.A., Pellow, D.N. and Schnaiberg, A. (2008) *Treadmill of Production: Injustice and Unsustainability in the Global Economy*. New York: Routledge.

Government of Alberta (2015) *Alberta's Oil Reserves Compared to Other Countries*. Available at: https://open.alberta.ca/dataset/4ad7b5c8-8fdf-42a4-bec4-e57fae9f058e/resource/e5dd5f00-5139-4b0d-ba45-49908 24b81af/download/did-you-know-fact-sheet-7-sept28.pdf (accessed 8 June 2022).

Government of Alberta (2020) *Alberta Mining and Oil and Gas Extraction Industry Profile, 2020*, p 12. Available at: https://open.alberta.ca/dataset/ f4f39b9e-48cb-4f6a-b491-25ee6f9c281e/resource/8db15e6c-5826-4ac5-b804-675e95867e9e/download/lbr-alberta-mining-and-oil-and-gas-ext raction-industry-profile-2020.pdf (accessed 8 June 2022).

Government of Alberta (2022) *Coal Policy*. Available at: www.alberta.ca/ coal-policy-guidelines.aspx (accessed 16 June 2022).

Government of Alberta (nd) *Cutting Red Tape*. Available at: www.alberta. ca/cut-red-tape.aspx (accessed 16 June 2022).

Graney, E. (2019) Swings and roundabouts: A timeline of conservatism in Alberta politics. *Edmonton Journal*. Available at: https://edmontonjournal. com/news/politics/swings-and-roundabouts-a-timeline-of-conservatism-in-alberta-politics (accessed 16 June 2022).

Griffin, S. (1978) *Woman and Nature: The Roaring Inside Her*. San Francisco: Sierra Club Books.

Hanvelt, J. (2004) *Cowboy Up: Gender and Sexuality in Calgary's 'Gay' and 'Straight' Rodeo*. Vancouver: University of British Columbia.

Haraway, D. (1988) Situated knowledges: The science question in feminism and the privilege of partial perspective. *Feminist Studies*, 14(3), 575–99.

Haraway, D. (2016) *Staying with the Trouble: Making Kin in the Chthulucene*. Durham, NC: Duke University Press.

Harding, S.G. (1991) *Whose Science? Whose Knowledge? Thinking from Women's Lives*. Ithaca: Cornell University Press.

Hayward, K.J. and Young, J. (2004) Cultural criminology: Some notes on the script. *Theoretical Criminology*, 8(3), 259–73.

Hearn, J., Nordberg, M., Andersson, K., Balkmar, D., Gottzen, L., Klinth, R., Pringle, K. and Sandberg, L. (2012) Hegemonic masculinity and beyond: 40 years of research in Sweden. *Men and Masculinities*, 15(1), 31–55.

Hochschild, A.R. and Machung, A. (1989) *The Second Shift*. New York: Avon Books.

Hultman, M. and Anshelm, J. (2017) Masculinities of global climate change: Exploring ecomodern, industrial and ecological masculinity. In M.G. Cohen (ed) *Climate Change and Gender in Rich Countries.* New York: Routledge, pp 19–34.

Hultman, M. and Pulé, P.M. (2018) *Ecological Masculinities: Theoretical Foundations and Practical Guidance.* New York: Routledge

Jaremko, D. (2022) A matter of fact: Canada's oil and gas workers don't need a forced 'just transition'. *Canadian Energy Centre*, 27 April. Available at: www.canadianenergycentre.ca/a-matter-of-fact-canadas-oil-and-gas-workers-dont-need-a-forced-just-transition/ (accessed 24 May 2023).

Jarrell, M.L. and Ozymy, J. (2010) Excessive air pollution and the oil industry: Fighting for our right to breathe clean air. *Environmental Justice*, 3(3), 111–15.

Keener, E. and Mehta, C. (2017) Sandra Bem: Revolutionary and generative feminist psychologist. *Sex Roles*, 76(9–10), 525–8.

Keith, T. (2017) *Masculinities in Contemporary American Culture: An Intersectional Approach to the Complexities and Challenges of Male Identity.* New York: Routledge.

Kenney, J. (2018) Twitter update, 30 November. Accessed October 23, 2021, Available at: https://twitter.com/jkenney/status/1068582203225890816 (accessed 16 June 2022).

Kenney, J. (2020) *20200225_bill-001.pdf, cC-32.7.* Available at: https://docs.assembly.ab.ca/LADDAR_files/docs/bills/bill/legislature_30/session_2/20200225_bill-001.pdf (accessed 16 June 2022).

Kenney, J. (2022) Twitter update, 28 March. Accessed May 8, 2022, Available at: https://twitter.com/jkenney/status/1508460397002571776?s=20&t=hI0kB3YYFxw6xj7WMyiGVw (accessed 16 June 2022).

Kramer, R.C. (2013) Carbon in the atmosphere and power in America: Climate change as state-corporate crime. *Journal of Crime and Justice*, 36(2), 153–70.

Krauss, C. (1993) Women and toxic waste protests: Race, class and gender as resources of resistance. *Qualitative Sociology*, 16(3), 247–62.

Leavitt, K. (2020) 'Not in our backyard': Alberta Wet'suwet'en rail blockade meets stiff opposition and shuts down after injunction granted. *The Toronto Star*, 19 February. Available at: www.thestar.com/news/canada/2020/02/19/new-wetsuweten-rail-blockade-pops-up-in-the-heart-of-canadas-oil-country.html (accessed 16 June 2022).

Leonard, E.B. (1982) *Women, Crime, and Society: A Critique of Theoretical Criminology.* New York: Longman.

Liboiron, M. (2021). *Pollution Is Colonialism.* Duke University Press.

Lowman, E.B. and Barker, A.J. (2015) *Settler: Identity and Colonialism in 21st Century Canada.* Halifax: Fernwood Publishing.

Lynch, M.J. and Stretesky, P. (2001) Toxic crimes: Examining corporate victimization of the general public employing medical and epidemiological evidence. *Critical Criminology: The Official Journal of the ASC Division of Critical Criminology*, 10(3), 153–72.

MacGregor, S. (2009) A stranger silence still: The need for feminist social research on climate change. *The Sociological Review*, 57(2_suppl), 124–40.

Martin, A., Myers, N. and Viseu, A. (2015) The politics of care in technoscience. *Social Studies of Science*, 45(5), 625–41.

McCright, A.M. and Dunlap, R.E. (2011) Cool dudes: The denial of climate change among conservative white males in the United States. *Global Environmental Change*, 21(4), 1163–72.

McKenzie, J., Beedell, E. and Corkall, V. (2022) *Blocking Ambition: Fossil Fuel Subsidies in Alberta, British Columbia, Saskatchewan, and Newfoundland and Labrador*. International Institute for Sustainable Development. Available at: www.iisd.org/system/files/2022-02/blocking-ambition-fossil-fuel-subsidies-canadian-provinces.pdf (accessed 22 May 2022).

Mellström, U. (1995) Engineering lives: Technology, time and space in a male-centred world. Available at: http://urn.kb.se/resolve?urn= urn:nbn:se:liu:diva-35073 (accessed 13 May 2022).

Merchant, C. (1980) *The Death of Nature: Women, Ecology, and the Scientific Revolution*. San Francisco: HarperOne.

Messerschmidt, J.W. (2018) *Hegemonic Masculinity: Formulation, Reformulation, and Amplification*. Lanham: Rowman & Littlefield.

Mies, M., Salleh, A. and Shiva, V. (2014) *Ecofeminism*, 2nd edn. London: Zed Books.

Miller, G.E. (2004) Frontier masculinity in the oil industry: The experience of women engineers. *Gender, Work & Organization*, 11(1), 47–73.

Naegler, L. and Salman, S. (2016) Cultural criminology and gender consciousness: Moving feminist theory from margin to center. *Feminist Criminology*, 11(4), 354–74.

National Inquiry into Missing and Murdered Indigenous Women and Girls (2019) *Reclaiming Power and Place: The Final Report of the National Inquiry into Missing and Murdered Indigenous Women and Girls*. Available at: www.mmiwg-ffada.ca/final-report/ (accessed 2 June 2022).

Nikiforuk, A. (2020) Threatened by coal, ranchers take the Kenney government to court. *The Tyee*. Available at: https://thetyee.ca/News/2020/12/07/Threatened-By-Coal-Ranchers-Take-Kenney-Court/ (accessed 16 June 2022).

O'Brien, M. (1981) *The Politics of Reproduction*. New York: Routledge & Kegan Paul.

OilfieldWifeSorority (@OWSorority) (2014) Twitter update, 15 August. Available at: https://twitter.com/owsorority/status/500331182530494464 (accessed 10 May 2022).

Pappano, G. (2022) Divestment push tone deaf amid surging global energy crisis. *Canadian Energy Centre*. Available at: www.canadianenergycentre.ca/pappano-divestment-push-tone-deaf-amid-surging-global-energy-crisis/ (accessed 21 June 2022).

Parlee, B.L. (2015) Avoiding the resource curse: Indigenous communities and Canada's oil sands. *World Development*, 74, 425–36.

Pearce, F. and Tombs, S. (1998) *Toxic Capitalism: Corporate Crime and the Chemical Industry*. Farnham: Ashgate.

Peeples, J.A. and DeLuca, K.M. (2006) The truth of the matter: Motherhood, community and environmental justice. *Women's Studies in Communication*, 29(1), 59–87.

Plumwood, V. (1993) The politics of reason: Towards a feminist logic. *Australasian Journal of Philosophy*, 71(4), 436–62.

Plumwood, V. (2018) Ecofeminist analysis and the culture of ecological denial. In L. Stevens, P. Tait and D. Varney (eds) *Feminist Ecologies: Changing Environments in the Anthropocene*. Cham: Springer International Publishing, pp 97–112.

Richards, K.S. (2019) Crude optimism: Romanticizing Alberta's oil frontier at the Calgary Stampede. *TDR*, 63(2 [T242]), 138–54.

Rodríguez Goyes, D. and South, N. (2016) Land grabs, bio-piracy and the inversion of justice in Colombia. *The British Journal of Criminology*, 56(3), 558–77.

Seiler, R. and Seiler, T. (2004) Managing contradictory visions of the west: The great Richardson/Weadick experiment. In L. Felske and B. Rasporich (eds) *Challenging Frontiers: The Canadian West*. Calgary: University of Calgary Press, pp 164–89.

Shriver, T., Adams, A. and Einwohner, R. (2013) Motherhood and opportunities for activism before and after the Czech Velvet Revolution. *Mobilization: An International Quarterly*, 18(3), 267–88.

Slovic, P., Peters, E., Finucane, M.L. and MacGregor, D.G. (2005) Affect, risk, and decision making. *Health Psychology*, 24(4S), S35.

Sollund, R. (2013) The victimization of women, children and non-human species through trafficking and trade: Crimes understood through an ecofeminist perspective. In N. South and A. Brisman (eds) *Routledge International Handbook of Green Criminology*. Oxon: Routledge, pp 317–30.

Sollund, R. (2017) Doing green, critical criminology with an auto-ethnographic, feminist approach. *Critical Criminology: The Official Journal of the ASC Division on Critical Criminology and the ACJS Section on Critical Criminology*, 25(2), 245–60.

Stretesky, P., Long, M.A. and Lynch, M.J. (2014) *The Treadmill of Crime: Political Economy and Green Criminology*. New York: Routledge.

Trembath, T. (2020) Alberta seeing 'one pandemic layered on top of another,' anti-domestic violence advocate says. *CBC*. Available at: www.cbc.ca/news/canada/calgary/domestic-violence-crystal-boys-jan-reimer-maryam-monsef-1.5551841 (accessed 2 June 2022).

Turnbull, S. and Aiello, R. (2020) 'Barricades must now come down,' PM Trudeau says of rail blockades. *CTV News*. Available at: www.ctvnews.ca/politics/barricades-must-now-come-down-pm-trudeau-says-of-rail-blockades-1.4821889?cache=sgjigezwskldpu%3FclipId%3D64268 (accessed 6 June 2022).

Vowles, K. and Hultman, M. (2021) Dead white men vs. Greta Thunberg: Nationalism, misogyny, and climate change denial in Swedish far-right digital media. *Australian Feminist Studies*, 36(110), 414–31.

Wachholz, S. (2007) At risk: Climate change and its bearing on women's vulnerability to male violence. In P. Beirne and N. South (eds) *Issues in Green Criminology: Confronting Harms against Environments, Humanity and Other Animals*. Cullompton: Willan, pp 161–85.

Weber, B. (2020) Alberta suspends environmental monitoring for oilpatch, citing COVID-19. *CBC*. Available at: www.cbc.ca/news/canada/edmonton/alberta-energy-regulator-suspends-environment-monitoring-for-oilpatch-over-covid-1.5578994 (accessed 16 June 2022).

White, R. (2011) *Transnational Environmental Crime: Toward an Eco-global Criminology*. Abingdon: Routledge.

Whitt, L. (2009) *Science, Colonialism, and Indigenous Peoples: The Cultural Politics of Law and Knowledge*. Cambridge: Cambridge University Press.

Williams, C. (1996) Environmental victims: An introduction. *Social Justice*, 23(4), 1–6.

Williams, K.A. (2021) *Stampede: Misogyny, White Supremacy, and Settler Colonialism*. Black Point: Fernwood Publishing.

Wright, W. (2001) *The Wild West: The Mythical Cowboy and Social Theory*. Abingdon: Sage.

15

David and Goliath:
Exploring the Male Burdens
of Patriarchal Capitalism

Rob White

Introduction

This chapter provides commentary on the burdens experienced by men in combating ecological destruction and in particular contesting the pressures and limits of patriarchal capitalism. A central focus is climate change and the responses of men to the challenges and burdens generated by global warming. There are rapid transformations occurring in the social, political and environmental landscape. This is a global process which is experienced differentially within the world's male population depending on social status, economic resources and geographical location. Precisely because material circumstances vary, the responses of men to climate crises likewise varies. Why and how this is the case informs the main discussions in this chapter.

At the heart of the chapter are two central conceptual considerations. The first relates to gender inequality and gender difference; the second to the materiality of masculinity as a lived practice. In relation to the first consideration, there is ample evidence that climate change is not gender-neutral. This is demonstrated by its effects on farming. For example, in the least affluent countries, in Africa and elsewhere, 79 per cent of women who are economically active report agriculture as their primary economic activity. Yet, only between 10 and 20 per cent of all landholders are women (UNDP, 2013). Women are especially vulnerable to the consequences of climate change, in part due to the prior disadvantages they suffer generally. For instance, compared with men, women have less access to land, financial services, livestock, social capital and technology. Impoverishment and lower social status weaken the ability of women to be resilient in the face of the burdens associated with climate change. These are compounded by the extra responsibilities associated with caring work for children, the ill and the elderly. Dramatic changes in temperature, climate and seasons exacerbate existing inequalities and hardships (United Nations Women Watch, 2009; UNDP, 2013).

Women who farm in the advanced capitalist countries likewise suffer from the consequences of entrenched gender inequalities. A major outcome of years of drought in Australia, for example, is an increase in women's labour on and off the farms. It is anticipated that with further global warming, the nature of agricultural labour will continue to change, including greater contributions being asked of farming women. Yet, women's labour remains largely taken-for-granted and is largely invisible in relation to policy development and national responses to climate change (Alston et al, 2018). Again, central to this, is the fact that, while women are active contributors as farmers, they rarely figure in agricultural land ownership. Climate change is leading to changed gender workloads and increased workloads for women. Yet, women's input is still being treated as a 'farm survival strategy' rather than a major personal economic contribution by women to the enterprise (Alston et al, 2018: 12). On the other hand, male suicide rates among farmers experiencing drought are high compared to female rates (Alston, 2012), again indicating a clear, but different, gendered response to crisis. In both instances, gender inequalities and gender differences carry with them immense social costs, especially for the health and wellbeing of farms and farmers under stress and for women as well as men. These kinds of issues are explored further in this chapter from the point of view of gender expectations and male lived experiences.

The chapter begins by introducing examples of how men are resisting and responding to ecological destruction, followed by a summary of the causes of contemporary ecocidal tendencies. The chapter then discusses which men are navigating the waters of social and ecological change, in which ways, and for which reasons, drawing upon the notions of masculinity and its intersections with class to explain varied forms of male agency. The burdens of patriarchal capitalism are experienced by many, but the solutions and responses very much depend upon the material location and existential realities of diverse groups of men.

Men resisting and/or responding to ecological destruction

This chapter was inspired by the biblical story of 'David and Goliath' in which a young man (barely a boy) stood up to and defeated a giant enemy warrior using a slingshot. The theme of the tale is how the apparently small and weak can defeat the evidently powerful. David succeeded because of his faith, strength of conviction and choice of weapon.

What triggered this recollection was the fact that two of my personal heroes of the climate justice movement are David Attenborough (British film maker, commentator and activist) and David Suzuki (Canadian scientist, broadcaster and activist) (see for example, Suzuki, 2010). They, too, are facing their Goliath, the immense political economic monster that is ravaging the planet.

While patriarchal capitalism entrenches male domination and the capitalist ruling class worldwide, it is neither monolithic nor static. It is possible to be challenged, and among those doing so are men and women from many backgrounds. Activists such as Attenborough and Suzuki are high-profile (male) experts who have devoted their lives to fighting against the forces behind climate change. In this process, they have also been very supportive of female voices (for example, Greta Thunberg). Other male activists, such as climate scientist Michael Mann (2021), are likewise engaged and supportive of figures like Thunberg, although resistant to the ideas of some women, such as Naomi Klein (2014), who take a more explicit class position in relation to capitalism and political action. The climate wars involve varying degrees of understanding of the underlying politics, resulting in differences in tactical and strategic vision. But they are nonetheless united in their fight for a fossil fuel free future.

The fight to stop the cranking up of the Earth's heat is not only symbolic and/or representational. It is also intensely physical and personal. Research by Global Witness has exposed the killing of environmental activists and local residents by perpetrators of harm such as private corporations and state officials. A recent report shows that in the past decade (to 2021), a land and environment defender was killed every two days, with more than 1,700 people killed. The deadliest countries include Brazil, Colombia, the Philippines and Mexico. The violence is linked to territorial conflicts and the pursuit of economic growth based on the extraction of natural resources from the land. The report highlighted that Indigenous communities in particular face a disproportionate level of attacks – nearly 40 per cent – even though they make up only 5 per cent of the world's population (Global Witness, 2022). It was reported that one in ten of the defenders killed in 2021 were women, nearly two-thirds of whom were Indigenous (Global Witness, 2022). Put another way, this means that nine out of ten defenders killed were men, fighting at the ground level against the exploitation of the environment.

Other men in other places have likewise stepped up to challenge the dominant discourses and failures of government policy that have allowed climate disasters to unfold. In Australia, for example, former fire chiefs from around the country have inserted themselves into the national political debates surrounding climate change. They have lobbied politicians and been publicly critical of 'business as usual'. Based on their experience and expertise, these men have stood tall in the midst of ignorance, denial and scorn from their elected political leaders. Yet, they have remained staunch in their critique and passionate about needed response and relief strategies. They have also been unafraid to take on the media moguls shaping the climate debates in this country, in particular the Murdoch media empire (Mullins, 2021). As will be seen shortly, responding to disasters resonates in particular ways for men compared to women. It frequently involves socially patterned roles in which men take specific kinds of risks, including the potential to lose their lives, in protecting the vulnerable.

Sourcing the problem

Global capitalism is the driving force behind ecological destruction, global warming and threats to biodiversity – it has a built-in 'growth imperative' (Kramer, 2020; van der Velden and White, 2021; Whyte, 2020). This is an historical process, evident in 500 years of imperialism and colonial rule, mainly emanating out of Europe, and involving land grabs, resource extraction and the subjugation of Indigenous peoples worldwide.

Class power intersects structurally with gender division, but neither is reducible to the other – male domination and domination by the (transnational) capitalist class is an historical configuration that is socially constructed but not intrinsic to the capitalist political economy as such. For example, there are nation states that have higher gender equality than others, but which simultaneously exhibit concentrations of wealth and power in private hands; moreover, some of this concentration of wealth also resides in the hands of women, such as mining magnate Gina Rinehart in Australia, and landowner the late Queen Elizabeth in the United Kingdom.

Men, as such, are not the enemy – patriarchal capitalism is. The key issues here relate to gender orders and class divisions, in that the main victims of environmental harm are working-class men and women, with women especially vulnerable on a global scale due to their structural location in economies and households. The dominance of a global hegemonic masculinity (Connell, 2000) also means that the leading role of women in peasant and Indigenous resistance movements worldwide is obscured in the same moment that traditional subsistence and ecologically benign forms of production are ignored.

Buchbinder (1998) makes the point that patriarchy is both remunerative *and* punitive for men. Similarly, Connell (1995, 2000) explains that there is a 'patriarchal dividend' for men in the form of higher status and greater control over societal resources at the social structural level. Fundamentally, though, 'being a man' is not reducible to an essential social characteristic, and men can likewise be socially positioned in extremely vulnerable and marginalised ways. Class matters. As does race, ethnicity, Indigeneity, age and ability.

Masculinity is multidimensional – it has 'heroic' and 'caring' aspects, as well as its destructive and oppressive dimensions. Men are both complicit in and opposed to the damage wrought by the state and corporate capitalist machine. With respect to this, it is important to acknowledge that masculinities are collective human projects, they are internally complex, they take myriad social forms, and they change.

Masculinity is a social process, and as such is bound up with issues of social structure and social resources. What this means at a practical level is that there are bound to be considerable variations in how specific groups of men respond to, and are affected by, social pressures and social change. In this realisation – that masculinity involves process, complexity and differentiation – lies the hope that gender can possibly be renegotiated, not within, but in opposition to, the dominant culture (White, 2002: 284).

Therefore, there is a need to account for the commonalities among men, as well as the differences in their lived experiences – the variety of ways in which masculinity is constructed in the course of day-to-day living.

Men responding to crisis

This section points to diverse situations and the different ways in which men are responding to the climate crisis. The urgency and realities of climate change are eloquently conveyed each time data is released on global heating trends. The effects of this global warming are manifest in climate disruption, involving high impact and extreme weather events such as heat and cold waves, precipitation, heavy rainfalls and floods, severe storms, and drought (World Meteorological Organization, 2020). These events and trends, in turn, present as a series of key risks.

The risks include increased damage from wildfires, heat-related human mortality and increased damage from river and coastal urban floods. They include a distributional shift and reduced fisheries catch potential at low latitudes; compounded stress on water resources; increased mass coral bleaching and mortality; reduced crop productivity and livelihood and food security; and the loss of livelihoods, settlements, infrastructure, ecosystem services and economic stability. Other risks include the spread of vector-borne diseases – the global COVID-19 pandemic illustrating just how quickly future risks can translate into present harms (IPCC, 2014).

Climate disruption means that there are sharp changes in weather, as longer-term weather and climate patterns shift outside previous norms and wreak havoc in relation to precipitation and the composition and direction of predominant wind and ocean currents. For climate-exposed sectors, regional economic effects to agriculture, forestry, fishery, energy and tourism have increasingly been identified (IPCC, 2022). However, current systems of natural resource extraction and use, and widespread contamination of the natural world, simultaneously undercut potential measures to bolster resilience. For example, mining (both legal and illegal) has impacts on water systems and land use that can be counter-productive in terms of climate responses (for example, because of deforestation and diminishment of the water table). Many of these activities take place in rural and remote locations which makes them even harder to monitor and regulate (Zabyelina and van Uhm, 2020). Meanwhile, climate change continues to radically alter the basis of world ecology while, simultaneously, 1 million species are considered at threat of extinction (IPBES, 2019; Portner et al, 2021). These trends manifest in events and problems that have specific gender dimensions, reflecting both gender difference and gender inequality.

Disaster relief: immediate material needs

As mentioned, global warming is a powerful driver of natural disasters as it generates greater propensity towards extreme weather events. These are projected to increase in intensity and frequency in the foreseeable future. They include phenomenon such as floods, cyclones and heat waves.

Disasters generate many kinds of vulnerabilities, including and especially those based upon gender and age. The very young and the very old are particularly vulnerable to disease, social predation and inability to get out of harm's way. Men and women face different kinds of pressures, risk and vulnerabilities (see Table 15.1). While it is true that women and children are especially vulnerable to violence, and specifically male violence, during disaster periods, not all men participate in this violence, and many are themselves victims of it.

For present purposes, however, the emphasis is on acknowledging the different roles and vulnerabilities of men compared to women during disasters. Many of these stem from the traditionally gendered manner in which men (and women) respond to the tasks and threats associated with disaster. Men are usually first responders, when danger is especially high, and usually the last to leave when evacuations are signalled. They risk their lives for others while receiving little formal assistance themselves. These are heroic attributes. They contrast sharply with images and behaviours in which men are perpetrators of violence and sexual abuse, looting and malicious damage. The majority of men involved in disaster response belong in the first category.

The discourses describing women in periods of disaster – as vulnerable or virtuous – also shape how women generally are viewed and portrayed in relation to climate change issues. Much the same applies to men who are seen as 'vulnerable', 'virtuous' or 'vile'. Too often this type of framing of the

Table 15.1: Gender-based vulnerabilities in disaster responses

Issues	Male/men	Female/women
Activity	First responders	Forced to stay at home
Risk taking	Risk lives for victims	Risk lives for children
Proximity	Last to evacuate	Forced to evacuate
Assistance	Less assistance	Greater family responsibility
Needs	Finding work/income	Finding food and shelter
Victimisation	Sexual and physical violence	Sexual and physical violence
Perceptions	Seen as threat or perpetrators	Seen as victims
Living arrangements	Living alone	Enforced communal living
Uncertainties	Head of household duties	Domestic duties

Source: Drawing from observations and data compiled by Heckenberg and Johnston (2012)

issues leaves begging fundamental questions regarding inequalities of power. They can also gloss over important specificities of context and circumstance:

> Marginality needs to be viewed through the power relations that produce the vulnerability in the first place. Different power relations are privileged in different situations and class, gender, ethnicity or nationality assume importance depending on the context. The specificity of vulnerability may differ. A generalized belief in women's vulnerability silences contextual differences. Gender gets treated not as a set of complex and intersecting power relations but as a binary phenomena carrying certain disadvantages for women and women alone. (Arora-Jonsson, 2011: 750)

How best to deal with existential challenges is therefore partly a matter of understanding the nuances of difference and the contours of inequality across diverse lived situations. Importantly, these observations apply to men as well as women. That is, different men are positioned differently regarding both disaster response (for example, as victims to be rescued or responders who do the rescuing) and disaster relief (for example, eligibility and availability depending on geopolitical location, status in household and political ties).

Changing social roles: farming and food

The contexualisation of male responses also needs to be front and centre in analyses of the impact of changing work and life circumstances. For instance, arable land and potable drinking water are already at a premium worldwide and this will be exacerbated under conditions of wide-scale climate change. Thus, '[r]ural areas are expected to experience major impacts on water availability and supply, food security, infrastructure and agricultural incomes, including shifts in the production areas of food and non-food crops around the world' (IPCC, 2014: 11). This is occurring in the context of the domination of global food production by transnational corporations for the purposes of profit not social need.

Indeed, transnational corporations, in conjunction with hegemonic nation states and local political elites, are implicated in many of the present changes occurring in global food production and consumption. The exploitation of the world's natural resources by the major transnational corporations occurs through the direct appropriation of lands, plants and animals as 'property' (including intellectual property as in the case of patents). It also occurs through the displacement of existing systems of production and consumption by those that require insertion into the cash–buyer nexus, in other words, the purchase of goods and services as commodities (Gray and Hinch, 2018; Schally, 2018). This has happened in food production as it has in other spheres of human life.

Subsistence and family farming are suffering due to overexploitation, corporatisation and climate change, and great shifts in human populations and in resource use are taking place worldwide. These farming communities, usually set in non-urban locations, are beset by intense weather events such as droughts, floods, storms, bushfires and cyclones but without the urban infrastructure of police and emergency services, hospitals and medical staff, rapid transit systems and logistical support. Prolonged drought may be linked to algae blooms that directly threaten irrigation systems and potable drinking water. Fires and fire smoke can kill stock as well as native endemic animals. Projections about human vulnerability include the observation that '[i]n rural areas vulnerability will be heightened by compounding processes including high emigration, reduced habitability and high reliance on climate-sensitive livelihoods' (IPCC, 2022: 12).

Farmers who are forced to leave land and home have several immediate issues to grapple with, such as where to live, what to do, how to earn an income, and with whom. For young men, particularly, this makes them vulnerable to being recruited into criminal organisations. As one commentator puts it:

> But what happens to young men who no longer have a productive path forward, whether it is on the farm or in the city? They are going to be ripe to be recruited into criminal organizations, whether cartels or human smuggling networks, that can pay them something and give them a sense of status and pride that they don't have starving on a

farm. (Oliver Leighton Barrett, retired US Navy lieutenant, quoted in Albaladejo and LaSusa, 2017)

In regions such as South America, Central America and the Horn of Africa, and countries such as Venezuela, Brazil and Somalia, these are present-day and persistent matters of concern. The conundrums of survival mean that at least for some, victimisation caused through climate change could well result in the 'victims' becoming 'offenders' due to the severe restrictions in life options accompanying displacement (Hall and Farrall, 2013).

Oppositions and social conflicts

Social inequality and environmental injustice will undoubtedly be the drivers of continuous conflicts for many years to come, as the most dispossessed and marginalised of the world's population suffer the brunt of food shortages, undrinkable water, climate-induced migration and general hardship in their daily lives (Crank and Jacoby, 2015; White, 2018; Heydon, 2020). These conflicts will likely become even more evident in relation to struggles related to the necessities of life (for example, food, water), territorial claims (including those related to receding coastlines and desertification), the exploitation of natural resources (for example, mining, agricultural) and the mass movements of people (for example, climate-induced migration, sea-rise refugees).

Exploitation of natural resources is a major cause of armed conflict within and between communities and nation states (Homer-Dixon, 1999; Klare, 2012; Le Billon, 2012). This is largely due to scarcity of resources, which can arise from depletion or degradation of the resource (supply), increased demand for it (demand), and unequal distribution and/or resource capture (structural scarcity) (Homer-Dixon, 1999). What humans do to the environment is directly implicated in the production of scarcity, and hence conflict. And armed men tend to feature predominantly although not exclusively in such conflicts.

Whether it is responding to disasters, adjusting to shifting work and life roles, or participating in the conflicts over natural resources, men are necessarily active players in shaping the future of their homes, communities, regions and states. Whether this is driven by survivalist imperatives and/or linked to broader political agendas is one of the major questions of our age. Posed in relation to patriarchal capitalism, the oppositions are clear.

- The 'Davids' are those fighting against the structural power and institutional weight of patriarchal capitalism. They include direct victims of the burdens imposed by gender and class structuring – such as poor people, Indigenous people, the working class and those who have been de-classed (for example, refugees). They also include those who are critical of the ecocidal tendencies and inequalities of present structures even though benefiting from the structurally provided patriarchal dividends – such as educated, informed scientists and middle-class activists (including academics).

- The 'Goliaths' are those at the helm of the existing global capitalist power structure (male and female) whose class interests necessarily involve exploitations of humans (men, women and children) and natural resources (minerals, trees, fish, seeds). The Goliath structure is comprised of patterns of behaviour, culture and institutional power, affecting all men and women, that variously benefit or disadvantage particular individuals (men and women) depending upon where they are located in the overarching power structure.

However, this bifurcated political divide needs further unpacking if we are to adequately capture the nuances and challenges of everyday life.

Key analytical and political issues

While many men experience the burdens of patriarchal capitalism, their responses to situations and predicaments are largely shaped by circumstance and opportunity. In other words, politics is forged in the exigencies of the moment, not simply in the words of the manifesto. Conversely, there is also a need to go beyond assertions that 'who men are' (for example, white, heterosexual, middle class, middle aged) somehow always and already determines their specific interests. In the case of our favourite 'Davids' (Attenborough and Suzuki), for instance, their public stance on climate change puts into question assumptions that structural privilege necessarily leads to specific political and social behaviour.

There is a need as well for considerations of 'what men do' (which is not reducible to ascribed characteristics), for example, manual labour on the farm and in the mines, and how the way of being as men translates into certain behaviours. Again, it is important to contextualise diverse expressions of masculinity rather than treating masculinity as universal, totalising or simply a variation of hegemonic masculinity. Segal (1990: 288) tells the story, for example, of an underground miner who very much valued the solidarity, comradeship and humour of his workmates, and who felt that the aggression and physicality of male-to-male relationships was integral to building deep friendships and general confidence in oneself. In the context of the tensions and boredom associated with mining, such qualities were highly valued. Interestingly, Segal (1990: 278–9) notes that while middle-class men's group members bemoaned the lonesomeness of their masculinity, the working-class perspective, and lived experience, was vastly different.

There are complex personal and social processes at play here. As indicated, a brutal environment (of poverty, unemployment, racism) may call forth an emphasis on 'toughness' and 'respect', of not displaying fear. Hegemonic masculinity is reinforced not simply by ideological process, but profoundly disadvantaged and threatening social environments. But life and work environments also provide the tools for forging solidarity, for engaging in resistance, and for sidestepping or dismantling the disempowering aspects of hegemonic masculinity (such as misogyny and homophobia). Much, however, depends upon how change affects individuals, families, neighbourhoods and collectivities.

For example, not all subjugated men (for example, Indigenous, low income, traditional land users) are prepared to 'fight for our future'. Some may sell-out or opt-out of the struggles (due to family concerns, fearfulness, threats from the powerful, personal gain or simply trying to make a living). The killings documented by Global Witness are real and terrifying.

These observations highlight the importance of *class* and *agency* in analysis of how patriarchy plays out as a social practice. This includes socially patterned forms of class exploitation, including exploitations of nature and which includes the use of working-class labour to carry out these exploitations (that is, wage labour is the source of livelihood in global capitalism, an inherently exploitive relationship between corporate employer and the employee who has nothing to sell except their labour power). Those employed in the resource extraction industries (coal, logging) may be drawn into a cross-class alliance with capitalist owners due precisely to their vulnerability in a precarious labour market (and demise of the welfare state). Discussion of 'just transitions' hinges precisely on this matter of precarity and vulnerability. Without policy and planning initiatives that make sense to those most affected, their future looks bleak. Under these conditions, the status quo seems attractive.

The limits and pressures on climate politics works in other ways as well. For instance, regardless of their personal feelings and possibly even progressive ideas about environment/society, the owners and managers of capital must perform certain roles or else they will lose their jobs and/or competitive advantage (Bakan, 2004). On the other hand, agency that is directed at social transformation is nonetheless exercised by those with enough privilege and power, individually and collectively (via unions and political organisation/social movements), that enables them to make a difference. Perversely, this includes billionaire owners of capital who also gain economically by pursuing 'green capitalist' alternatives to existing fossil fuel industries.

Political resolve and activism are also exercised by those who have 'nothing left to lose', such as environmental victims, and those who consciously forsake career/occupation in favour of their commitment to environmental justice movements (in which case, coming from a 'privileged' background may assist in terms of income and family support). In other words, those who do not have resources or who use their existing resources to turn their backs on the dominant political economy, are among the leaders in the struggles against patriarchal capitalism.

Social and ecological crisis can generate profound ontological insecurities and fears. It can also underpin a drift towards radicalisation, reactionary or progressive. The typical characteristics of radicalisation include a sense of injustice or humiliation; response to perceived injustice against a group of people, who are not necessarily related to the protagonists; direct experience of disrespect and oppressive state intervention; a need for identity; and search for defining purpose or goal in life (Walklate and Mythen, 2015). It is also linked to large-scale transformations such as climate change. According to Mary Robinson, President of the Mary Robinson Foundation-Climate Change, and former Irish President and United Nations Commissioner for Human Rights, '[c]limate change is a threat multiplier – it exacerbates poverty and water scarcity, it compounds food and nutrition insecurity and it makes it even harder for poor households to secure their rights', and moreover, '[i]n a world where climate change exacerbates the stresses of daily life on people already disenfranchised by poverty or social standing, radicalisation is very likely' (Robinson, 2015). Men and boys suffering from material deprivations and status insecurity are especially prone to radicalisation. Which way they turn is uncertain and yet is integral to contemporary political struggles.

Conclusion

This chapter has provided a brief review of how men are responding to the climate crisis, through the lens of gender, class and patriarchal capitalism. It has been argued that 'politics' cannot be read off from class position or the position of men in the gender order – it is forged in experience, education and direct action. Much depends upon the material circumstances within which men find themselves. For many, gender and class combine in forms of masculinity that manifest as ways to cope with precarious jobs, engagement in disaster relief, negotiating social conflict and ensuring basic survival.

The burdens of patriarchal capitalism are both structural (in the sense of narrowly shaping human experience insofar as it is informed by specific notions of 'hegemonic masculinity' and 'economic imperatives') and oppressive (in the sense that pushing back against the power structure inevitably comes at a cost – financial, reputational, employment and, in some instances, lives).

Thus, the politics of climate justice involving men is grounded in diversity of experience, opportunity and material circumstance. The bottom line, however, is that combating the Goliaths of late capitalism demands alliances and solidarity. This, in turn, rests upon a critique of gender essentialism (for example, 'women are always nurturing; men are always domineering') and class reductionism (for example, 'middle-class men are always conservative; working-class men are always duped'). Fundamentally, significant social transformation requires the fostering of social and individual agency directed towards communal empowerment. In a nutshell, we need more Davids.

References

Albaladejo, A. and LaSusa, M. (2017) The perfect storm: How climate change exacerbates crime and insecurity in LatAm. *Insight Crime*, 25 September. Available at: www.insightcrime.org/news/analysis/perfect-storm-climate-change-exacerbates-crime-insecurity-latinamerica-caribbean/ (accessed 30 November 2017).

Alston, M. (2012) Rural male suicide in Australia. *Social Science and Medicine*, 74, 515–22.

Alston, M., Clarke, J. and Whittenbury, K. (2018) Contemporary feminist analysis of Australian farm women in the context of climate changes. *Social Sciences*, 7(2), 16, doi:10.3390/socsci7020016

Arora-Jonsson, S. (2011) Virtue and vulnerability: Discourses on women, gender and climate change. *Global Environmental Change*, 21, 744–51.

Bakan, J. (2004) *The Corporation: The Pathological Pursuit of Profit and Power*. London: Hachette.

Buchbinder, D. (1998) *Performance Anxieties: Re-producing Masculinity*. Sydney: Allen & Unwin.

Connell, R. (1995) *Masculinities*. Sydney: Allen & Unwin.

Connell, R. (2000) *The Men and the Boys*. Sydney: Allen & Unwin.

Crank, J. and Jacoby, L. (2015) *Crime, Violence, and Global Warming*. London: Routledge.

Global Witness (2022) Decade of defiance: Ten years of reporting land and environmental activism worldwide. Available at: www.globalwitness.org/en/campaigns/environmental-activists/decade-defiance/ (accessed 21 October 2022).

Gray, A. and Hinch, R. (eds) (2018) *A Handbook of Food Crime: Immoral and Illegal Practices in the Food Industry and What to Do About Them*. Bristol: Policy Press.

Hall, M. and Farrall, S. (2013) The criminogenic consequences of climate change: Blurring the boundaries between offenders and victims. In N. South and A. Brisman (eds) *Routledge International Handbook of Green Criminology*. London: Routledge, pp 120–33.

Heckenberg, D. and Johnston, I. (2012) Climate change, gender and natural disasters: Social differences and environment-related victimisation. In R. White (ed) *Climate Change from a Criminological Perspective*. New York: Springer, pp 149–72.

Heydon, J. (2020) *Sustainable Development as Environmental Harm: Rights, Regulation, and Injustice in the Canadian Oil Sands*. London: Routledge.

Homer-Dixon, T. (1999) *Environment, Scarcity, and Violence*. Princeton: Princeton University Press.

IPBES (Intergovernmental Science-Policy Platform on Biodiversity and Ecosystem Services) (2019) *The IPBES Global Assessment Report on Biodiversity and Ecosystem Services*. Paris: IPBES.

IPCC (Intergovernmental Panel on Climate Change) (2014) *Climate Change 2014 Synthesis Report, Approved Summary for Policymakers*. Geneva: UN.

IPCC (Intergovernmental Panel on Climate Change) (2022) *Climate Change 2022: Impacts, Adaptation and Vulnerability. Summary for Policymakers*. Geneva: UN.

Klare, M. (2012) *The Race for What's Left: The Global Scramble for the World's Last Resources*. New York: Metropolitan Books, Henry Holt and Company.

Klein, N. (2014) *This Changes Everything: Capitalism versus the Climate*. New York: Simon & Schuster.

Kramer, R. (2020) *Carbon Criminals, Climate Crimes*. New Brunswick: Rutgers University Press.

Le Billon, P. (2012) *Wars of Plunder: Conflicts, Profits and the Politics of Resources*. New York: Columbia University Press.

Mann, M. (2021) *The New Climate War: The Fight to Take Back Our Planet*. Melbourne: Scribe.

Mullins, G. (2021) *Firestorm: Battling Super-Charged Natural Disasters*. Sydney: Viking.

Portner, H., Scholes, R., Agard, J., Archer, E., Ameth, A., Bai, X. and Ngo, H. (2021) *IPBES-IPCC Co-sponsored Workshop Report on Biodiversity and Climate Change*. IPBES and IPCC. DOI: 10.5281/zenodo.4782538.

Robinson, M. (2015) Mary Robinson: Climate change 'very likely' to increase radicalisation. *The Conversation*, 7 December. Available at: http://theconversation.com/mary-robinson-climate-change-very-likelyto-increase-radicalisation-51508 (accessed 15 December 2015).

Schally, J. (2018) *Legitimizing Corporate Harm: The Discourse of Contemporary Agribusiness*. London: Palgrave Pivot.

Segal, L. (1990) *Slow Motion: Changing Masculinities, Changing Men*. London: Virago.

Suzuki, D. (2010) *The Legacy: An Elder's Vision for Our Sustainable Future*. Sydney: Allen & Unwin.

UNDP (United Nations Development Programme) (2013) Overview of linkages between gender and climate change. *Gender and Climate Change: Asia and the Pacific, Policy Brief 1*. New York: UNDP.

United Nations Women Watch (2009) *Fact Sheet: Women, Gender Equality and Climate Change*. Available at: www.un.org/womenwatch/feature/cli mate_change/downloads/Women_and_Climate_Change_Factsheet.pdf (accessed 17 May 2023).

van der Velden, J. and White, R. (2021) *The Extinction Curve: Growth and Globalisation in the Climate Endgame*. Bingley: Emerald Publishing.

Walklate, S. and Mythen, G. (2015) *Contradictions of Terrorism: Security, Risk and Resilience*. London: Routledge.

White, R. (2002) Social and political aspects of men's health. *Health: An Interdisciplinary Journal for the Social Study of Health, Illness and Medicine*, 6(3), 267–85.

White, R. (2018) *Climate Change Criminology*. Bristol: Policy Press.

Whyte, D. (2020) *Ecocide: Kill the Corporation Before It Kills Us*. Manchester: Manchester University Press.

World Meteorological Organization (2020) *State of the Global Climate 2020*. Geneva: WMO.

Zabyelina, Y. and van Uhm, D. (eds) (2020) *Illegal Mining: Organized Crime, Corruption and Ecocide in a Resource-Scarce World*. London: Palgrave.

Index

A

ability *see* ableness
ableness 28, 37, 62, 131, 138, 293
abolitionist theory 129, 260, 261
Abolitionist Vegan Society 260, 261
activism 26–7, 35, 44–5, 63, 199, 220, 229,
   232, 236, 241–3, 252–3, 256–7, 259,
   260–2, 299
 anti-speciesist 256–7
 community 262
 environmental 62, 63, 235, 259,
   272
 feminist 241
 nonhuman animal rights 252–3, 257,
   262
 political 20, 22, 191, 243, 259
activist criminology 45
additionality 214
afro-descendent populations 229–43
agency 11, 54, 98, 149, 159, 231, 237, 240,
   243, 290, 299, 301
Alberta, Canada 11, 267, 272, 273–80
animal, nonhuman 2, 4, 6, 17, 18, 19, 20,
   23, 24, 25, 34–46, 97–113, 123, 124,
   132, 190, 191, 192, 212, 252–6, 258–9,
   261, 262–3
animal abuse, nonhuman 3, 39–41, 212;
   see also nonhuman victimisation and
   nonhuman violence
animal liberation 112, 254
Animal Liberation Victoria 257
animal rights, nonhuman 24, 40, 99, 251–61
animal suffering, nonhuman 97–8, 100–1
animal welfare, nonhuman 9–10, 24, 97, 99
   100, 104, 106, 109–13, 252, 256
anthropocentric 2, 3, 9, 11, 28, 34, 36, 38,
   41, 43, 44, 45, 210, 211, 216, 263
anthropocentrism 219
anti-speciesist 252–3, 256–7, 260, 261,
   262
anti-vivisection 254–7
assembly line workers 10, 149, 152, 156,
   158, 160, 161
Attenborough, David 291, 298

B

Bali 216
Bangladesh 10, 149–50, 152–7, 160–1
Basel Convention 173, 174, 175, 177
binary *see* gender binary
black feminism 229, 231, 239, 240, 242, 243
Black Sea 10, 187, 193–4
Brown Dog Affair, Battersea 254–6

C

Calgary Stampede 273–5
capitalism 7, 11, 58, 126, 128, 130, 132, 133,
   134, 149, 188, 211, 241, 271, 272, 292,
   299, 300
 necro 10, 128, 130, 134, 136, 162
 patriarchal 11, 193, 289–92, 297, 298, 300
carbon emissions 6, 55–7, 59, 154, 163, 214
cisgender 55, 62, 103, 104, 105–7, 109, 111,
   112, 126, 130, 132, 136
class 7, 22, 27–8, 37, 55, 62, 63, 123, 125,
   127, 130, 138, 161, 162, 231, 254, 255,
   263, 268, 269, 272, 290, 291, 292–3, 295,
   297–301
climate change 3–4, 6, 25–7, 43, 53, 55, 57,
   59–61, 64–6, 79, 121, 123–5, 127–9, 134,
   135, 149, 150, 153, 208, 209, 213, 241,
   272, 289–91, 293–4, 296–8, 300
 denial 60–1, 292
climate justice 56, 65, 188, 291, 300
Cobbe, Frances Power 256, 261
collective trauma 133
Colombia 11, 229–43, 291
colonialism 23, 28, 58, 137, 139, 151, 172,
   211, 230, 232–3, 237, 240, 271, 273, 280
community activism 262
companion animals 6, 99–102, 104, 106–9
consumption 37, 42, 44, 57, 58, 62, 66, 77,
   78, 80, 89, 101, 134, 151–4, 177, 180,
   188, 199, 259, 296
Convention on Biological Diversity (Rio
   1992) 80
Convention on the Elimination of All
   Forms of Discrimination against
   Women 88, 217

Convention on International Trade in Endangered Species of Wild Fauna and Flora (CITES) 81, 88
corporate social responsibility 156, 159
Covid-19 54, 59, 66–7, 122, 124, 125, 126–7, 130, 132–4, 135, 136–7, 138, 139–40, 149, 152, 155, 158, 159, 160, 161, 162–3, 217, 278, 293
cultural green criminology 267–9
cultural harm 179, 234
cultural narratives 269, 273, 275
cyber ethnography 191–2

D

debt regimes 218–19
debt-for-nature (DFN) swaps 205, 212–14
decolonial feminism 229, 239, 240, 242, 243
deforestation 125, 190, 193, 234, 294
Despard, Charlotte 252, 253–6
direct action 256, 257, 259, 300
disability 27, 28, 131, 138
disasters *see* industrial disasters or natural disasters
discrimination 7, 27, 37, 38, 101, 128
  employment 126–7, 130, 137
  housing 126–8, 130, 137
dispossession 150, 218, 230, 232–3, 236–7, 238–9, 241, 275
dog fighting 99, 100, 102, 104–5, 106–9, 114
domestic abuse 26, 54, 63; *see also* domestic violence
domestic violence 23, 98, 101, 191, 218, 219; *see also* domestic abuse

E

East Africa 79, 80, 81, 90
eco-colonialism 214
eco-feminism 7, 17–29, 58, 121, 128, 131–2, 134, 137–8, 139, 188, 189, 192, 205, 209, 211–12, 215, 218–20, 252–3, 262, 263, 271
eco-feminist *see* eco-feminism
eco-imperialism 214
ecological damage *see* ecological destruction
ecological degradation *see* ecological destruction
ecological destruction 11, 22, 24, 128, 170, 188, 189, 206, 208, 259, 289–90, 292
ecological disorganisation 2, 128, 188, 198
ecological harm 43, 191, 263, 268
ecological justice (including eco-justice) 17, 19, 25, 170
ecological masculinities 64
ecology 20, 21, 22, 40, 59, 278, 294
economic harm 170, 193
employment discrimination 126–7, 130, 137
enforceability 214

environmental activism 62, 63, 235, 259, 272
environmental damage *see* ecological destruction
environmental defenders 230, 236, 242
  environmental harm 1, 2, 3, 4, 8, 9, 12, 24, 25, 27, 28, 29, 43, 45, 61, 64, 66, 123, 128, 138, 154, 155, 158, 162, 164, 193, 211, 212, 229, 230–2, 236–7, 241, 242, 243, 263, 267, 268, 269, 272, 280, 292
environmental justice 17, 18, 29, 40, 128, 129, 137, 150, 199, 206, 211, 219, 230, 231, 262, 263, 297, 299
environmental security 205, 206, 207, 208–9, 216, 218, 219
environmental threats 28, 199
environmental violence 61–2, 64, 150
erosion 135, 150, 188
ethic of care 64, 66
ethical veganism 34, 41, 44–6
ethics 3, 19, 22, 42, 75, 76, 86, 164, 190, 252, 256
e-waste 171, 175, 177–9, 181
extractive industries 24, 56, 135, 188, 189, 190, 234, 235–6, 238, 239, 240, 242, 243, 273, 276
extractivism 230, 231, 235, 237, 239, 242, 243

F

FARC (Revolutionary Armed Forces of Colombia) 234, 236
farmers 163, 234, 259, 290, 296
fast fashion 148, 151, 153, 154, 155, 157, 158, 161, 162, 163–4
feminine principle 211
femininity 61, 66, 75, 103, 238, 270, 281
feminism 4, 22, 38, 190
  black 229, 231, 239, 240, 242, 243
  decolonial 229, 239, 240, 242, 243
  eco 7, 17–29, 58, 121, 128, 131–2, 134, 137–8, 139, 188, 189, 192, 205, 209, 211–12, 215, 218–20, 252–3, 262, 263, 271
  indigenous 229, 239, 240, 241, 242, 243
  vegan 251–63
feminist 3, 6, 7, 17, 20, 21, 22, 23, 28, 29, 37, 55, 137, 149, 190, 193, 198, 211, 212, 218, 229, 230, 231–2, 236, 240, 241, 243, 252, 259, 267, 271, 281
  criminology 1–2, 5–7, 8, 18, 25, 34, 37, 38, 45, 187, 190, 191, 199, 200, 252, 262
  green criminology 187–8, 189, 190–1, 199, 200
  vegan 253
financial harm *see* economic harm
food justice 260, 261, 262
forced displacement 230, 234, 235, 238
forced marriage 219; *see also* underage marriage

forest loss 188; *see also* deforestation
fossil fuel hegemony 281
fossil fuels 53, 57, 60, 61, 62, 66, 134, 151,
    278, 279, 280, 291, 299
frontier masculinity 273, 275, 278, 279, 280–1

G

garment industry 152, 153–4, 156–8, 160, 161
gender-based violence 54, 151, 205–7,
    231, 241
gender-based vulnerability 295
gender binary 97, 98–9, 100, 102, 104, 107,
    109, 111, 136, 137, 277, 295
gender difference 100, 177, 289, 290, 294
gendered dimensions of illegal wildlife
    trade 73, 74, 75, 76, 78–9, 80, 82, 87–8,
    90, 91
gender identity 27, 86, 98–9, 100–12
gender inequality 22, 53, 86, 88, 187, 191,
    198, 199, 207, 219, 256, 289, 290, 294
gender narratives 91, 243, 269, 271, 272,
    280
gender power hierarchies 101, 109, 148, 269,
    270, 271
gender roles 98–102, 106–12, 114–15, 137,
    171, 242, 273
Global North 58, 59, 61, 132, 134, 148,
    153, 155, 156, 163, 170–3, 175, 179, 180,
    231–2, 243
Global South 59, 148, 151, 154, 156–63,
    170–3, 175–6, 179–80
Goliath 298
Green Road Project 187–200
green victimisation 123, 170–1, 187, 191,
    192–3, 195, 199, 212
green victimology 25, 170

H

habitat fragmentation 188
harm 2, 4, 6, 7, 8, 18, 23, 24, 25, 27, 34, 36,
    39, 40, 41, 44–5, 53, 54, 60, 61, 62, 63,
    65, 123, 126, 129, 130, 131, 132, 133–4,
    135, 136, 138, 139, 176, 177, 178, 179,
    188, 189, 190, 191, 197, 206, 210, 212,
    229, 231, 236, 237, 238, 242, 243, 256,
    261, 267, 268–9, 272, 280, 281, 291, 293
  cultural 179, 234
  ecological 43, 191, 263, 268
  economic 170, 193
  environmental 1, 2, 3, 4, 8, 9, 12, 24, 25,
    27, 28, 29, 43, 45, 61, 64, 66, 123, 128,
    138, 154, 155, 158, 162, 164, 193, 211,
    212, 229, 230–2, 236–7, 241, 242, 243,
    263, 267, 268, 269, 272, 280, 292
  green *see* environmental
  financial *see* economic harm
  social 19, 25, 56, 121, 128, 139, 153, 163,
    171, 175, 179–80, 267, 272
  socioenvironmental 122, 133

hegemonic masculinity 56, 62–3, 269–73,
    292, 298–300
heteronormativity 23, 130, 132, 134,
    137–9, 271
HIV/AIDS 128, 130, 132–3, 135–6, 139
HIV criminalization 135
housing discrimination 126–8, 130, 137
human security 206, 208–9, 217, 219
human species 18, 19, 36

I

identity 7, 27, 28, 37, 43, 57, 60, 110, 123,
    137, 198, 233, 237, 252, 253, 268, 269,
    270, 280–1, 300
  gender 27, 86, 98–9, 100–12
  queer 122, 138, 195
illegal wildlife trade 72–91
incarceration 63, 126, 128–30, 137, 139
indigeneity 293
indigenous feminism 229, 239, 240, 241,
    242, 243
Indonesia 151, 160, 205–20
industrial disasters 151, 152, 156, 158
inequality 21, 37, 67, 121, 125, 138, 139,
    179, 198, 214, 235, 269, 295, 297
  gender 22, 53, 86, 88, 187, 191, 198, 199,
    207, 219, 256, 289, 290, 294
International Consortium on Combatting
    Wildlife Crime (ICCWC) 86, 88–9
International Union for the Conservation of
    Nature (IUCN) 207, 220
Intersectionalities *see* intersectionality
intersectionality 7, 18, 23, 27–8, 34–5, 38,
    41, 45, 113, 130, 133, 137, 205, 212, 229,
    243, 252–3, 256, 260, 261, 263, 268, 269,
    271, 272
Irish Women's Franchise League 253
isolation 66, 98, 126, 132–4, 137, 139, 234

J

justice 2, 17, 25, 29, 36, 45, 59, 206, 219,
    230, 232, 243, 252, 253, 255, 260, 300
  climate 56, 65, 188, 291, 300
  ecological (including eco-justice) 17, 19,
    25, 170
  environmental 17, 18, 29, 40, 128, 129,
    137, 150, 199, 206, 211, 219, 230, 231,
    262, 263, 297, 299
  food 260, 261, 262
  social 7, 45, 170, 219, 243, 252, 253, 254,
    256, 260, 261, 262, 263
  species 17, 19, 40, 44, 170
Justice and Development Party (AKP) 189

K

Kistle, Sarah 252, 260–2

L

logic of domination 205–11, 215–16, 218–19

M

male domination 6, 59, 60, 63, 66, 110, 270, 272, 291, 292
malestream norms 5, 269–72
Mark, Patty 252, 256–7
masculinity 5, 6, 55–67, 75, 76, 101, 103, 148, 267, 270, 271, 273, 275, 276, 281, 289, 290, 293, 298, 300
 frontier 273, 275, 278, 279, 280–1
 hegemonic 56, 62–3, 269–73, 292, 298–300
meat 39, 56, 77, 78, 80
meat-eating 38, 56, 101
Mellor, Mary 20–1, 22, 23
military 62, 153, 161, 208, 209
money marriages 216
monitoring 214

N

natural disasters 25, 54, 63, 121–8, 132, 133, 134, 135, 137, 138, 139, 140, 149, 217, 291, 292, 294–5, 297, 300
necrocapitalism 10, 128, 130, 134, 136, 162
nikah siri 217
non-binary (NB) 98–9, 101–5, 109–12
nonhuman 3, 18, 19, 20, 22, 23, 25, 66
 animal rights 24, 40, 99, 251–61
 animals 2, 4, 6, 17, 18, 19, 20, 23, 24, 25, 34–46, 97–113, 123, 124, 132, 190, 191, 192, 212, 252–6, 258–9, 261, 262–3
 victimisation 2, 3, 34
 violence 3, 39

O

oil and gas industry 57, 267, 272, 273, 276–7, 278–80
open rescue 257–9, 262
oppressive conceptual framework 209–10, 215, 218

P

patriarchal capitalism 11, 193, 289–92, 297, 298, 300
patriarchy 6, 7, 21, 56, 58, 60, 61, 62, 66, 134, 148, 150, 151, 189, 191, 194, 198, 199, 210, 211, 218, 219, 231, 236, 238, 239, 241, 242, 243, 251, 252, 262, 271, 292, 298, 299
permanence 214
pets 36, 101, 104, 107, 113, 258, 262; see also companion animals
political activism 20, 22, 191, 243, 259
politics of disposability 131, 132, 133, 134, 140
pollution 22, 154, 162, 173, 175, 176, 177, 179, 181, 188, 190, 193, 208, 240, 268
 stream 188
poorcide 148, 150, 156, 158, 162

poverty 3, 74, 131, 149, 150, 179, 214, 215, 216, 217, 238, 240, 299, 300
privatisation 211, 212, 215

Q

quantitative methods 111
queer 98, 99, 103, 105, 121–40
queer green criminology 121, 123, 138, 139

R

race 7, 22, 27, 28, 37, 123, 125, 127, 130, 131, 138, 150, 162, 211, 231, 232, 253, 272, 293
Rana Plaza 149, 155, 156–8, 163
resistance 2, 21, 22, 66, 113, 189–90, 191, 192, 193, 195, 196–200, 230, 231, 235–6, 240–3, 252, 261, 262, 292, 299

S

security
 environmental 205, 206, 207, 208–9, 216, 218, 219
 human 206, 208–9, 217, 219
sexism 3, 22, 23, 36, 38, 101, 269, 271
sexual orientation 111, 127, 136
sexual violence 54, 55, 89, 129, 151, 206, 207, 231, 238, 242, 276, 294, 295
sexuality 27, 28, 38, 103, 122, 137, 138, 231; see also sexual orientation
ship breaking yards 171, 175, 176–7, 178, 181
slow violence 61, 148, 150, 157, 162
social conflict
social harm 19, 25, 56, 121, 128, 139, 153, 163, 171, 175, 179–80, 267, 272
social justice 7, 45, 170, 219, 243, 252, 253, 254, 256, 260, 261, 262, 263
species justice 17, 19, 40, 44, 170
speciesism 3, 23, 34, 35–41, 43–5, 99, 101, 251, 252, 256, 257, 258, 260, 261, 262
speciesist 34, 38, 39, 41, 45, 99, 252, 258, 259, 260, 262; see also anti-speciesist
Sri Lanka 149, 150, 152–62, 163
status dogs 99, 100, 102, 104, 105, 106, 107–9, 110, 114
stream pollution 188
structural violence 8, 61, 135, 148, 149, 150, 151, 229
Sub-Saharan Africa 4, 72–91
suffering 25, 26, 27, 37, 38, 39, 43, 44, 63, 97–8, 100–2, 109, 148, 151, 153, 157, 158, 161, 162, 163, 177, 181, 192, 206, 234, 240, 254, 255, 256, 259; see also nonhuman animal suffering
Suffragettes 253, 255, 259
Sumatra 213, 216, 218
sustainable wildlife management 73, 82, 88, 91
Suzuki, David 291, 298

T

territory 89, 148, 172, 230, 232, 233, 234,
    235, 237, 238–43, 280, 291, 297
theriocide 39, 44, 212
Turkey 175, 176, 187–200

U

underage marriage 205–6, 209, 212, 216–19
United Nations Convention against
    Transnational Organized Crime 81
urban 26, 40, 151, 152, 158, 196, 197, 231,
    232, 239, 293, 296

V

vegan 34–46, 56, 252, 253, 257, 258, 260,
    261, 262, 263
vegan feminism 251–63
veganising 34–5, 44–6
veganism 34, 41–6, 98, 252
victimisation 1–5, 7, 18, 19, 23, 24, 25, 26,
    27, 45, 123, 125, 126, 127, 129, 130, 132,
    133, 134, 135, 137, 139, 149, 170–1, 177,
    178, 181, 187, 188, 189, 190, 191, 192,
    193, 197, 198, 269, 295, 297
  green 123, 170–1, 187, 191, 192–3, 195,
    199, 212
  nonhuman 2, 3, 34

victimology
  green 25, 170
violence
  domestic 23, 98, 101, 191, 218, 219;
    see also domestic abuse
  gender-based 54, 151, 205–7, 231,
    241
  nonhuman animal 3, 39
  sexual 54, 55, 89, 129, 151, 206, 207, 231,
    238, 242, 276, 294, 295
  slow 61, 148, 150, 157, 162
  structural 8, 61, 135, 148, 149, 150,
    151, 229
violence prevention 64

W

waste crime 170–81
Wildlife and Forest Crime Analytical
    Toolkit 88–9
Women's Freedom League 253
women's green victimisation 187, 191,
    193, 199
World Health Organization 207

Z

zemiology 24
zoonotic disease 66, 121–40